DYNAMICS OF COMETS: THEIR ORIGIN AND EVOLUTION

ASTROPHYSICS AND SPACE SCIENCE LIBRARY

A SERIES OF BOOKS ON THE RECENT DEVELOPMENTS
OF SPACE SCIENCE AND OF GENERAL GEOPHYSICS AND ASTROPHYSICS
PUBLISHED IN CONNECTION WITH THE JOURNAL
SPACE SCIENCE REVIEWS

VOLUME 115
PROCEEDINGS

DYNAMICS OF COMETS: THEIR ORIGIN AND EVOLUTION

PROCEEDINGS OF THE 83rd COLLOQUIUM
OF THE INTERNATIONAL ASTRONOMICAL UNION,
HELD IN ROME, ITALY, 11–15 JUNE 1984

Edited by

ANDREA CARUSI

and

GIOVANNI B. VALSECCHI

*Istituto di Astrofisico Spaziale – Consiglio Nazionale delle Ricerche,
Reparto Planetologia, Rome, Italy*

SPRINGER-SCIENCE+BUSINESS MEDIA, B.V.

Library of Congress Cataloging in Publication Data

International Astronomical Union. Colloquium (83rd: 1984: Rome, Italy)
    Dynamics of comets.

    (Astrophysics and space science library; v. 115)
    Includes indexes.
    1. Comets–Congresses. I. Carusi, Andrea, 1946–
II. Valsecchi, Giovanni B. (Giovanni Battista), 1951–   . III.   Title.
IV. Series.
QB717.I58   1984   523.6   85–11767
ISBN 978-94-010-8884-8   ISBN 978-94-009-5400-7 (eBook)
DOI 10.1007/978-94-009-5400-7

TABLE OF CONTENTS

PREFACE

The Colloquium

The IAU Colloquium n. 83 "Dynamics of Comets: Their Origin and Evolution"
(Rome, Italy, 11-15 June 1984) took place well over a decade after the
first IAU meeting devoted to essentially the same subject, i.e. IAU Symposium n. 45 "The Motion, Evolution of Orbits and Origin of Comets"
(Leningrad, USSR, 4-11 August 1970). During the time interval separating the two meetings cometary astronomy has made big steps forward from
the point of view of both the physics and the dynamics, and further
progress is expected in the near future, with the coming of the many
space missions aimed to P/Halley. However, the scientific meetings totally devoted to comets held in the seventies and early eighties (IAU
Colloquium 25 "The Study of Comets", Greenbelt, U.S.A., 1974; IAU Colloquium 61 "Comets: Gases, Ices, Grains and Plasma", Tucson, U.S.A.,
1981) emphasized the physical aspects, and did not cover satisfactorily
matters related to dynamics, origin and early evolution. These were
confined to individual sessions in meetings on the minor bodies of the
solar system (IAU Colloquium 22 "Asteroids, Comets, Meteoric Matter",
Nice, France, 1972; IAU Colloquium 39 "Relationships between Comets,
Minor Planets and Meteorites", Lyon, France, 1976; "Asteroids, Comets,
Meteors", Uppsala, Sweden, 1983). It was therefore felt necessary to
organize a meeting centred on the dynamics, a field which still comprises such a large fraction of all what is known about comets. This
idea was extensively discussed with L. Kresák, B.G. Marsden and E. Everhart, who strongly supported it.

Sponsored by Commission 20, and co-sponsored by Commission 7, the
meeting was approved by the Executive Committee of the International
Astronomical Union as IAU Colloquium 83. The Scientific Organizing
Committee consisted of P. Babadzhanov (U.S.S.R.), N.A. Belyaev (U.S.S.R.),
A. Carusi (Italy, Chairman), E. Everhart (U.S.A.), J. Kovalevsky (France),
Y. Kozai (Japan), L. Kresák (Czechoslovakia), B.G. Marsden (U.S.A.), H.
Rickman (Sweden), E. Roemer (U.S.A.), G.B. Valsecchi (Italy, Secretary),
and the Local Organizing Committee of A. Carusi, E. Perozzi (Secretary),
G.B. Valsecchi (Chairman).

The Colloquium was financially supported by the Consiglio Nazionale
delle Ricerche, by the Provincia di Roma, Assessorato alla Cultura, and
of course by IAU; the Accademia Nazionale dei Lincei hosted the meeting
in its beautiful historical complex of buildings in the centre of Rome;
the last session was held in CNR headquarters.

A total of 64 scientists from 16 countries attended the meeting,
giving 9 Invited Reviews and 31 Contributed Papers. Another three In-

vited Reviews and six Contributed Papers were initially to be given
by Soviet scientists, but later they could not participate to the Colloquium; the abstracts of some of these presentations were read during
the corresponding sessions by other participants. It is sad that, independently of their will such a large fraction of the invited reviewers,
not mentioning the authors of Contributed Papers, were unable to come
from the Country that more than many others has given significant contributions to the birth of modern cometary dynamics.

The Colloquium has been structured in seven sessions: Origin of
Comets, The Oort Cloud of Comets, Meteor Streams and Interrelations
with Minor Planets, Dynamics of Comets: Numerical Modelling, Dynamics
of Comets, Nongravitational Forces, Comet P/Halley and Future Missions
to Comets, whose Chairmen were respectively A.H. Delsemme and J.M. Greenberg, C. Froeschlé, B.A. Lindblad, I.P. Williams, B.G. Marsden, J.C.
Brandt, P.R. Weissman.

Half day of the meeting has been devoted to a visit to the Vatican
Observatory in Castelgandolfo and its beautiful collection of meteorites – one of the richest in the world. We want to thank very much
Father G.V. Coyne S.J., the Director of the Specola Vaticana, for his
warm hospitality.

A special thank also to Dr. E. Perozzi, currently at ESOC, Darm-
stadt (W. Germany), for the valuable help as a member of the Local
Organizing Committee, and to Dr. S. Pozio, to J. Vannozzi and to G.
Sabatino for their assistance at the Registration Desk and at the slide
projector.

The Proceedings

This book contains papers presented at the IAU Colloquium 83 "Dynamics
of Comets: Their Origin and Evolution"; each paper, either invited or
contributed, has been submitted to two referees. This choice has been
made to increase the value of the book itself, especially considering
that a comparable collection of papers may not appear for another decade, given past experience.

We therefore thank the referees M.E. Bailey, S.V.M. Clube, A. Coradini, A.H. Delsemme, E. Everhart, P. Farinella, J.A. Fernández, G. Forti, C. Froeschlé, M. Fulchignoni, R. Greenberg, A. Hajduk, L. Kresák,
M. Kresáková, B.A. Lindblad, Rh. Lüst, A. Manara, B.G. Marsden, F. Mignard, A. Milani, W.M. Napier, E.M. Pittich, H. Rickman, V.S. Safronov,
H. Scholl, P.R. Weissman, I.P. Williams, S. Yabushita, D.K. Yeomans, P.
Zadunaisky for their cooperation; we are of course responsible for the
final appearance of the book, since in some cases we have decided to
include papers not too warmly recommended by the referees, in order to

avoid an excess of orthodoxy.   Special thanks are due to the British
and American referees, who substantially improved many papers from the
linguistic point of view.

The material is divided in seven sections, corresponding to the
seven sessions of the Colloquium; some of them, like for instance those
on the Origin and on the Oort Cloud, are deeply related, and the inclusion of some papers in either of them would have probably been equally
reasonable.

The book should have contained papers presented at the Colloquium,
possibly in complete form; however the authors of two communications
have preferred to publish them in the form of extended abstracts.   Another exception has been made for those Soviet colleagues who could not
attend, but were able to send their papers early enough (in fact, before the end of May 1984) to allow their refereeing together with all
the others.   Two papers were added at the end of the session on nongravitational forces, and two papers at the end of that on the numerical modelling of cometary dynamics.   Of these, one is of a particular
nature, being centred on the outstanding contributions of E.I. Kazimirchak-Polonskaya to this field of studies; the second, by herself,
is a review of her work on the role of Neptune in the dynamical evolution of comets.   We have been very pleased by the opportunity to publish these papers since, as already said, E.I. Kazimirchak-Polonskaya
has been a pioneer in the subject, and IAU Colloquium 83 has followed
the footsteps of IAU Symposium 45, whose organization was due in great
part to her efforts.

Last, but not least, we want to say that, throughout all the phases
of our work, from the organization of the colloquium up to the editing
of these proceedings, we have been continuously encouraged and helped
in many ways by Lubor Kresák.   Our most sincere thanks to him.

Rome, 29 April, 1985   Andrea Carusi
                                            Giovanni Battista Valsecchi

SECTION I

ORIGIN OF COMETS

THE ORIGIN OF COMETS AMONG THE ACCRETING OUTER PLANETS

Richard Greenberg
Planetary Science Institute
2030 E. Speedway, Suite 201
Tucson, Arizona 85719
USA

ABSTRACT. The hypothesis of formation of comets as an accompaniment to formation of Uranus and Neptune from icy planetesimals is attractive for several reasons, but has suffered from long-standing problems regarding formation of the planets themselves. The history of this problem is reviewed, and recent results are described that may help solve it. Numerical simulations of planet growth show that when the system of planetesimals is no longer artificially constrained to a power-law size distribution, growth of planets may occur in reasonable time. An adequate number of comet-sized bodies to populate the Oort cloud is not produced as collisional debris during the planet-building process. Rather, the comets are probably a remnant of the original planetesimal "building blocks" from which the planets grew.

The origin of the Oort Cloud of comets was likely to have been connected with the formation of planets in the solar system. Nebular densities beyond the planetary system were probably too low to have permitted accretion of comet-sized bodies (Öpik 1973, Safronov 1977a). But closer to the sun, planet formation was apparently accompanied by production of smaller bodies, some of which would necessarily be perturbed by planetary encounters into orbits in the Oort Cloud. Thus, comets are a plausible by-product of planetary formation.

In the context of the planetesimal hypothesis of planet formation, it seems plausible that comets are planetesimals that were removed to the Oort Cloud by close encounters with growing (or nearly grown) planetary embryos before they could be accreted. For a number of reasons, the most promising candidate region for cometary origin is the Uranus-Neptune zone. Uranus and Neptune are quite likely to have been formed from icy planetesimals. Moreover, Uranus and Neptune's sizes and positions are appropriate for having scattered residual planetesimals out to the Oort Cloud with reasonable (~10%) efficiency (Fernandez and Ip 1981, Safronov 1969). From closer to the sun, it was much harder to scatter planetesimals out that far. After Jupiter's sudden increase in mass with gas accretion around its solid core

3

A. Carusi and G. B. Valsecchi (eds.), Dynamics of Comets: Their Origin and Evolution, 3–10.
© *1985 by D. Reidel Publishing Company.*

(Safronov and Ruskol 1982), it became <u>too</u> effective at scattering planetesimals; most were ejected from the solar system on strongly hyperbolic trajectories, with only a very small fraction contributing to the Oort Cloud region. Closer to the sun, planetesimals were rocky, not icy, and hence not the source population for comets.

While the evidence has pointed to cometary origin near Uranus and Neptune, quantitative analysis has awaited resolution of a fundamental problem regarding formation of the planets themselves: Accretion models (e.g., Safronov 1969) generally required $\sim 10^{11}$ yr for outer planet growth, assuming a plausible surface density of the planetesimal swarm of $\sigma \sim 0.3$ gm/cm^2. The slow growth was due to the increase in relative velocities among planetesimals believed to accompany growth of the planetary embryos, which kept gravitational cross-sections small.

Attempts to modify the theory to accommodate the actual existence of the outer planets involved <u>ad hoc</u> assumptions of either very high surface density of the planetesimal swarm or lower values of relative velocities among planetesimals. Levin (1972) considered the implications of increasing σ one-hundred-fold to 30 gm/cm^2. Availability of so much mass increased accretion rates so as to give growth in $<10^9$ yr. But the excess material needed to be removed, and to eject so much material would require great loss of angular momentum from the planets. Levin pointed out that an implication is that Uranus and Neptune would have had to have formed ten times farther from the sun than their present orbits. With σ thus ~ 30 gm/cm^2 at >200 AU, the total nebular mass would have had to have been ~ 2 M$_\oplus$, which as Levin concluded is much too large to be consistent with the planetesimal model of planet growth.

Safronov considered the possibility that growth rates were enhanced by a combination of high σ and low velocities. The latter help by increasing gravitational cross-sections and thus speeding accretion. With $\sigma \sim 3$ gm/cm^2, the extreme problems noted by Levin are avoided. Safronov offered speculative suggestions as to why relative velocitkes might have been lower than for his nominal model, which was based on an assumed equilibrium between collisional damping and gravitational stirring by mutual encounters and which gave relative velocities on the order of the escape velocities of the larger bodies. Those suggestions included the following: (a) Relative velocities were distributed over some range of values. The segment of the population with higher velocities was preferentially ejected from the system, leaving only the low velocity portion of the population (Safronov 1969). (b) The low strength of icy planetesimals might have given a steep size distribution which yields lower relative velocities (Safronov 1972).

There are problems with both those ideas. Suggestion (a) raises questions about other planets' growth. For example, for the Earth, would such a low velocity component speed growth relative to the growth rate computed by Safronov based on the average velocity? Suggestion

(b) is contradicted by experimental evidence (e.g. Hartmann 1969) which indicates that weak materials do not have such steep distributions; they simply break up more easily. Safronov (1977b) later suggested that gravitational instabilities directly produced large embryos, thus by-passing much of the evolutionary time required for collisional accretion. However, as described below, it is implausible that the gravitational instability could have produced such large bodies.

Levin (1978) suggested that relative velocities may have been lower than in Safronov's nominal growth models for another reason. He invoked Safronov's own dynamical theory in pointing out that velocities would be low compared with the escape velocity of the largest body, when in the late stages the planetary embryos "ran away" in terms of growth from the remaining planetesimal distribution in its zone. Once an embryo becomes detached from the continuous part of the size distribution, relative velocities no longer increase with the embryo's size.

In fact, more recent numerical simulations (Greenberg et al. 1978) of planet growth show that the size distribution may have been very different than assumed in Safronov's theory. For the terrestrial planets, most of the mass remained in small planetesimals (original building blocks plus a power law distribution of smaller debris), which damped velocities as the embryo grew. Velocities did not increase directly with embryo size. Growth of a substantial embryo was $\sim 10^{2-3}$ times faster than in Safronov's model. Qualitatively, such simulations, applied to the outer solar system, were expected to solve two problems, yielding (a) planets in reasonable time, and (b) a large reservoir of small bodies available for removal to the Oort Cloud.

In order to apply such simulations to the outer solar system we first needed to select plausible initial conditions. The conventional theory of gravitational instability in a flat dust disk (Safronov 1969, Goldreich and Ward 1973) predicts that the first generation of planetesimals at a given heliocentric distance is characterized by sizes proportional to σ, yielding radii >60 km. It seemed reasonable, based on the numerical results for terrestrial planet growth, that with this initial size the outer planets could have grown quickly, and that the comet-size bodies (1 to 10 km) would be produced as collisional debris.

Numerical simulations have now been applied to outer planet growth (Greenberg et al. 1984). We modeled accretion of solid icy material in Neptune's zone for cases with σ in the range of 0.3 gm/cm^2 (near the minimum to form the planet) to 3 gm/cm^2. Initial planetesimals were given the characteristic size, produced by gravitational instability, corresponding to the value of σ, with initial relative velocities on the order of their escape velocities. In these simulations, the Neptune embryo grew rapidly, reaching 10% of its final mass in $\sim 10^8$ yr, at which time it is growing at a rate such that full size would be reached in $< 10^9$ yr. Most of the mass remained in bodies of the original size, but collisional debris extended down through the cometary size range. The quantity of comet-sized debris is comparable to the estimated number of Oort Cloud comets ($\sim 10^{10}$ of ~ 10 km, 3×10^{12} of ~ 1 km), but not enough to account for the order of magnitude loss in transporting them to the Cloud. After 10^8 yr, the number of comet-sized bodies decreased as they were rapidly broken into even smaller pieces. This problem remained even when we modeled the bodies as being as strong as solid rock (impact strength 10^8 ergs/cm^3).

Even if the initial population is taken to include in addition the required number of comet-size bodies, the presence of a comparable mass of 100 km bodies is sufficient to raise relative velocities enough to destroy the comet-size bodies before Neptune grows large enough to scatter material to the Oort Cloud. Neptune does grow rapidly, however, because, as in the earlier experiments, relative velocities are much less than the embryo's escape velocity.

The implication of our numerical experiments is that an adequate comet-size population can exist long enough for the Neptune embryo to reach scattering size only if such a distribution exists from the beginning and if there is initially a negligible mass contribution from bodies $\gtrsim 10$ km. Such an initial population consists of smaller bodies than predicted by the conventional gravitational instability models, even for the minimum σ needed to make Neptune. However, such instability models assume that σ refers to a dust layer of uniform density settling homologously to the plane of the nebula. In a non-uniform layer, gravitational instability occurs in regions that exceed a critical density. Thus clumping into planetesimals may begin even before all material has settled to the midplane.

We have modeled the earlier settling process, and find that if coagulation among dust grains occurs, larger grain aggregates experience runaway growth and rapid settling, forming a dense sub-layer in the central plane. This sub-layer may reach the critical density for instability while containing $\lesssim 1\%$ of the total mass of solids. The resulting planetesimals are correspondingly small; their actual sizes depend on the rate at which mass arrives at the central plane relative to the growth time of instabilities ($\sim$ the orbital period). Gravitational encounters among this first generation of bodies stir them out of the plane on the same time scale, but their perturbations do not affect later settling dust which is damped by gas drag. The process may repeat for several generations, while collisional accretion proceeds. A comet-like size distribution, rather than bodies $\gtrsim 60$ km, is a reasonable outcome of the gravitational instability process.

Numerical simulation of planet growth with this comet-size initial population in the outer solar system shows that growth of substantial planetary embryos occurs in very short time (see Fig. 1). A sufficient population of the comet-sized bodies remains to account for population of the Oort Cloud by scattering as the planetary embryos approach full size. The embryo is sufficiently detached from the size distribution that subsequent final accretion should be fairly fast. This model seems to satisfy all of our requirements.

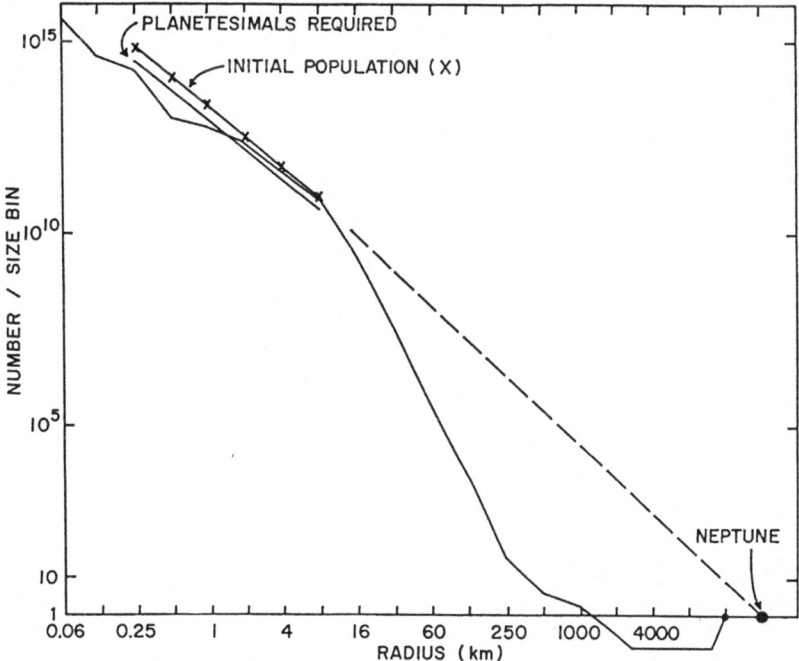

<u>Figure 1</u>: Evolution of a population of initially
comet-sized bodies, shown by x's. The solid curve
shows the population after 1.4 x 10^5 yr. For
reference, the dashed line represents the slope for
equal mass per size bin (factor of 2 in radius),
and the uncompressed size of Neptune is shown. The
line labelled "Planetesimals Required" shows the
number of comet-sized bodies needed to account for
populating the Oort Cloud with 10% efficiency (Fig.
from Greenberg <u>et</u> <u>al</u>. 1984).

    However, evolution beyond the stage shown cannot be modeled
adequately by our numerical simulation in its present form, because a
number of late-stage effects are not readily incorporated into our
particle-in-a-box statistical approach. The dominance of a single body
would make certain regions (e.g. the neighborhood of its own orbit)
special. Also in the late stage questions arise as to the validity of
computing gravitational cross-sections using the two-body encounter
model. Because our model is not applicable to the late stage, there
remain important questions about late-stage growth. Do the first-
formed embryos accrete or scatter the small bodies between their
orbits, or, alternatively, do many additional embryos grow among the
first-formed ones, only later to be consolidated into a few planets?
Similar questions remain regarding late-stage planet growth in the
inner solar system.

Nevertheless, combining the results of our models for mid-plane settling and for outer planet growth, strongly suggests that comets are a representative residue of the initial population of planetesimals in the outer solar system, not fragments of larger bodies. At the very least, these results demonstrate that the long-standing problems with time required for outer-planet growth are not as serious as previously thought.

ACKNOWLEDGMENTS

This paper describes research carried out with the collaboration of S.J. Weidenschilling, C.R. Chapman, and D.R. Davis. I am grateful to M.E. Bailey, J.A. Fernandez, W.K. Hartmann, and G.W. Wetherill for helpful comments. The Planetary Science Institute is a division of Science Applications, Inc.

REFERENCES

Fernandez, J.A., and W.H. Ip (1981). 'Dynamical evolution of a cometary swarm in the outer planetary region', Icarus 47, 470-479.
Goldreich, P., and W.R. Ward (1973). 'The formation of planetesimals'. Astrophys. J. 183, 1051.
Greenberg, R., J.F. Wacker, W.K. Hartmann, and C.R. Chapman (1978). 'Planetesimals to planets: Numerical simulation of collisional evolution', Icarus 35, 1-26.
Greenberg, R., Weidenschilling, S.J., Chapman, C.R. (1984). 'From icy planetesimals to outer planets and comets'. Icarus 59, 87-113.
Hartmann, W.K. (1969). 'Terrestrial, lunar, and interplanetary fragmentation', Icarus 10, 201-213.
Levin, B.J. (1972). 'Revision of initial size, mass, and angular momentum of the solar nebula and the problems of its origin', in The Origin of the Solar System (H. Reeves, Ed.), CNRS, Paris.
Levin, B.J. (1978). 'Relative velocities of planetesimals and the early accumulation of planets', Moon and Planets 19, 289-296.
Opik, E.J. (1973). 'Comets and the formation of planets', Astrophys. Space Sci. 21, 307-398.
Safronov, V.S. (1969). Evolution of the Protoplanetary Cloud and Formation of the Earth and the Planets, Nauka Publ., Moscow, English translation: NASA TT F-677 (1972).
Safronov, V.S. (1972). 'Accumulation of the planets', in Origin of the Solar System (H. Reeves, Ed.), CNRS, Paris, 89-113.
Safronov, V.S. (1977a). 'Oort's cometary cloud in the light of modern cosmogony', in Comets, Asteroids, and Meteorites (A. Delsemme, Ed.), Univ. of Toledo Press, 483-484.
Safronov, V.S. (1977b). 'Time scale for the formation of the Earth and planets and its role in their geochemical evolution', in Soviet-American Conference on Cosmochemistry of the Moon and

Planets, NASA SP-370.
Safronov, V.S., and Ruskol, E.L. (1982). 'On the origin and initial
    temperature of Jupiter and Saturn', Icarus 49, 284-296.

DISCUSSION

P. Farinella: How are your collisional accretion models affected by
changing parameters like impact strength, fraction of kinetic energy
going into fragments, etc.?

R. Greenberg: The basic conclusions are not affected by such
choices. We have experimented with a wide range of assumed impact
parameters. Interestingly, we find the greatest success in producing
and preserving comet-size planetesimals when the material is assumed to
be very strong. At first I was surprised by that result because I had
expected that weaker material would tend to produce comet-size fragments more easily. From our numerical experiments, though, I learned
that weak material tends in fact to be easily ground down to sub-comet-
sized bodies; the runs which produced the most comets assumed strengths
equal to that of strong rock or even iron!

J. Lissauer: Could you get more comet-sized bodies by ejecting them
before they can get "ground down"?

R. Greenberg: Velocities are too small for ejection during the
stages modeled in our numerical simulations; however, this process may
be relevant later on.

B.A. Lindblad: Does it follow from your work that a significant
amount of comets would at the present time be moving in nearly circular
orbits between Uranus and Neptune?

R. Greenberg: That does not follow from our work, because we have
not studied the late stage of planet growth. However, there may be
some small zones of stability between the planetary orbits.

A. Fernandez: Have you considered how the possible migration of the
accreting proto-Uranus and proto-Neptune, due to the exchange of angular momentum with planetesimals, might affect your model parameters;
for instance, the surface density σ, the size range of formed planetesimals, and the total mass required to form Uranus and Neptune?

R. Greenberg: This may be a way for the planetary embryos (or proto-
planets) to move through the zones that otherwise would be relatively
isolated between the embryos' orbits. It could be a solution to some
of the questions about late-stage evolution that I have raised.
However, we have not yet examined the late stage in any detail.

M.E. Bailey: You mentioned that Levin's suggestion of increasing σ by increasing the mass of the solar nebula went agaiost the spirit of the planetesimal hypothesis. What is the largest mass nebula that is consistent with this scheme of things?

R. Greenberg: Usually the planetesimal hypothesis assumes a nebula mass ~5% of the sun's mass, while the "giant gaseous protoplanet" hypothesis assumes the total mass (sun + nebula) to be ~2$M_\odot$. One could imagine intermediate cases, which might have hybrid processes, but such models have not been studied.

P. Weissman: I have two comments: first, in answer to Dr. Lindblad's question, we would expect 0.5 - 1.0% of Uranus-Neptune zone planetesimals to be surviving in Uranus-Neptune-crossing orbits. Objects such as Chiron are almost certainly Uranus-Neptune plane-tesimals evolving dynamically out of that zone, as shown by Scholl.
      Secondly, the distribution of cometary magnitudes found by Everhart after corrections for observational selection has a knee in the curve at about $H_{10} \approx 6$. If one goes through the steps of converting the distribution to a distribution of cometary masses, we find that for a reasonable albedo, say 0.3, the size of the comets at the knee is about 8 km. Thus, we may ave observational confirmation of the type of size distribution you are talking about here.

STATISTICAL TEST OF THE DISTRIBUTION OF PERIHELION POINTS
AND ITS IMPLICATION FOR COMETARY ORIGIN

S. Yabushita
Department of Applied Mathematics and Physics,
Kyoto University, Kyoto 606, JAPAN

ABSTRACT. The distribution of perihelion points of long-period comets
is known to cluster towards the solar apex, and some authors ascribe it
to north-south asymmetry in the distribution of observers. Validity or
otherwise of this alleged selection effect is tested by randomly picking up the same number of perihelia in the southern ($\delta < 0$) as those in
the northern ($\delta > 0$) hemisphere. It is shown that the observed clustering cannot be ascribed to the asymmetry of observers. Further, 67
comets which are *new* in Oort's sense are tested similary. The character
of their distribution is similar to that of all the known comets. It
appears difficult to interpret the clustering in terms of a recent
stellar disturbance of the Oort cloud.

1. Introduction
     Whether the comets originated in the primitive solar nebula and
hence represent the most primitive form of the solar system material or
they originated in the interstellar medium is one of the most important
problems in cometary research. The most direct approach to this problem
would be to measure the isotope abundance ratio of one of the comets,
which may be achieved by the GIOTTO mission. However, careful analysis
of the observational data as well as their proper interpretation by
taking into account the dynamical evolution of the cometary orbits can
provide an answer. I have recently (Yabushita 1983) reviewed the processes of dynamical evolution, and it need not be repeated here.
Relevant problems are;
     1. Is the distribution of original 1/a values as calculated by
         Marsden, Sekanina & Everhart (1978) consistent with the assumption
         of steady state Oort cloud ?
     2. How does the planetary perturbation change the distribution of
         1/a values ?
I have shown (Yabushita 1983) that the present cometary population
cannot be regarded as being in a steady state. There is too much
excess of new comets in Oort's sense. Everhart (1979) earlier pointed
out that three out of four *new* comets are not observed, if the cloud is
to be in a steady state. On the other hand, it has been shown that if

11

A. Carusi and G. B. Valsecchi (eds.), Dynamics of Comets: Their Origin and Evolution, 11–17.

comets are injected into the observable region 6 ~ 9 million years ago,
the planetary perturbation can bring the 1/a distribution to what it is
observed now (Yabushita 1979a). Napier & Clube (1979) argue that there
was a comet capture event 10 million years ago, as the solar system
passed through the Gould belt.

        If the cometary capture from interstellar medium is a correct
theory, then there should be peculiarities arising from the solar system
motion relative to the nearby stars.

2. Statistical test.

        It has been known (see for instance, Hasegawa 1976) that the
perihelion points are not distributed randomly, but clustered toward the
solar apex. Let $(\bar{\ell}, \bar{m}, \bar{n})$ be the direction cosines given by

$$\bar{\ell} = \frac{1}{N} \Sigma \cos L_i \cos B_i , \qquad \bar{m} = \frac{1}{N} \Sigma \sin L_i \cos B_i$$

$$\bar{n} = \frac{1}{N} \Sigma \sin B_i$$

where (L_i, B_i) are the ecliptic longitude and latitude of the perihelion
point of the i-th comet. We write (λ, β) to denote the longitude and
latitude specified by $(\bar{\ell}, \bar{m}, \bar{n})$.

        Tyror (1957) obtained $\lambda = 261°$, $\beta = 71°$, while Yabushita (1979b)
obtained $\lambda = 259°.7$, $\beta = 66°$ from Marsden's (1972) catalogue.

        On the other hand, the ecliptic coordinates of the solar apex can
be calculated from the data given by Allen (1976);
        L = 271°.5,   B = 53°.4 .
Thus, it is apparent that the direction of the solar apex is away from
the direction of the clustering of the perihelia by less than 20 degrees,
and the closeness of the two directions have been noted by all of the
authors referred to above. On the assumption that the Oort cloud is
primordial, one would expect a uniform distribution of perihelia over
the sky. The probability of the two directions being separated by less
than 20 degrees is $(1-\cos 20°)/2 \sim 0.03$ on the nul hypothesis of uniform
distribution.

        Some authors (cf. Kresak 1975) argue that since there are more
observers in the northern hemisphere and since comets are bright while
close to perihelia and are more likely to be discovered, the closeness
of the two directions merely reflects the asymmetry of the distribution
of observers over the globe; that β is positive might reflect the
asymmetry of the distribution of the observers.

        In order to investigate if the asymmetry gives any bias concerning
the calculated values of λ and β, the present author (Yabushita 1979b)
proposed the following test.

        A larger number of comets have perihelia in the north of the
ecliptic (B > 0) than in the south. For instance, in Marsden's (1972)
catalogue, there are 307 comets with B > 0, while there are 196 comets
with perihelia in the south of the ecliptic. In order to eliminate the
alleged bias which might arise from the larger number of comets with
B > 0, take equal numbers of comets with $\underline{B \geq 0}$ and those with B < 0, and
calculate the average direction cosines $(\bar{\ell}, \bar{m}, \bar{n})$. The direction so

obtained should be free from the effect which might arise from the inequality in the numbers of comets with B > 0 and those with B < 0. In case of 503 comets contained in Marsden's catalogue (1972), it would be appropriate to take 196 comets among those with B > 0, since there are 196 comets such that B < 0. However, since there are many ways of picking 196 comets among the 307 comets with B > 0, the direction cosines $(\bar{\ell}, \bar{m}, \bar{n})$ are not uniquely calculated. The 196 comets may be randomly chosen among the 307 comets by a Monte Carlo method. One will then have a distribution of $(\bar{\ell}, \bar{m}, \bar{n})$, or a distribution of (λ, β). (see Figs. 5,6,7 of Yabushita 1979b)

If the direction of the solar apex is not far from a region where the calculated points (λ, β) are densely distributed, the alleged effect due to the asymmetry of the distribution of the observers cannot be accepted, and the closeness of the two directions (solar apex and the direction (λ, β)) will be real. In the earlier paper, I have calculated the directions by

$$\bar{\ell} = \frac{1}{2} (-0.1232 + \Sigma \ \ell_i/196), \qquad \bar{m} = \frac{1}{2} (-0.0331 + \Sigma \ m_i/196)$$

$$\bar{n} = \frac{1}{2} (-0.4366 + \Sigma \ n_i/196) \qquad\qquad\qquad\qquad\qquad (2.1)$$

where (ℓ_i, m_i, n_i) are the direction cosines of 196 randomly chosen comets among 307 comets with B > 0, and (−0.1232, −0.0331, −0.4366) are the average direction cosines of 196 comets with B < 0.

Now, the ecliptic is inclined to the equator so that the north-south asymmetry over the globe is not the same as the asymmetry with respect to the ecliptic. So, it will be more appropriate to divide the comets according as $\delta > 0$ or $\delta < 0$, where δ is the declination of a cometary perihelion. The following table gives the average direction cosines of 503 comets in the Marsden catalogue.

Table 1. Numerical values of $(\bar{\ell}, \bar{m}, \bar{n})$ calculated from Marsden's (1972) catalogue of cometary orbits.

N	$\bar{\ell}$	$\bar{m}$	$\bar{n}$	classification
293	0.04534	0.1568	0.4872	$\delta > 0$
210	−0.08896	−0.3643	−0.3481	$\delta < 0$
503	−0.01073	−0.06076	0.1385	$\delta \gtrless 0$
174	0.03151	0.1359	0.5026	$q < 1$ AU $\delta > 0$
134	−0.05956	−0.3374	−0.4012	$q < 1$ AU $\delta < 0$
119	0.06556	0.1875	0.4647	$q > 1$ AU $\delta > 0$
76	0.1408	−0.4119	−0.2546	$q > 1$ AU $\delta < 0$

Figs. 1 ~ 3 show the distribution of (λ, β) which are calculated from equations similar to (2.1). Since a set of (λ, β) is obtained from one random sampling, we get a distribution of (λ, β) and it is possible to judge if the solar motion is related to the clustering direction of perihelion points. Figs. 1 ~ 3 show the distribution of (λ, β) so obtained.

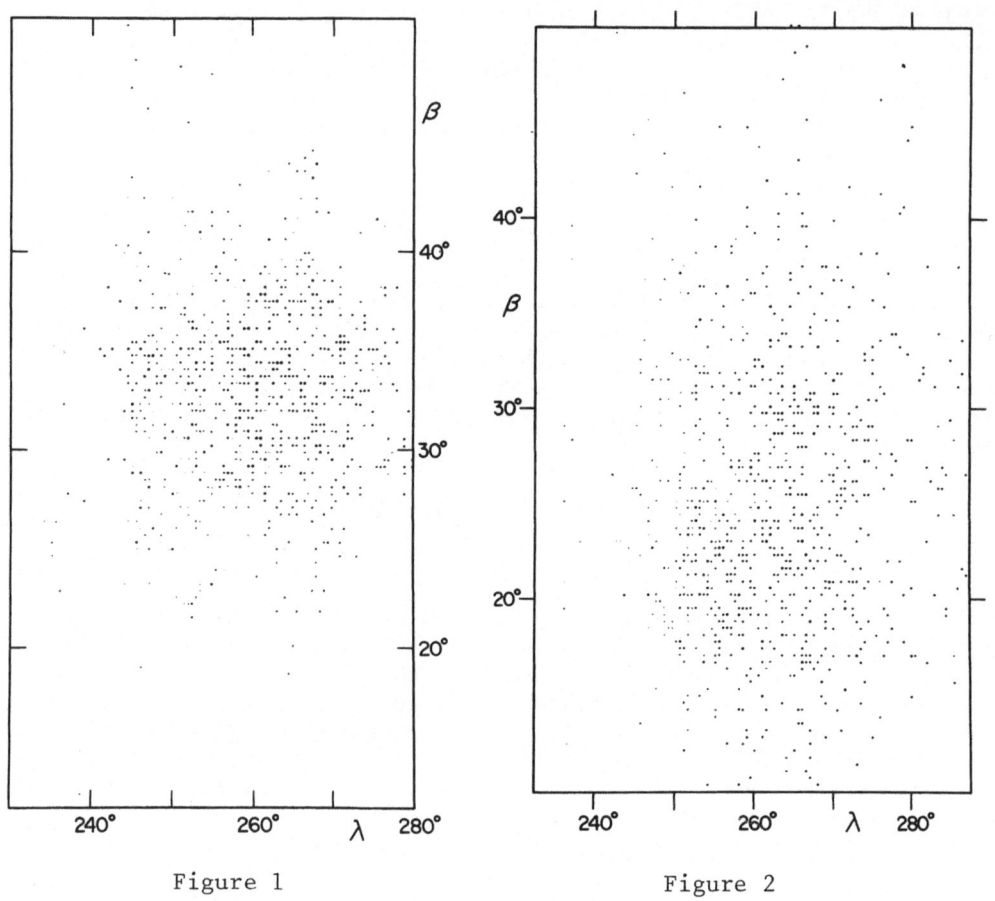

Figure 1   Figure 2

Fig.1. Distribution of directions calculated by adopting the same
        number of comets with perihelion points in the north ($\delta > 0$)
        and those in the south ($\delta < 0$). The comets are not classified
        according to the perihelion distance, q. 1000 random directions
        have been generated.

Fig.2. Distribution of average directions calculated by adopting the
        same number of comets with perihelion points in the north as
        those in the south. Those comets such that q < 1 a.u. are taken
        into account.

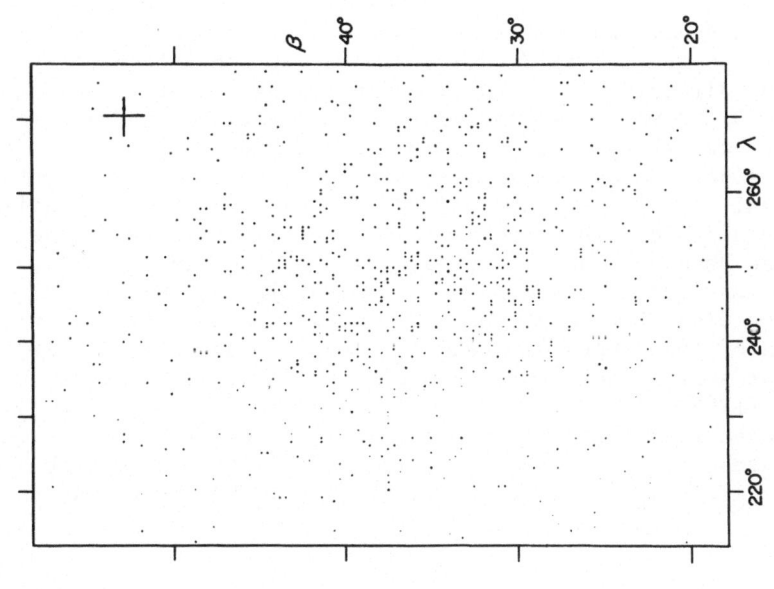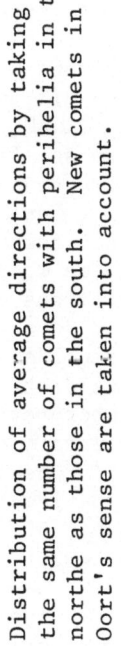

Figure 4

Distribution of average directions by taking the same number of comets with perihelia in the northe as those in the south. New comets in Oort's sense are taken into account.

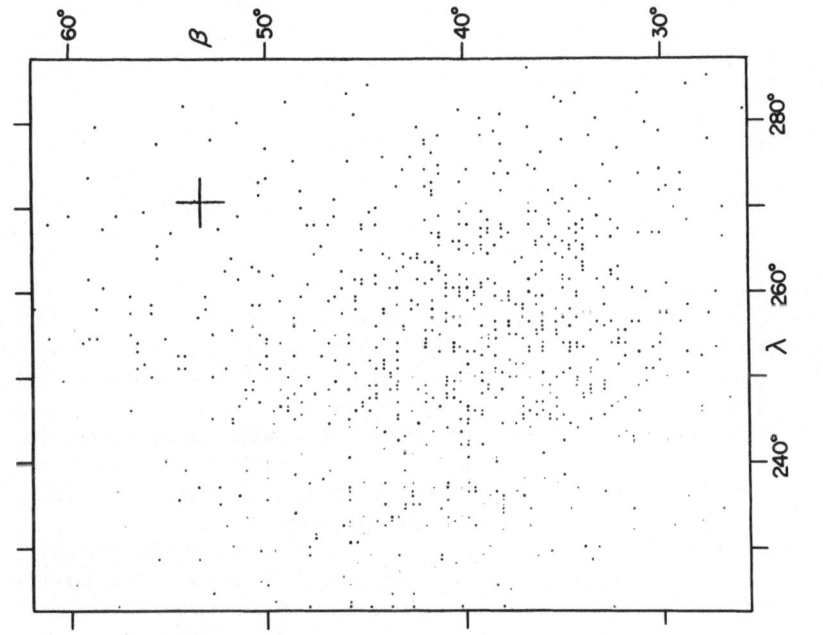

Figure 3

The same as Fig.2, except that comets with q > 1 a.u. are taken into account. The cross denotes the solar apex.

One may note that the direction of the solar apex, (L = 271.5, B = 53.4), is not far from the direction where (λ, β) points are densely distributed. In Fig.1, which include comets with q(perihelion distance) $\geqq$ 1 AU as well as, q < 1, the distance between the two is some 18 degrees or so. In Fig.2, which corresponds to comets such that q < 1 AU, the distance between the two is 25 degrees. On the other hand for q > 1 AU, the two directions are separated by 15 degrees or so, as one can see from Fig.3. Almost the same was obtained earlier, where comets were divided into two groups according to the latitude (B).

Next we consider *new* comets in Oort's sense. There are 67 comets such that original 1/a is less than $5\times10^{-5}\mathrm{AU}^{-1}$ (a $\gtrsim 2\times10^{4}\mathrm{AU}$) including those whose original 1/a values are negative. A test similar to the previous ones has been done and the result is shown in Fig.4. Here, one does not see any marked difference from Figs. 1 ~ 3. We may conclude that the distribution of the *new* comets is almost similar to the distribution of the comets as a whole. Thus, it seems difficult to interpret them as star stracks through the Oort cloud, as suggested by Biermann *et al.* (1983).

One also notes that among the *new* comets, the number of those with q(perihelion distance) greater than 1 AU is greater than those with q < 1 AU. This reflects that they are intrinsically brighter than the others.

Table 2. Average direction cosines of 67 comets such that original 1/a as calculated by Marsden, Sekanina & Everhart is less than $5\times10^{-5}\mathrm{AU}^{-1}$.

N	$\bar{\ell}$	$\bar{m}$	$\bar{n}$	classification
36	−0.01064	0.07554	0.5387	$\delta > 0$
31	−0.1151	−0.3606	−0.3026	$\delta < 0$
67	−0.05898	−0.1262	0.1495	$\delta \gtrless 0$
47				q > 1 AU
20				q < 1 AU

3.   Summary and discussions.

It is known that perihelion points of long-period comets cluster toward the solar apex. Some authors criticize that this could be due to the north-south asymmetry in the distribution of observers over the globe. A statistical test has been made in order to see the validity or otherwise of the criticism. It has been shown that wnen equal numbers of comets with perihelia in the north and in the south of the equator are adopted, they still cluster toward the solar apex. This remains true even if comets are divided into two groups, namely q > 1 AU and q < 1 AU. Further, the same is true with *new* comets in Oort's sense.

That the preferred direction of cometary perihelia is close to the solar apex is difficult to be reconciled with the intra-solar system origin of comets, Biermann *et al.* (1983) regard this trend of the new comets as the result of recent stellar encounters, but the present study shows that the statistical character of the distribution of perihelia of the new comets is not different from that of the rest. Thus, a capture of interstellar comets (by the action of Jupiter, or by direct interaction with the molecular clouds which contain comets (Clube & Napier (1984)) appears to be in better accord with the present study.

References
Allen, C.W., 1976. *Astrophysical Quantities*, Athlone Press, London.
Biermann, L., Heubner, W.F. & Lust, Rh., 1983. *Prof. Natl. Acad. Sci.*, 80, 5151.
Clube, S.V.M. & Napier, W.M., 1984. *Mon. Not. R. astr. Soc.*, 208, 575.
Everhart, E., 1979. In *The Dynamics of the Solar System*, p.273, ed. Duncombe, R.L., Reidel, Dordrecht, Holland.
Hasegawa, I., 1976. *Publ. Astron. Soc. Japan*, 28, 259.
Kresak, L., 1975. *Bull. Astr. Insts. Czech.*, 26, 92.
Marsden, B.G., 1972. *Catalogue of Cometary Orbits*, Smithsonian Astrophysical Observatory, Cambridge, Massachusetts.
Marsden, B.G., Sekanina, Z. & Everhart, E., 1978. *Astr. J.*, 83, 64.
Napier, W.M. & Clube, S.V.M., 1979. *Nature*, 282, 455.
Tyror, J.G., 1957. *Mon. Not. R. astr. Soc.*, 117, 370.
Yabushita, S., 1979a. *Mon. Not. R. astr. Soc.*, 187, 445.
Yabushita, S., 1979b. *Mon. Not. R. astr. Soc.*, 189, 45.
Yabushita, S., 1983. *Quart. J. R. astr. Soc.*, 24, 430.

DISCUSSION

Kresák: Did you also consider the longitude-dependent biases ? For example, the Holetschek effect, when combined with the discovery opportunities varying as a function of solar longitude (length of the night, seasonal variations of cloudiness at the contributing observatories), should make comet discoveries more probable for particular perihelion longitudes.

Yabushita: I agree with you in that such an effect may exist. This problem is now under investigation by I. Hasegawa and myself using a most recent database.

Lüst: 1) I would like to underline the comment just made by Dr. Weissman. It should be interesting to compare the small arc around the apex with the overall distribution on the complete sky. There may be other clusters of comparable density.

2) The results depend on the material. Dr. Yabushita worked with a comet sample different from that of our investigation in which we found a somewhat different result. It further more depends on the exact position adopted for the apex. If we shift that position for some degrees, it is also located in a dense region of perihelia. This is, however, not meaningful because there are about 5 such dense "groups" distributed over the whole sky.

MOLECULAR CLOUDS: COMET FACTORIES?

S.V.M. Clube
Department of Astrophysics,
South Parks Road, Oxford OX1 3RQ

ABSTRACT. Recent discoveries seem to indicate a catastrophic history of terrestrial evolution, explicable in terms of Oort cloud disturbance by molecular clouds in the Galactic disc. The problem of Oort cloud replenishment thus assumes considerable significance and reasons are given for supposing comet exchange takes place during actual penetration of molecular clouds. The number density of comets in molecular clouds, thereby implied, seems to suggest primary condensations of $\lesssim 10^3$km in a dense precursor state of spiral arms. If chemical and/or isotopic signatures of comets should indicate an extra-Solar System source, the theory of terrestrial catastrophism may place new constraints on our understanding of the origin of molecular clouds.

INTRODUCTION

That the Galaxy may influence the Earth is a comparatively old idea. Cycles of some 200-250 Myr and some 30 Myr for example were noted in the terrestrial record at least fifty years ago (e.g. Holmes 1927) and because no earthbound engine operating at these frequencies was known, the possibility of Galactic control became a subject for speculation. Thus, the Sun's motion relative to the underlying Galactic substratum, $[(\Pi_0 - \Pi_G), (\Theta_0 - \Theta_G), (Z_0 - Z_G)] \simeq [10, 15, 7]$ kms^{-1}, happens to combine with the above periods to produce characteristic displacements parallel and perpendicular to the plane of ~ 5000 pc and ~ 200 pc respectively. These are not very different from the separation and thickness of spiral arms, the separation here being a rough average of the radial and tangential values. That something in spiral arms might affect the Earth seemed a reasonable possibility therefore, and it is understandable that speculations should have arisen linking interstellar clouds with ice ages for example, or supernovae with extinctions (McCrea 1981). About ten years ago however, the Galaxy's massive molecular cloud system was detected (Gordon and Burton 1980, Solomon and Sanders 1980) and planetary scientists were meanwhile demonstrating a likely connection between the Apollo asteroid system and the continuous

19

A. Carusi and G. B. Valsecchi (eds.), Dynamics of Comets: Their Origin and Evolution, 19–30.
© *1985 by D. Reidel Publishing Company.*

cratering record (Shoemaker et al 1979). Now, it had already been
recognized that long period comets that were not deflected by
planetary perturbations out of the Solar System altogether may be
the principal source of short period comets and thus, in
devolatilised form, of Apollo asteroids (Opik 1961), so it was
possible that the Oort cloud was being regularly perturbed by
molecular clouds in spiral arms and that corresponding enhancements
in the flux of live and dead comets were directly responsible for
the episodic terrestrial record, controlling both ice-ages and
extinctions, as well as major orogenic cycles and the fluctuating
pattern of magnetic reversals (Clube 1978; Napier and Clube 1979;
cf Clube and Napier 1984a,b and references therein). In adopting
this picture, the assumption that the asteroid belt was a principal
source of Apollos had of course to be abandoned, there being no
known mechanism in this case permitting Galactic periodicities.

Until this time, the Oort cloud had been considered to play a
comparatively passive role in Solar System history, being supposedly
perturbed only rather gently by passing stars, but it had also
become clear that this picture might be in need of revision. Thus,
neither the perihelion distribution of the long-period comet system
nor its orbital energy distribution appeared to be in a relaxed
state; moreover, dynamical simulations of Jupiter's role in
transferring long-period comets to the short-period population
seemed to suggest the latter was too large, indicating perhaps a
recent surge in the long-period comet flux. Admittedly, passing
stars might explain the perihelion distribution (but see Yabushita :
these proceedings), fading might explain the orbital energy
distribution (see Bailey: these proceedings), and an arbitrary
comet cloud beyond Saturn might be invoked to explain the short-
period comets (e.g. Fernandez & Jockers 1983), but there was also
obvious attraction in a proposal that attributed to all three the
same cause. Thus, given (1) the relative proximity of the molecular
clouds in Gould's Belt through which the Sun had just passed, and
(2) the many indications that the Earth is currently immersed in an
extended period ($\gtrsim 3$ Myr) of generally high glacial, magnetic and
orogenic activity, these signs of recent Oort cloud disturbance were
not inconsistent with the general proposition.

The credibility of theories involving comet-induced terrestrial
catastrophism was considerably enhanced by the discovery of excess
iridium at the Cretaceous-Tertiary boundary suggestive of extra-
terrestrial input (Alvarez 1983) but later work showed the
concentration of such material was too large to be explicable in
terms of impact alone (Kyte and Wasson 1984) even though quartz
crystals with impact signatures were evidently present (Bohor et al
1984). In addition, a similar iridium enhanced layer at the Eocene-
Oligocene boundary did not coincide exactly with the microtektite
layer which is thought to be impact induced and to correlate with
the corresponding extinction (Kyte et al 1981); thus, Δt was found
to be ~ 0.03 Myr. If such theories were to be relied upon therefore,
it was apparent that processes of some complexity were at play
involving the direct deposition of cosmic dust as well as impacts.

To meet the facts therefore, we may envisage the following sequence of events: during its journey through the Galaxy, the Solar System experiences successive close passages by molecular clouds, of relatively short duration (~ 0.25 Myr), the effect on the orbit of a typical comet in the Oort cloud each time being that of a simple random impulse (Clube & Napier 1983). Each passage results in a cascade of comets into the Solar System which lasts typically half an orbital period (i.e ~ 2 Myr) thereby replenishing the short and intermediate period comet population from which a small fraction diffuses over the subsequent ~ 5 Myr into orbits of the Apollo-Amor kind (Rickman and Froeschlé 1980). On entering these relatively more stable orbits, the physical evolution and disintegration of each successive comet proceeds rather rapidly, probably with a spell of continuous devolatilisation that is strongest during the first $\sim 10^2 - 10^3$ perihelion passages, followed by a period of more intermittent fragmentation, each producing either continuous or intermittent dust in orbit that feeds into a contemporary zodiacal cloud. Since the mass spectrum of comets (eg Hughes and Daniels 1982) is such that the very largest bodies contain most of the mass, we can expect the evolution of Chiron-like bodies ($\gtrsim 100$ km) arriving at ~ 0.1 Myr intervals to dominate the inner region of the Solar System i.e. the zone bounded by the Jovian orbit. How the evolution proceeds obviously depends on the structure and composition of large comets : 'dirty snowballs' accreted in a cold environment would probably simply disintegrate into dust and gas but if gravitational or radioactive heating occurs during formation, with differentiation we can also expect more solid remnants. The stream of disintegration products (dust, boulders, Apollo asteroids) associated with each of these particular bodies is thus largely responsible for the course of terrestrial evolution. Indeed, each giant comet tends to produce a recurring dust veil on Earth of generally diminishing intensity, resulting probably in a major glaciation followed by a sequence of climatic recessions interspersed with the effects of an occasional asteroidal impact, all within a period of $\sim 0.01 - 0.1$ Myr (Clube and Napier 1984b). The effect of a single molecular cloud passage is thus an episode of glaciations possibly correlated with magnetic reversals (Doake 1978), at $\sim 0.1 - 1.0$ Myr intervals lasting $\gtrsim 5$ Myr during which the largest asteroid encounter can result in a single major extinction. Now, at the present time, the terrestrial planets appear to be swept twice a year by the broad Taurid-Arietid streams of dust (Stohl 1983) and boulders (Dorman et al 1978) produced during past fragmentations of the Comet Encke progenitor (Whipple and Hamid 1952), whilst cosmic dust extracted from the late Pleistocene ice (LaViolette 1983, 1984) and debris scattered by the more recent Tunguska missile (Ganapathy 1983; Golenetskii et al 1981) appear to have remarkably similar, yet anomalous, chemical signatures, iridium in particular being well above the normal terrestrial level (see Figures 1 and 2). The evidence therefore suggests a single disintegrating giant comet which first gave rise to the last ice-age on entry into the sub-Jovian region ~ 0.02 Myr ago, and then later

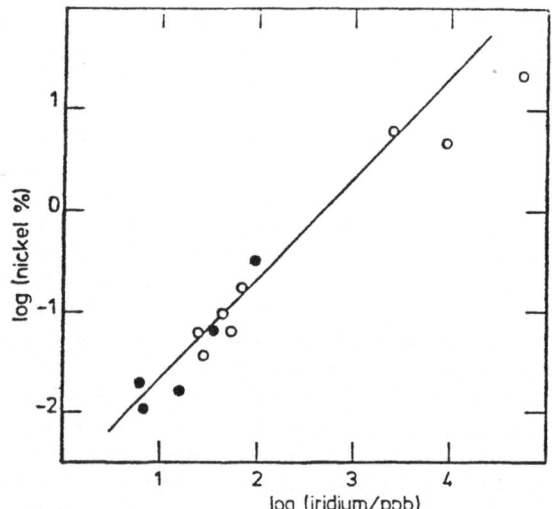

Figure 1
The nickel-iridium ratio for dust in the Camp Century ice-core
covering the period 19,700-14,200 BP (filled circles) and from
microparticles discovered at the site of the 1908 AD Tunguska
explosion. The solid line represents a 'cosmic' abundance ratio but
according to Ganapathy (1983), randomly selected meteor ablation
products show a very large scatter (~1 dex) about this line. The good
fit of both sources to this line seems therefore to imply a single
progenitor body, especially if the common anomalies are also taken
into consideration (see Figure 2).

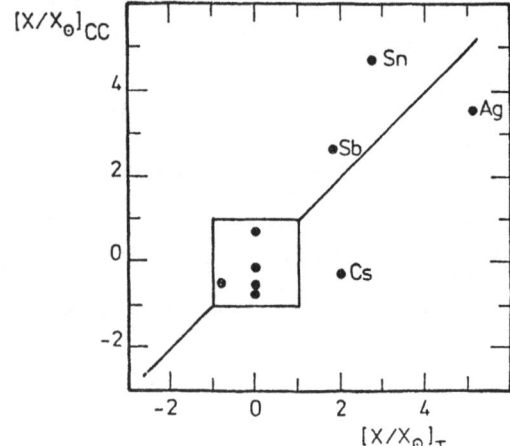

Figure 2
Log-log plot of abundances relative to CI for various elements
detected in Camp Century polar ice and the Tunguska fallout. The
error box centred on 0,0 is expected to take in all elements
displaying normal chondritic abundance while the 45° line is that
corresponding to identical anomalies. There appears to be a common
trend in the anomalies, tin and silver being significantly
overabundant by a factor $\sim 10^4$.

left a devolatilised core which is probably still circulating in the
Taurid stream, possibly the source of the prominent boulder swarm
encountered by the Moon in June 1975 (Dorman et al 1978) as well as
the Tunguska missile and the current zodiacal cloud. The existence
of the devolatilised body remains of course to be verified but it
seems reasonable to assume the recent events and the immediate state
of the sub-Jovian region closely replicate the circumstances
producing the most active periods in the terrestrial record i.e. the
effects that have followed our recent passage through Gould's Belt
simply typify the Solar System's past interaction with molecular
clouds in the Galactic disc.

    Although, on the above chain of argument, molecular clouds are
believed to play a dominant role in Earth history, many aspects of
the thesis remain to be examined in detail and one cannot yet be
certain as to its validity. Nevertheless, it is difficult on
present evidence to see how it can be excluded; past encounters
with dead comets and their debris may therefore be leaving a
significant record of interaction with the Galaxy on Earth which can
have new, far-reaching consequences for our understanding of the
Galaxy. In this paper, the immediate aim is to review some possible
implications as far as the molecular cloud system is concerned.

INNER CLOUD OR CAPTURE?

Although the current long-period and short-period comet systems
together with the terrestrial record through the last few million
years are now taken to characterise transient states of the Solar
System following molecular cloud passages, it is clear in principle
that the $\sim$10-20 most energetic encounters during the lifetime of the
Solar System, possibly involving actual penetration of GMCs, should
have dispersed the Oort cloud altogether (Clube and Napier 1983). It
is necessary therefore to consider how the latter may have been
replenished. Two main suggestions have been made: (1) that there
exists an inner cloud $\sim$10^3 a.u. in size releasing comets into the
Oort cloud by the action of the very same molecular cloud
perturbations that cause the latter to be removed and (2) that
comets are occasionally captured from dense star-forming regions in
the molecular clouds themselves.

    The extent of any postulated inner cloud could reflect
primordial conditions (e.g. Bailey 1983, Hills 1981) or it may be
that it is regularly replenished itself by planetary perturbations
from an even more tightly bound Uranus-Neptune cloud (Shoemaker and
Wolfe 1984). In either case, however, there is a potential
difficulty. Thus, any process replenishing the Oort cloud from
further in is likely to be inefficient (say a few percent) since a
relatively small excess energy imparted by molecular clouds will
take comet orbits not only beyond the Oort cloud but outside the
solar sphere of influence altogether. It follows that a typical
Oort cloud replenishment involving $\sim$10^{11} comets may require $\sim$10^{13}
comets to be drawn from the inner cloud. And if there are $\sim$10
typical penetrations of GMCs during the lifetime of the Solar System

producing major replenishments of the Oort cloud <u>without</u> significant depletion of the parent source (which is necessary to be consistent with the approximately constant cratering rate), then an inner cloud, or Uranus-Neptune cloud, of $\gtrsim 10^{15}$ comets may have to be envisaged. There has as yet been no detailed modelling of this process but with typical cometary masses $\sim 10^{17}$ gm, it is possible that an inner cloud of $\gtrsim 0.1 M_\odot$ may be required. Indeed, if the formative material is drawn from a medium of $\sim$ solar abundance, a primordial Solar System of $\gtrsim 10 M_\odot$ could well be necessary, not in very satisfactory agreement with the rough balance of mass between recently formed stars and the amount of material in molecular clouds.

Although these arguments are not definitive, the inner cloud hypothesis appears to have drawbacks: it is necessary therefore to consider the capture hypothesis (Clube and Napier 1984a). In the past, comet capture from the interstellar medium has usually been regarded as inherently unlikely due to the relatively high velocity (~ 20 kms^{-1}) of the Sun and such limits as are placed on the space density of interstellar comets by the lack of observed hyperbolic candidates travelling through the Solar System. The problem cannot be resolved by simply postulating a high space density of comets in molecular clouds for such comets would, like stars, eventually dissipate into the Galactic disc, and the detection limit for hyperbolic comets also places a constraint on the possible number of comets in previous generations of molecular clouds - always assuming of course that comets survive virtually intact at their expected average distance from radiating sources. However, one possible combination of factors has been overlooked, namely the fluctuating gravitational potential inherent in the hierarchical structure of molecular clouds, and the gradual enlargement of the solar sphere of influence as the Sun climbs out of the potential well associated with a GMC. The result under these circumstances is that a comet entering the solar sphere of influence at above the escape velocity during molecular cloud passage can subsequently experience a random set of perturbations which cause it later to arrive at the enlarged sphere of influence with less than the escape velocity and thus be captured. Quantitatively, this implies that a much larger volume of velocity space relative to the Sun than has been previously considered possible is in fact available for capture. The corollary is that a typical family of $\sim 10^{11}$ comets in orbits like those associated with the Oort cloud can be captured by the Sun during GMC penetrations if a comet population of mean density ~ 0.1 a.u.$^{-3}$ is assumed to be present (Clube and Napier 1984a). There is perhaps order of magnitude uncertainty in this figure depending on the precise motion of the Sun through the GMC and the distribution of material throughout the cloud. Nevertheless, if such a figure be adopted generally for GMCs throughout Galactic history, the implicit number of comets in the disc is then consistent with the observed limit based on the lack of hyperbolic comets in the Solar System (Sekanina 1976). It thus seems that terrestrial history and the current state of the comet population attached to the Sun are in

principle explained if GMCs generally contain comets at the above
space density. Even though such comets will probably be
undetectable in situ, the inference can hardly be disallowed a
priori since GMCs house star-forming regions and comets are usually
assumed to be formed along with stars. The space density is at
first sight a rather inconsequential result, but there are further
possible implications concerning the properties of molecular clouds,
particularly their role as comet factories.

THE HIGH NUMBER DENSITY OF COMETS IN GMCs

The number density implies the observed molecular hydrogen in a GMC
in combination with its comets gives a roughly solar mix, whilst the
amount of high Z material in comets (i.e. $Z>2$) is evidently
comparable to that in stars, even to the extent that most of the
high Z material in GMCs may at some stage reside in comets. An
extremely efficient mode of comet formation seems to be indicated
therefore, but if it occurs more or less randomly throughout
molecular clouds by sedimentation say (McCrea and Williams 1965) or
by the action of differential radiation pressure for example (Hills
1982), some additional (unknown) method of gathering together
preformed planetesimals with the appropriate proportion of gas would
have to be invoked to make stars. In practice, random comets would
be expected to disperse into the Galactic disc, as do stars, thus
comets captured from GMCs seem more likely to have emerged from
preformed dense star-forming regions, consistent with generally
accepted ideas regarding Sun and planet cosmogony. However, the
very large mass of comets escaping from such regions within GMCs
would have to imply compact but nevertheless very loosely bound (i.e
non-virialised) aggregates, not unlike the assemblies of supersonic
floccules proposed by McCrea (1978) as the starting point of star
formation. Under these circumstances, with comets preformed, we
might expect to see element deficiencies relative to solar abundance
in the interstellar medium which increase strongly with hydrogen
column density in star-forming regions, and such an effect is in
fact observed (Cardelli and Bohm-Vitense 1982; Phillips et al 1982;
Mentense 1982; Tarafdar 1983). These large Z depletions are
already a problem for current theories of stellar collapse for if
hydrodynamic or magnetohydrodynamic forces control the gas and
purely ballistic ones control the comets, it is not clear how stars
of roughly normal abundance can arise from a gas that is strongly
depleted of its high Z component. Again, it seems more reasonable
to treat the dense concentrations as dissipating regions with comets
and gas emerging together from a pre-stellar compact state of near
zero energy. The assumption that comets are stellar precursors is
however contrary to the current orthodoxy.
     Both the large Z depletions in the densest regions of molecular
clouds and the possible existence of captured comets seem then to be
indicating a very dense starting point for star formation which is
not initially self-gravitating (i.e. not virialised) and whose

initial condensations are ultimately of cometary dimensions.
Conceivably though, most observed comets are fragments of these
primary condensations, in which case the initial bodies would
correspond to the largest observed comets or their 'volatilised'
counterparts of approximately solar abundance. Such bodies would
have mass $\lesssim 10^{24}$ gm (dimensions $\lesssim 10^3$ km.) and are likely to have
cores that have experienced some degree of differentiation. Thus,
dense non-virialized systems have short survival times and one
expects the primary bodies to have undergone some internal heating
if the final stage of their formation involves relatively rapid
gravitational collapse. Meteoriticists often favour primary bodies
of this character (Grossman and Larimer 1974) and it has been known
for some time of course that there is advantage in making
planetesimals first since this would in principle resolve the
angular momentum problem of Solar System (and star) formation
(McCrea 1978; Woolfson 1979). Moreover, if aggregates of such
bodies are supersonic as well as being dense, as McCrea has
envisaged, we might expect any virialised disc that forms along with
a centrally condensed star will be associated with the escape of a
comparable mass of material which is likely to be most clearly seen
where it is unimpeded by the primordial disc. Roughly symmetrical
bipolar outflows are thus a reasonable expectation, and it may be
significant that bipolar flows are probably associated with the
making of all stars (Lada 1983). Now, it would be premature to
suggest the argument developed here is necessarily unique but there
does seem to be a reasonably supported line of evidence based on the
'observed' number density of comets in GMCs, indicating that the
first stage in making stars might be the formation of $\sim 10^{24}$ gm
bodies in a medium of $\sim$ solar abundance which is necessarily very
dense. Neither the process of coagulation nor the pressure under
which such bodies would form are of course known at present, but if
the trigger is pressure alone, it is even possible the primary
condensations precipitate from the compressed medium as simple
gravitational instabilities. Thus, the picture that seems to emerge
is of the interstellar medium being rather rapidly compressed to
produce the primary condensations, half then combining to produce
stars and planets, whilst the remainder devolatilise to produce
comets and molecular clouds. Any gain in simplicity such a picture
might bring would clearly be offset by the problem of identifying
the apparently non-gravitational process that induces the necessary
compression. Nevertheless, if future studies of cometary material
should indicate that it is indeed captured from outside the Solar
System, significant new insights into the processes leading to the
formation of comets and molecular clouds in spiral arms may be at
hand: understanding terrestrial catastrophism can thus in principle
influence our understanding of the way the Galaxy works.

INTERSTELLAR COMETS?

Cometary material, as distinct from meteorites which have been more
extensively studied, currently deposits on Earth in the form of

micrometeorites. These are irregularly shaped interplanetary dust
particles with dimensions up to a few hundred microns, many in the
form of low density, so-called Brownlee particles composed of
submicron size dust grains embedded together in a fragile, porous
matrix with flat 'plate-like' or 'whisker-like' crystals of larger
size which probably formed by gas-to-particle condensation from a
low-pressure hot vapour phase (Bradley et al 1983). Included also
in the extra-terrestrial input are small spherical bodies which seem
to be micrometeorites that have been heated to a molten state before
entering the Earth's atmosphere (Parkin et al 1977): they have been
detected in the stratosphere, ocean sediments, glacial ice, lunar
soil and as 'chondrules' in chondritic meteorites. Besides
exhibiting normal chondritic elemental signatures, many of these
microspheres have highly anomalous compositions. There are
instances for example where calcium, aluminium and titanium are
exceptionally concentrated but in other cases, other elements may be
greatly enriched. In general then, the micrometeoritic material
deposited on the Earth during the last 20,000 years may be
attributed to a relatively steady flux of disintegrating comets that
underwent various degrees of differentiation during formation.

Many of the centimetre-sized primitive chondritic inclusions in
the Allende meteorite are also rich in Ca-Al-Ti and it has been
suggested that such excesses may have a similar explanation to that
of micrometeorites (LaViolette 1983). So far as the Allende
meteorite is concerned, the inclusions are thought also to contain
material preserved from the earliest stages of Solar System
condensation, revealing contamination from a (possibly nearby)
exotic source such as a supernova (Wasserburg and Papanastassiou
1982; however cf Clayton 1984). Since comets are generally
supposed to be among the most primitive of bodies, it would perhaps
not be unexpected if they too displayed contamination of a similar
kind. The cosmic dust deposited in glacial ice does indeed carry
plate-like particles of anomalous composition which physically
resemble the (enstatite) crystal platelets found in Brownlee
particles (LaViolette 1983). The possibility arises therefore that
different comets bear different signatures reflecting their
particular condensation and contamination histories. It follows
that material from comets from inside or outside the Solar System
will not necessarily display identical chemical signatures. The
similarly tin-rich chondritic dust discovered in abundance in the
Camp Century late Pleistocene ice and in the Tunguska fallout is
thus probably indicative of a single common source (ie a giant
comet) which may or may not be from outside the Solar System.

It has been shown by Kyte et al (1983) that rather similar
considerations apply to cosmic material deposited at the Cretaceous-
Tertiary boundary. The abundance pattern in this case resembles a
Solar System source very closely but it does not fit known meteorite
types: thus, there is significant enrichment of Os, Re and other
elements relative to chondritic meteorites or irons. Isotope
anomalies moreover have been detected in the material at this
boundary, such as $^{187}Os/^{186}Os$ (Luck and Turekian 1983) and $^{12}C/^{13}C$

(Bontê et al 1984) and these are at least suggestive of a non-Solar
System source. Whilst the origin of the chemical and isotopic
anomalies in cometary material has not been settled, and until
suitable isotope age indicators have been developed, one cannot say
with certainty whether these data imply a non-Solar System source
for (live or dead) comets striking the Earth. Nevertheless, the
balance of evidence may now be favouring an interstellar origin.

CONCLUSIONS

It is possible the discovery of molecular clouds in the Galaxy,
combined with recent cometary studies and the terrestrial record,
leads now to a very specific picture of star formation in which the
primary process is one of very rapid compression of the interstellar
medium. Spiral arms initially composed of comets are thus implied
which, contrary to a common view, are themselves the dissipating
source of molecular clouds. Molecular clouds are then unlikely to
be comet factories.

REFERENCES

Alvarez, L.W. 1983 Proc. Natl. Acad. Sci USA 80, 627.
Bailey, M.E. 1983 Mon. Not. R. astr. Soc. 202, 603.
Bohor, B.F., Foord, E.E., Modreski, P.J. & Triplehorn, D.M. 1984
    Science 224, 867.
Bontê, Ph., Delacotte, O., Renard, M., Laj, C., Boclet, D.,
    Jehanno, C., & Rocchia, R. 1984 Geophys. Res. Lettr. 11, 473.
Bradley, J.P., Brownlee, D.E. & Veblen, D.R. 1983. Nature 301, 473.
Cardelli, J. & Böhm-Vitense, E. 1982. Astrophys J. 262, 213.
Clayton, D.D. 1984 Astrophys. J. 280, 144.
Clube, S.V.M. 1978 Vistas in Astronomy 22, 77.
Clube, S.V.M. & Napier, W.M. 1983 Highlights of Astronomy 6, 355.
Clube, S.V.M. & Napier, W.M. 1984a. Mon.Not.R.astr.Soc. 208, 575.
Clube, S.V.M. & Napier, W.M. 1984b. Mon. Not. R. astr. Soc.
    in press.
Doake, C.S.M. 1978 Earth Planet. Sci. Lett. 38, 313.
Dorman, J., Evans, S., Nakamura, Y. & Latham, G. 1978 Proc. Lunar
Planet Sci Conf 9, 3615.

Fernandez, J.A. & Jockers, K.   1983   Rep. Prog. Phys. 46, 665.
Ganapathy, R. 1983   Science 220, 1158.
Golenetskii, S.P., Stepanok, V.V. & Murashov, D.A. 1982
   Translated from Astr. Vestu. 15, 167, 1981.
Gordon, M.A. and Burton, W.B. 1980   Proc. Third Gregynog
   Astrophysics Workshop on Giant Molecular Clouds (ed. Solomon
   et al) 25.
Grossman, L. & Larimer, L. 1974   Rev. Geophys. Space Phys. 12, 71.
Hills, G.J. 1981   Astron. J. 86, 1730.
Hills, G.J. 1982   Astron. J. 87, 906.
Holmes, A. 1927   The age of the Earth - An Introduction to
Geological
   Ideas (Benn, London).
Hughes, D.W. & Daniels, P.A. 1982 Mon. Not. R. astr. Soc. 198, 573.
Kyte, F.T. & Zhou, Z., 1980   Meteoritics 15, 320.
Kyte, F.T., Zhou, Z. & Wasson, J.T. 1981   Nature 292, 417.
Kyte, F.T. & Wasson, J.T. 1984   Meteoritics 19, 332.
Lada, C.J. 1983   Star Formation Workshop (Royal Observatory,
   Edinburgh), in press.
LaViolette, P. 1983   Thesis, Portland State University.
LaViolette, P. 1984   Evidence of High Cosmic Dust Concentrations
   in Late Pleistocene Polar Ice (20,000-14,000 years B.P.),
submitted.
Luck, J.M. & Turekian, K.K. 1983   Science 219, 613.
McCrea W.H. 1978 In 'The Origin of the Solar System' (ed. Dermott),
   75.
McCrea, W.H. 1981   Proc. R. Soc. London A375, 1.
McCrea, W.H. & Williams, I.P. 1965   Proc. R. Soc. London A287, 143.
Mentense, H.H. 1982   Astrophys Space Sci 82, 173.
Napier, W.M. & Clube, S.V.M. 1979   Nature 282, 455.
Opik, E.J. 1961   Adv. Astro. Astrophys. 2, 219.
Parkin, D.W., Sullivan, R.A.L. & Andrews, J.N. 1977 Nature 266, 515.
Phillips, A.P., Gondhalekar, P.M. & Pettini, M. 1982   Mon. Not. R.
   astr. Soc. 200, 687.
Rickman, H. & Froeschlé, C. 1980   Moon and Planets 22, 125.
Sekanina, Z.   1976   Icarus 27, 123.
Shoemaker, E.M., Williams, J.G. Helin, E.F. and Wolfe, R.F. 1979
   In Asteroids (ed: Gehrels), 253.
Shoemaker, E.M. & Wolfe, R.F. 1984   Meteoritical Society,
   abstract submitted.
Solomon, P.M. & Sanders, S.B. 1980   In Proc. Third Gregynog
   Astrophysics Workshop on Giant Molecular Clouds (ed: Solomon), 41.
Stohl, J. 1983   In Asteroids, Comets, Meteors (ed: Lagerkvist &
   Richman, Uppsala), 419.
Tarafdar, S.P., Prasad, S.S. & Huntress, W.T. 1983   Astrophys. J.
   267, 156.
Wasserburg, G.J. & Papanastassiou, D.A. 1982   In 'Essays in Nuclear
   Astrophysics" (ed: Barnes), 77.
Whipple, F.L. & Hamid, S.E. 1952   Hellwan Obs. Bull. 41, 1.
Woolfson, M.N. 1979   Quart. J.R. astr. Soc. 20, 97.

DISCUSSION

Yabushita: You mention Jeans' gravitational instability as a
    mechanism of cometary formation. I think sedimentation is a
    more natural process.
Clube: You may be correct though on the picture I have been
    discussing, the formation of regions of high density would
    still be expected to precede the sedimentation.
Fernandez: For a capture event ~5 Myr ago, you might expect an
    increase of the comet bombardment of the terrestrial planets
    during the last few million years. Is there any evidence in
    the cratering record of the Earth in support of this view?
Clube: Precipitation may cause large craters formed more than about
    1 Myr ago to be deeply eroded or filled by sediments whilst in
    areas of glaciation, they may be completely obliterated.
    Detailed geological mapping of the Earth's surface is also
    incomplete, so it is unlikely that all young continental impact
    structures greater than, say, 18 km have been recognized
    (Shoemaker, Ann. Rev. Earth Planet Sci. 11, 461, 1983).
    Nevertheless, based on the current asteroid population, which
    the theory requires to be on the order of the time-averaged
    population through the Phanerozoic, there is a 99% probability
    of at least one 18 km crater within the last 3.5 Myr. One such
    crater is known (Lake Elgygytgyn, Eastern Siberia) with a K/Ar
    age of 3.5±0.5 Myr.
Delsemme: The periodicity of mass extinctions of species on Earth
    (~30 Myr) does not seem to fit in with the periodicity of
    molecular cloud collisions with the Solar System. What do we
    know about molecular cloud statistics?
Clube: Although a very unlikely companion star hypothesis has been
    put forward, the ~30 Myr cycle probably relates to the Sun's
    galactic z-motion and implies a rather high frequency of
    molecular cloud encounters. Our knowlege of the statistics and
    the Sun's past motion is not sufficient at present to exclude
    the z-motion hypothesis.

DYNAMICAL INTERACTIONS OF THE SOLAR SYSTEM WITH MASSIVE NEBULAE

W.M. Napier,
Royal Observatory, Edinburgh, U.K.

ABSTRACT

The effects of encounters with massive nebulae on the long-period comet population are examined, paying particular attention to the uncertainties in the data. An earlier conclusion, that the long-period comet system is dynamically unstable, is upheld. Whether replenishment by unbinding from a dense inner comet cloud is a viable hypothesis awaits detailed modelling, but a qualitative discussion is given which argues tentatively against it. If comets occur in molecular clouds, however, their capture into temporarily bound Solar System orbits is a natural consequence of close encounters for realistic velocities and potentials. A large disturbance or capture may have occurred a few Myr ago as the Sun emerged from the Orion spiral arm.

INTRODUCTION

In a series of papers (Napier & Clube 1979; Napier & Staniucha 1982; Clube & Napier 1982a, 1983, 1984a; Bailey 1983) it has been argued that the long-period comet system is likely to be dynamically unstable, because of the tidal action of molecular clouds, and must therefore be replenished from elsewhere. Such replenishment might come from a hypothetical dense inner cloud of comets, by capture of comets from the molecular clouds themselves, or from some combination. It has been stated (but not so far demonstrated: Weissman 1983, Dr. Fernandez, these proceedings) that the properties of molecular clouds are so uncertain that the issue of Oort cloud stability cannot be advanced beyond the speculative stage; therefore in this contribution I examine the question taking account as far as possible of the assumptions and uncertainties in the data. In particular the problem is formulated in such a way as to bring out the dependence on such factors as the adopted mass distribution of the nebulae, the local column density, gravitational focussing and so on.

A. Carusi and G. B. Valsecchi (eds.), Dynamics of Comets: Their Origin and Evolution, 31–41.
© *1985 by D. Reidel Publishing Company.*

ADOPTED PROPERTIES OF THE MOLECULAR CLOUD SYSTEM

Early surveys of the CO emission in the galactic plane (Solomon &
Sanders 1975, Gordon & Burton 1976) led to the discovery that much
of the mass of the interstellar medium is in the form of molecular
hydrogen concentrated into cold, massive nebulae. Initially there
was much uncertainty about the mass and structure of the system,
arising partly from uncertainties in the scaling factor between CO
and H_2, partly from crowding effects in the inner regions of the
Galaxy, and partly because of ambiguities in the kinematic distances
of the nebulae. Thus while Solomon, Sanders & Scoville (S^3 1979)
considered that most of the CO emission derived from giant molecular
clouds (GMCs) of radii ∽20 pc and masses ∽5 x 10^5 M_Θ, Gordon &
Burton (1980) modelled the CO emission by a more numerous population
of clouds with masses ⪴ a few $10^4 M_\Theta$ and radii ∽2 - 10 pc. However
the latter authors found a striking tendency for these clouds to
occur in clusters, and the GMCs of S^3 show considerable
substructure. It appears now that the difference was semantic
(Liszt et al. 1981) and that there is essentially agreement between
all groups that most of the mass and volume of H_2 does reside in
very large sources.

More recently, CO observations covering galactic longitude -4° to
170° have been carried out by S^3 (1984), who also discuss the
conversion factor between integrated CO intensity W_{co} and the mass
column density $N(H_2)$ of molecular hydrogen. They adopt $N(H_2)/W_{co}$ =
3.6 x 10^{20} cm^{-2}/(K cm s^{-1}) and point out that all values now in use
are within a factor 2 of this. Empirical measures of the factor
yield ∽4±2 x 10^{20} cm^{-2}/(K cm s^{-1}). From this it is found that
within 16 kpc of the galactic centre, the total mass of H_2 is 3.5 x
10^9 M_Θ. Their study confirms that the emission is concentrated into
GMCs with diameters 20-80 pc and masses 10^5 - 3 x 10^6 M_Θ. They find
also that at the solar distance the molecular cloud surface density
is 5.2 M_Θ pc^{-2} measured in a column perpendicular to the galactic
plane. The scale height of the system appears to be $Z_{1/2}$ ∽75 pc as
measured by CO emission, or ∽50-60 pc if it is delineated by OB
associations.

A mass spectrum has been derived for molecular clouds, of the form

$$n(m) = km^{-\alpha} \tag{1}$$

where n(m) dm represents the number of molecular clouds per cubic
kiloparsec in the mass range (m,m+dm) solar masses. The population
index α = 1.45±0.08 according to Thaddeus & Dame (1983), and that
found by Xiang et al. (1984) can be fitted by α = 1.65±0.2 or so.
The distribution applies at least to masses ⪴ few 10^3 M_Θ, and is
such that the bulk of the mass resides in the few largest clouds.
Bhatt et al (1984), in a discussion of Lynds dark nebulae in the
general field and the Taurus cloud complex, find α = 1.5±0.15, but
it is interesting that within the Orion and ρ Ophiuchus complexes

the index is near unity. Drapatz & Zinnecker (1984) find a somewhat more complex mass distribution which, however, is in general accord with the power law spectrum for masses $\gtrsim 3 \times 10^3$ M_Θ.

According to Xiang et al (1984) molecular cloud radii are in the range $8 \lesssim R \lesssim 30$ pc, with mean ~ 13 pc. However the maximum contribution to the mass of the system comes from clouds with a mass of 8×10^5 M_Θ whose radii are 20 pc.

Whether molecular clouds delineate spiral structure in the inner Galaxy is a controversial question but not crucial to the present analysis. Towards the outer Galaxy, where individual clouds are more easily defined, the Local, Perseus and Sagittarius arms are delineated by molecular clouds (Thaddeus & Dame 1984), Cohen et al (1980) finding that the arm/interarm contrast is $\gtrsim 5:1$ around the Perseus Arm. It is suggested (Drapatz & Zinnecker 1984) that clouds with $R < 15$ pc occur in arm and interarm regions, larger ones outlining, more or less, spiral structure.

PERTURBATION OF THE OORT CLOUD

It is convenient to divide encounters of the Solar System with molecular clouds into those with impact parameters $p > 20$ pc and those with $p < 20$ pc, referring to the latter as 'close encounters'. Many of the passages with $p > 20$ pc will be 'flybys', whereas many close encounters will involve actual penetration of GMCs. Assuming the mean peculiar velocity of the Sun over its history has been $V = 20$ km s^{-1}, the volume swept out within 20 pc of the Sun over 4500 Myr is ~ 0.11 kpc^3, whence the number of close encounters with GMCs having masses $\geqslant m$ is

$$N(\geqslant m) = 0.11 \, \nu(\geqslant m) \qquad (2)$$

$\nu(\geqslant m)$ representing the number density of molecular clouds with masses $\geqslant m$ solar masses. It is measured in numbers kpc^{-3} and is given by

$$\nu(\geqslant m) = (\frac{2-\alpha}{\alpha-1}) \; (f^{\alpha-1}-1) \; \frac{\rho}{m_2} \qquad (3)$$

defining $f = m_2/m$. It is assumed that the mass spectrum (1) applies between lower and upper limits (m_1, m_2) respectively. ρ (M_Θ kpc^{-3}) is the density of molecular cloud material at the solar distance. Adopting $\alpha = 1.5$, we have that

$$\nu(\geqslant m) = \frac{\rho}{m_2} \; (f^{1/2} - 1) \qquad (4)$$

If instead of a finite mass distribution it is assumed that all GMCs are of the same mass $\bar{m}$, one finds

$$\nu = \rho / \bar{m} \qquad (5)$$

The density $\rho \, (M_\Theta \, kpc^{-3})$ can be found in terms of the surface density $s \, (M_\Theta \, pc^{-2})$ and total effective disc thickness $h(pc)$ through

$$\rho \;=\; 10^9 \;\; \frac{s}{h} \tag{6}$$

Hence for $s = 5 \, M_\Theta \, pc^{-2}$, $h = 100$ pc, $m_2 = 3 \times 10^6 \, M_\Theta$, one finds $\rho = 5 \times 10^7 \, M_\Theta \, kpc^{-3}$ and

$$\nu \; (\geqslant m) \;=\; 16.7 \;\; (f^{1/2} - 1) \tag{7}$$

which, with (2), yields

$$N(\geqslant m) \;=\; 1.8 \; (f^{1/2} - 1) \tag{8}$$

With these formulae, one find that there are

$\nu = 511 \, kpc^{-3}$ clouds of mass $m_1 \geqslant 3 \times 10^3 \, M_\Theta$ yielding $N = 56$
      close encounters,
  75 kpc^{-3} clouds of mass $m_1 \geqslant 10^5 \, M_\Theta$ yielding 8.2
      close encounters, and
  24 kpc^{-3} clouds of mass $m_1 \geqslant 5 \times 10^5 \, M_\Theta$ yielding 2.7
      close encounters.

Gravitational focussing increases these encounter rates at the massive end of the distribution, the ratio of gravitational to geometric target areas being

$$r \;=\; 1 + (\frac{V_e^{\,2}}{V}) \tag{9}$$

where V_e represents the escape velocity from the surface of the GMC of radius R, and we take $V = 20 \, kms^{-1}$. Thus for a GMC of mass 10^6 M_Θ and radius 20 pc, $V_e = 31$ km s^{-1} and $r = 3.2$. For radius 40 pc, $V_e = 22 \, kms^{-1}$ and $r = 2.2$. The number of close encounters with massive GMCs (say $\geqslant 5 \times 10^5 \, M_\Theta$) is therefore at least doubled to ~ 5. A 'best estimate' for the number of close encounters with GMCs of mass $\geqslant 10^5 \, M_\Theta$ then becomes ~ 10, half of which are with GMCs of mass $\geqslant 5 \times 10^5 M_\Theta$. It seems unlikely that the encounter rate would be less than half of this, on present assumptions.

Bailey (1983) has pointed out that encounter rates are lowered by a factor $\sim 1.5-2$ when a simple slab model for the vertical distribution of GMCs is replaced by an exponential one; on the other hand stellar kinematic evidence shows that the Galaxy has been 'rougher' in the past, by a factor up to ~ 10 (Lacey 1984), whence the mean number density of GMCs over the history of the Solar System has probably been about double the current value. I shall assume that these factors cancel out.

Consider first only encounters with $p > p_1 = 20$ pc, out to a large distance p_2 pc, and assume that these are all flyby. Each passage produces a velocity change of a comet relative to the Sun given, on the impulse approximation, by

$$\delta V = \frac{2Gm}{pV} \frac{d}{p} \tag{10}$$

G the gravitational constant and d the Sun-comet projected distance. In time t the Solar System interacts with $n(m)dm \times Vt \times 2\pi$ $(1+[\frac{2G}{pV^2}]m)$ pdp nebulae in the range (m,m+dm) and (p,p+dp), where the enhancement of rate due to gravitational focussing has been allowed for. Assuming the velocity increments add randomly, the r.m.s. velocity σ_c induced in comets by passages >20pc is given by

$$\sigma_c^2 = 2\pi Vt \int_{p_1}^{p_2} \int_{m_1}^{m_2} (\frac{2Gm}{pV} \frac{d}{p})^2 (1 + [\frac{2G}{pV^2}] m) n(m)dp \, dm \tag{11}$$

or, with (1),

$$\sigma_c^2 = \sigma_{nf}^2 + \sigma_f^2 \tag{12}$$

where

$$\sigma_{nf}^2 = 4\pi \frac{G^2}{V} t (\frac{d}{p_1})^2 (\frac{2-\alpha}{3-\alpha}) m_2 \rho \tag{13}$$

and

$$\sigma_f^2 = \frac{16\pi}{3} \frac{G^3}{V^3} t (\frac{d}{p_1})^2 (\frac{2-\alpha}{4-\alpha}) \frac{m_2^2 \rho}{p_1} \tag{14}$$

(σ_{nf}, σ_f) representing the dispersion components due to the non-focussing and focussing components respectively of (11). Their ratio is

$$\frac{\sigma_f}{\sigma_{nf}} = \left[\frac{4}{3} \frac{G}{V^2} (\frac{3-\alpha}{4-\alpha}) \frac{m_2}{p_1} \right]^{1/2} \tag{15}$$

which, for $V = 20$ km s^{-1}, $\alpha = 1.5$, $m_2 = 3 \times 10^6$ $M_\odot$, $p_1 = 20$ pc, is 1.15, yielding $\sigma_c = 1.53 \sigma_{nf}$.

Assuming d = 20,000 a.u. and t = 4.5 Byr in (13), one finds

$$\sigma_{nf}^2 = 4.8 \times 10^{-6} m_2 \rho \text{ (cm s}^{-1})^2$$

which, with $m_2 = 3 \times 10^6$ $M_\odot$ and $\rho = 5 \times 10^7$ $M_\odot$ kpc^{-3} as before, gives

$$\sigma_{nf} = 0.27 \text{ kms}^{-1}$$

and

$$\sigma_c = 0.41 \text{ kms}^{-1}$$

It is clear from (13) that the bulk of the power input to the Oort
cloud comes from the closest encounters.   Within 20 pc, we may
assume that the most significant close encounters will involve
actual passage of the Solar System through a GMC.   It may be shown
(CN 1983, Bailey 1983) that the cumulative effect of penetrating
encounters contributes 1.6 times as much energy as the non-focussing
component of flyby encounters.   Thus the velocity dispersion induced
by the penetrating encounters is roughly $\sigma_p = 0.27\sqrt{1.6} = 0.34$ kms^{-1}
and the overall effect of the molecular clouds is to introduce a
dispersion

$$\sigma = (\sigma_p^2 + \sigma_c^2)^{1/2}$$

or

$$\sigma = 0.53 \text{ km s}^{-1}$$

If the r.m.s. velocity induced by stars (~ 0.14 km s^{-1}) is added,
this increases to ~ 0.55 km s^{-1}: stellar perturbations are a minor
contributor to Oort cloud disturbance.   The escape velocity at
40,000 a.u. being ~ 0.2 km s^{-1}, the energy injected into the long-
period comets is about an order of magnitude in excess of that
required to disperse it even assuming it has no initial kinetic
energy.

Two important considerations have been omitted from the analysis.
First, the energy input is a random walk, but with an outward drift:
there is a systematic unbinding imposed on the random energy
changes.   This systematic component is comparable to the random one
already discussed (CN 1982a).   Second, real GMCs are not homogeneous
spheres but are highly structured objects comprising many clouds
within them;   a typical filling factor is ~ 0.05.   It is readily
shown, by application of these formulae suitably modified, that for
reasonable mass distributions of internal structure subject to the
filling factor = 0.05 constraint,the effect of clumpy internal
structure is generally larger than that of the GMC as a whole.   And
probably substantially larger: e.g. in the molecular cloud complex
towards M17 about a third of the mass is in four fragments with
radii 3-6 pc (Elmegreen & Lada 1976);   but a high resolution study
of one fragment (M17SW) of mass 20,000 M_θ reveals it to comprise
substructure of masses $\sim 50 M_\theta$ and radii $\sim 16,000$ a.u.: Martin et
al.(1984).   A single penetrating encounter with such a nebula would
probably remove comets $\gtrsim 10^3$ a.u. from the Sun, and there have
probably been ~ 5-10 such encounters.

There are uncertainties, eg. the impulse approximation may be very
poor for GMCs, and the past history of the solar orbit is uncertain.
Nevertheless it seems very likely that the long-period comet system
must be replenished from some other reservoir. The same conclusion
may be reached from numerical work (e.g. Napier & Staniucha 1982),
or semi-empirically from the distribution of semi-major axes of
binaries (Napier & Staniucha, unpublished), or from the observed
kinematic heating of stars in the disc.

THE OORT CLOUD: AN OPEN OR CLOSED SYSTEM?

The passage of the Solar System through a large GMC is likely to cause the ejection of comets orbiting more than a few 10^3 a.u. from the Sun and possibly even less.   There may have been ~10 such penetrations.   In addition the more numerous flyby encounters themselves substantially energise the Oort cloud.   Evidently replenishment must be taking place, and two conceivable reservoirs are a dense, compact inner cloud or the nebulae themselves.   There appear to be no <u>prima facie</u> astrophysical objections to the occurrence of comets in either source.   The question of provenance then reduces to that of mass and depletion timescale, which in turn depends on the celestial mechanics of unbinding or capture.

It now seems from the ~30 Myr periodicity in the cratering record (Seyfert & Sirkin 1979; Rampino & Stothers 1984; Alvarez & Muller 1984) that comets arriving from a disturbed Oort cloud are the major source of large impact cratering in the inner Solar System.   (If asteroids perturbed from the main belt were the main source of large craters, it is difficult to see how such sharp, quasi-periodic impact episodes would result.   Further the occurrence of such episodes was <u>predicted</u> on the basis of periodic comet disturbance by molecular clouds: NC 1979, CN 1982b, 1984b).   However the observed lunar cratering rate shows no sign of having decayed over the past ~3 Byr, whence the comet reservoir must have a timescale long compared with this.   The half-life of comets orbiting in the Uranus-Neptune region is ~1 Byr, and so a massive comet cloud in the region of these planets would appear to be ruled out.   One might have a cloud just beyond this, with orbits of half-lives say $\gtrsim$5 or 10 Byr (and therefore formed <u>in situ</u>), receiving just enough energisation to feed its members slowly into a zone where molecular cloud perturbations begin to take over.   But in that case, because of the dominance of a few large discrete energy inputs, when a comet does begin to move outwards it is more likely to be ejected than attain zero energy, the efficiency of transfer into the long-period system being perhaps ~1-10%.   Thus the replenishment of a cloud of ~10^{11}-10^{12} comets requires the unbinding of ~10^{12}-10^{14} which, with ~10 such replenishments in the history of the Solar System, implies that 10^{13}-10^{15} comets must have been thrown out altogether from the inner reservoir.   To achieve this without a discernible cratering decline probably implies an inner reservoir population at least an order of magnitude greater than this, say ~10^{14}-10^{16} comets.   For a mean comet mass $\gtrsim 10^{17}$ gm this implies an inner cloud mass $\gtrsim 10^{31}$-10^{33} gm, or $\gtrsim$1.7 10^3 - 1.7 x 10^5 times the mass of the Earth!   There are severe astrophysical objections to the existence of so massive an inner cloud not least of which is that the energy transferred to it by Uranus and Neptune would be substantially greater than the orbital energy of these planets.   For comparison the IR emission around Vega has been interpreted by Weissman (1984) as a comet cloud of radius 85 a.u. and mass 15 $M_{\oplus}$ (~10^{12} comets).

The constancy of the cratering record, in spite of the frequency with which the Oort cloud is emptied, indicates that a very large reservoir is being tapped, and the Galaxy would appear par excellence a natural place to look for it (Clube 1978; NC 1979). The molecular cloud system may be the specific reservoir we seek, since star formation occurs in the dense cores of molecular clouds, and comet formation is likely to be an adjunct to the process. The depletion of heavy elements in the denser regions of molecular clouds is consistent with a comet number density, averaged over a GMC, of $\nu_c \sim 10^{-1\pm1.5}$ a.u.$^{-3}$ (CN 1984a; Dr Clube, these proceedings; CN and Napier & Humphries, submitted). Allowing for filling factors and density contrasts, substructures with $\nu_c > 10-10^2$ a.u.$^{-3}$ are expected to be common. The question then is, would an Oort cloud of the 'observed' population and dimensions be captured during passage through a GMC? To retain the cloud, capture would have to occur just as the Solar System was climbing out of the GMC.

Classically, the problem of capturing comets has been seen as a severe one for an interstellar cosmogony, Whipple (1975), Noerdlinger (1977), Sekanina (1976) and others finding that, to capture an Oort cloud, the Sun has to be co-moving ($\lesssim 2$ kms^{-1}) with a comet cloud of very small velocity dispersion ($\lesssim 1$ kms^{-1}). However in such studies the ambient potential was taken to be static, the Sun capturing only those comets which crossed a fixed sphere of influence (radius $\sim 40,000$ a.u.) at $\lesssim 0.2$ kms^{-1}. The real situation in a GMC involves a dynamic, fluctuating potential with a solar sphere of influence which, on climbing out of a GMC potential well, may expand from $\sim 10^4$ a.u. to $\sim 10^5$ a.u. at a rate of ~ 1 kms^{-1} (the Sun moving 20 pc in 1 My, say). The problem of capture in these circumstances is discussed in CN (1984a), but a simple argument is presented here to illustrate the principles.

Neglecting factors of order unity, the number of comets entering the sphere of influence of radius r_0 in a time t at less than the escape velocity v_e is given by $N_c \sim v_e t \times 4\pi r_0^2 \times f\nu_c$, where ν_c represents the number density of comets and a fraction f of them are co-moving with the Sun to within $\pm v_e$ kms^{-1}. This fraction is given approximately by $f \sim (h/\sqrt{\pi})^3 \exp(-h^2 v_0^2) d^3u$ with $du \sim 2v_e$ and $h^2 = 1/2\sigma^2$. Assume for example that the Sun, in climbing out of a potential well, passes at $V_\odot = 20$ kms^{-1} through a dense structure with an internal velocity dispersion such that $h^{-1} = 10$ kms^{-1}. If the chord length of passage is 2 pc, $t = 10^5$ yr and $N_c \sim 3.9 \times 10^3 r_0^2 v_e^4 \nu_c$. Since $v_e^4 \alpha r_0^{-2}$ the radius of the sphere of influence cancels out and hence N_c is not sensitive to the detailed geometry and masses involved. One finds $N_c \sim 2.3 \times 10^7 \nu_c$ resulting in the capture of $\sim 10^9$ comets if $\nu_c \sim 10^2$ a.u.$^{-3}$. However, one expects that, since the velocities of bodies leaving the GMC decline relative to it, they will, statistically, also decline relative to each other. In particular, the sphere of influence is a purely formal concept and the GMC potential continues to affect the motions of comets not too far inside the sphere. In essence comets entering the solar potential well at hyperbolic speed may still be trapped by the increase in the height of the well caused by the recession of the GMC. This decline in velocity dispersion may be given, approximately, by $\delta\sigma \sim \delta V_{\odot}\, r_O/R$ where $\delta V_{\odot}$ represents the decline in the velocity of the Sun as it climbs out of the potential well of a structure of scale length R. It is easily found that the 'capture window' $\pm v_e$ may be increased by a factor 2 or 3 due to this if the substructure is a few pc across and has a few $10^4 M_{\odot}$ mass, in effect increasing $N_c \, \alpha \, v_e^4$ by one or two powers of ten. Accordingly, it seems that to order of magnitude an adequate capture mechanism may exist: the Sun merely has to pass through a moderately dense substructure as it leaves a GMC.

CONCLUSIONS

1. There may have been ~ 10 actual penetrations of GMCs during the course of Solar System history.

2. Within the acceptable ranges of molecular cloud parameters it is to be expected that the Oort cloud has been removed beyond a few 10^3 a.u. perhaps ~ 10 times in the past 4.5 Byr. Stars are a relatively minor contributor to perturbations of the Oort cloud.

3. Replenishment by unbinding from a very dense inner cloud has yet to be modelled quantitatively, taking full account of GMC perturbations, but a qualitative discussion is given which suggests that the hypothesis may not work.

4. On the other hand, capture from GMCs with a reasonable comet number density may take place and is consistent with the steady cratering rate.

5. The dynamical picture described here forms the basis for a theory of terrestrial catastrophism first described in NC(1979) and developed in CN 1982b, 1984b (and see Dr. Clube's contribution to these proceedings). Amongst the predictable (and predicted) consequences of the theory are mass extinctions of species and geophysically disturbed epochs. These are caused by episodes of bombardment which recur with galactic periodicities and are dominated by the disintegration products of the very largest comets.

References

Alvarez, W. & Muller, R.A., 1984.   Nature 308, 718.
Bailey, M.B., 1983. Mon. Not. R. astr. Soc. 202, 603.
Bhatt, H.C., Rowse, D.P. & Williams, I.P., 1984 Mon. Not.
  R. astr. Soc. 209, 69.
Clube, S.V.M. & Napier, W.M., 1982a. Q. Jl. R. astr. Soc. 23, 45.
Clube, S.V.M. & Napier, W.M. 1982b. Earth Planet. Sci. Lett. 57, 251.
Clube, S.V.M. & Napier, W.M., 1983. In Highlights of Astronomy,
  Vol. 6, (ed. R.M. West), 355-362, D. Reidel.
Clube, S.V.M. & Napier, W.M., 1984a. Mon. Not. R. astr. Soc. 208, 575.
Clube, S.V.M. & Napier, W.M., 1984b. Mon. Not. R. astr. Soc.
   In press.
Cohen, R.S., Cong, H., Dame, T.M. & Thaddeus, P., 1980. Astrophys. J.
  (Lett.) 239, L53.
Drapatz, S. & Zinnecker, H., 1984. In press.
Elmegreen, B.G. & Lada, C.J., 1976. Astron. J. 81, 1089.
Gordon, M.A. & Burton, W.B., 1976. Astrophys. J. 208, 346.
Gordon, M.A. & Burton, W.B., 1980. In Giant Molecular Clouds in the
  Galaxy, eds. M.G. Edmunds & P.M. Solomon, (Oxford : Perganon),
  p.41.
Lacey, C.G., 1984. Mon. Not. R. astr. Soc. 208, 687.
Liszt, H.S., Xiang, D. & Burton, W.B., 1981. Astrophys. J.
  249, 532.
Martin, H.M., Sanders, D.B. & Hills, R.E., 1984. In press.
Napier, W.M. & Clube, S.V.M., 1979. Nature 282, 455.
Napier, W.M., & Staniucha, M., 1982. Mon. Not. R. astr. Soc.
  198, 723.
Noerdlinger, P.D., 1977. Icarus 30, 566.
Rampino, M.R. & Stothers, R.B., 1984. Nature 308, 709.
Sanders, D.B., Solomon, P.M. & Scoville, N.Z., 1984.
  Astrophys. J. 276, 182.
Scoville, N.Z., & Solomon, P.M., 1974. Astrophys. J. (Lett.)
  187, L67.
Sekanina, Z., 1976. Icarus 27, 123.
Seyfert, C.K. & Sirkin, L.A., 1979. Earth History and Plate Tectonics
  (New York: Harper Row).
Solomon, P.M., Sanders, D.B. & Scoville, N.Z., 1979. In IAU Symp. 84,
  The Large Scale Characteristics of the Galaxy, W.B. Burton
  (Dordrecht : Reidel), p.35.
Thaddeus, P. & Dame, T.M., 1983. In Star Formation Workshop,
  Royal Observatory, Edinburgh.
Weissman, P.R., 1983. In Highlights of Astronomy, Vol. 6
  (ed. R.M. West, 363, D. Reidel.
Weissman, P.R., 1984. Science 224, 987.
Whipple, F.L., 1975. Astron J. 80, 525.
Xiang, D., Liszt, H.S. & Burton, W.B., 1984. Chin. Astron.
  Astrophys. 8, 195.

DISCUSSION

J.A. Fernandez: In my opinion our current knowledge of GMCs is too
poor to make strong statements on the capture-disruption of
cometary clouds. We still do not know very well the general
properties of GMCs such as their size, mass and frequency of
encounters with the solar system, let alone the problem of
discussing their fine structure. I think that changes in the
values of the parameters within the currently accepted ranges
will lead to quite different conclusions regarding the disruption
of cometary clouds and the capture of new ones by the solar
system.

W.M. Napier: Dr Fernandez's opinion should be demonstrated using
recently published work on GMCs. In any case the analysis is
based not just on the measured molecular cloud properties but
also on a comparison with the observed heating of the stellar
disc: the conclusion, that the energy injected into the long-
period comet system over 4.5 Byr is one or two powers of ten
greater than its binding energy, can be reached without reference
to GMCs. Further evidence on the dynamics and origin of the Oort
cloud should be forthcoming from the cratering record and the
geochemistry of iridium - enhanced layers of probably cometary
origin (cf the talk by Dr Clube, this volume).

SECTION II

THE OORT CLOUD OF COMETS

THE FORMATION AND DYNAMICAL SURVIVAL OF THE COMET CLOUD

Julio A. Fernández
Observatório do Valongo, U.F.R.J.
Ladeira do Pedro Antonio, 43
20.080 Rio de Janeiro
Brazil

ABSTRACT. The theory of a huge reservoir of comets (the "comet cloud") extending to almost interstellar distances is analyzed, paying special attention to its dynamical stability, formation process and orbital properties of the incoming cloud comets. The perturbing influence of passing stars and giant molecular clouds is considered. Giant molecular clouds may be an important perturbing element of the comet cloud, although they do not seem to change drastically former studies including only stellar perturbations. The more tightly bound inner portions of the comet cloud, say within 10^4 AU, would have withstood the disrupting forces over the age of the solar system. The theory of a primordial comet origin in the outer planetary region close to Neptune's orbit is specially analyzed. A primordial comet origin is consistent with the cosmogonic view that a large amount of residual material was ejected during the last stage in the formation of the Jovian planets. The smooth diffusion in the energy space of bodies scattered by Neptune guarantees that most of them will fall in the narrow range of energies close to zero (near-parabolic orbits) where passing stars and GMCs can act effectively on them. The long time scales of ~10^9 yr required for bodies scattered by Neptune to reach near-parabolic orbits would indicate that the buildup of the comet cloud was an event that took place long after the planets formed. Depending on the field of perturbing galactic objects, it is possible to conceive that most scattered comets were stored in rather tightly bound orbits (a ~10^4 AU), favoring the concept of their dynamical survival over several billion yr. Alternative theories of comet cloud formation, e.g. in-situ origin or interstellar capture, are also discussed. The main difficulty of the in-situ theory is to explain how comets could accumulate at large heliocentric distances where the density of the nebular material was presumably very low. The interstellar capture theory also meets severe dynamical objections as, for instance, the lack of observed comets with original strongly hyperbolic orbits and the extremely low probability of capture under most plausible conditions. Since our knowledge of the structure of giant molecular clouds and their frequency of encounters with the solar system is still very uncertain, the concept of capture of transient comet clouds during such encounters can be advanced very little beyond the speculative stage. Some other

A. Carusi and G. B. Valsecchi (eds.), Dynamics of Comets: Their Origin and Evolution, 45–70.
© *1985 by D. Reidel Publishing Company.*

dynamical properties of relevance to theories of origin and structure of
the comet cloud are also reviewed. We mention, for instance, the
distribution of perihelion points on the celestial sphere. There seems
to be here a well established deviation from randomness, although the
debate on whether or not there is a preference of the perihelion
clustering for the vicinity of the apex of the solar motion is still
unsettled. The alleged correlation with the solar apex may be biased by
the preference of comet discoveries in the northern hemisphere. Deviations from randomness might be caused by very close stellar passages in
the recent past. The excess of retrograde orbits among the observed
"new" and young comets – mainly those with q $\gtrsim$ 2 AU – is another well
known dynamical feature. Such an excess may probably be accounted for
by the combined action of planetary and stellar perturbations. Because
of the decreasing action of planetary perturbations with increasing
heliocentric distances, a significant increase in the rate of passages
of long-period comets is predicted for the outer planetary region.

1. OBSERVATIONAL BACKGROUND

The theory of a huge reservoir of comets surrounding the solar system
(the "comet cloud") was developed by Oort (1950). Assuming that the
comet cloud was thermalized by stellar perturbations, Oort was able to
derive a cloud population of 2 x 10^{11} comets by comparing the expected
theoretical influx rate of near-parabolic comets with the observed one.
A somewhat smaller cloud population of a few times 10^{10} comets has been
given by Fernández (1982) by arguing that cloud comets have kept rather
eccentric orbits up to the present. On the other hand, Weissman (1983)
obtains a much higher population of 1.4 x 10^{12} comets in correspondence
with his assumption of a higher influx rate of near-parabolic comets.
Given the different estimates of the cloud population and the average
comet mass, large uncertainties are involved in any estimate of the
cloud mass although several authors seem to agree that it should be of
the order of a few $M_\oplus$ (Oort 1950, Öpik 1973, Fernández 1982, Weissman
1983). Given the uncertainties, this should be taken as a very preliminary result. Although the theory of the comet cloud has been questioned
by some authors (e.g. Lyttleton 1974, Yabushita 1979), their criticisms
have not weakened its appeal arisen from its capability for explaining
some rather puzzling properties of long-period comets as we shall see
below.
     Comets with orbital periods P > 200 yr are usually referred to as
long-period (LP) comets. One of their most important orbital parameters
is certainly the "original" semimajor axis, a_{orig}, i.e. before being
perturbed by the planets, that gives information about the place where
LP comets come from. The orbital energy per unit mass, ε, is proportional
to the reciprocal semimajor axis, (-1/a), where a binding (negative)
energy corresponds to an elliptc (positive a) orbit. Choosing convenien-
tly the units we can thus set $\varepsilon = -1/a$. As seen in Fig. 1, the energy
distribution of the original orbits of LP comets shows a strong concentration in the narrow range - 10^{-4} AU^{-1} < ε < 0, that corresponds to
near-parabolic comets reaching heliocentric distances greater than

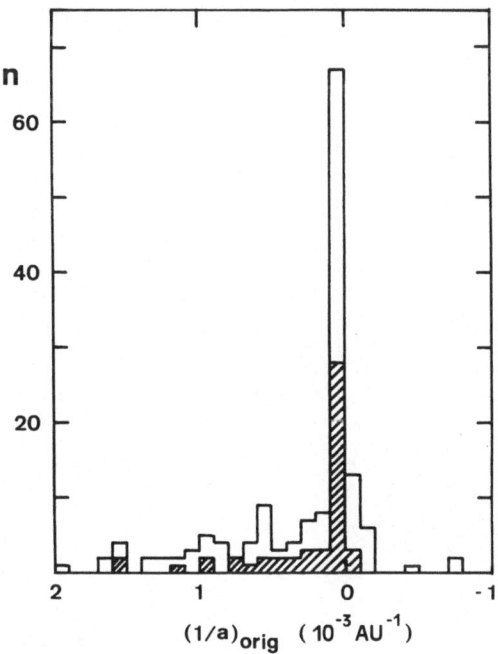

Figure 1. Distribution of the original reciprocal semimajor axes of LP comets with $(1/a)_{orig} < 2\times10^{-3}AU^{-1}$ as computed by Marsden et al. (1978) and Everhart and Marsden (1983). The shaded histogram is for those comets with q > 2 AU.

$\sim2\times10^{4}$AU. Oort argued that these comets were driven into the planetary region by stellar perturbations. For this reason he called them "new" on the belief that they were passing through the planetary region for the first time. We will show that most of the so-called new comets have probably passed before through the planetary region, so to avoid confusions we shall use the term "cloud comets" for those with $\varepsilon > -10^{-4}AU^{-1}$ ($a_{orig}>10^{4}$AU). Marsden and Sekanina (1973) noted a clustering of aphelion distances around 5×10^{4}AU which defines the "radius" of the comet cloud.

We also note in Fig. 1 the lack of comets with strongly hyperbolic orbits ($1/a_{orig} << 0$) which seems to rule out an interstellar origin. This conclusion is strengthened when only comets with perihelion distances q>2 AU are considered (shaded histogram). Very few comets remain with $1/a_{orig} < 0$ which suggests that negative values of $1/a_{orig}$ are probably caused by nongravitational forces (cf. Marsden et al. 1978).

In contrast to the rest of the known solar system bodies, LP comets show an almost random distribution of their orbital inclinations i (Fig. 2). Some departure from randomness is clearly indicated by the observed excess of retrograde orbits. This effect becomes more pronounced when only comets with q>2 AU are considered (shaded histogram).

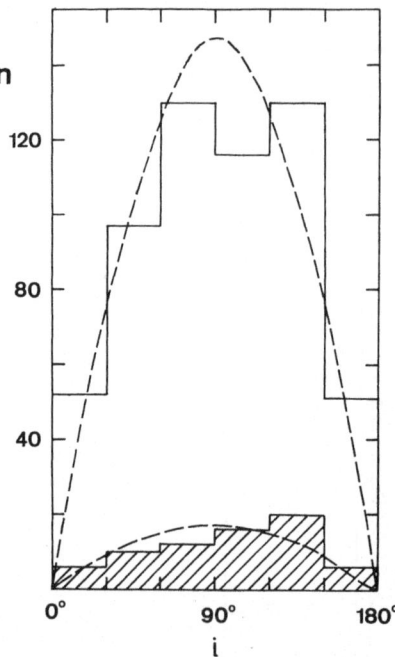

Figure 2. Inclination distribution of the LP comets appearing in Marsden's (1982) catalogue. The nine members of the Kreutz family have been considered as a single comet. The shaded histogram corresponds to the partial sample of LP comets with q > 2 AU. The (dashed) sine-curves correspond to random i-distributions.

      The q-distribution of LP comets with q ≲ 1.1 AU and discovered after 1800 is shown in Fig. 3. The choice of this rather restricted sample has the purpose of avoiding as much as possible the bias against the discovery of comets with larger q. LP comets show a steady decrease toward smaller perihelion distances. However, when only dynamically young comets with $a_{orig} > 10^3$ AU are considered, the q-distribution turns out to be rather uniform. The decrease in the number of LP comets whith decreasing q is probably an evolutionary effect due to physical decay of comets in one or a few perihelion passages (Fernández 1981a).
      A departure from randomness has long been noted in the distribution of perihelion points on the celestial sphere (e. g. Tyror 1957, Hurnik 1959). The crucial point is whether there is a correlation between the direction of the solar apex and the preferred direction of the perihelia. Oja (1975) found the preferred perihelia direction to be only 7° from the solar apex. Fernández and Jockers (1982) argue however that the predominance of comet discovery in the northern hemisphere introduces a bias favoring such correlation. In this respect, Bogart and Noerdlinger (1982) find that when they only consider LP comets discovered in the 20th century – which are presumably less biased to the north – the preferred perihelia direction moves to about 50° from the solar apex.

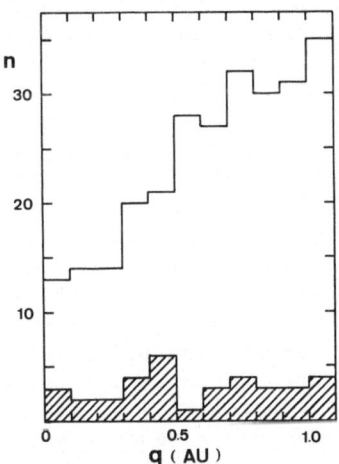

Figure 3. Distribution of the perihelion distances of LP comets with q<1.1 AU observed after 1800 as recorded in Marsden's (1982) catalogue. The shaded histogram corresponds to "new" and "young" comets with $a_{orig} > 10^3 AU$.

2. FORMATION OF THE COMET CLOUD

Most authors seem to agree at present about the existence of an extended cloud of comets surrounding the solar system. Yet various views on how and when such a cloud formed appear in the literature. We can consider three fundamental currents of opinion:

a) Primordial origin. The basic idea is that comets are a byproduct of the planet formation. Some authors take this as equivalent to saying that the comet cloud is nearly as old as the solar system (cf. Oort 1950). The implicit argument is that diffusion time scales from the planetary region to the Oort region are very short as compared to the solar system age. We shall see that this might not be the case.

The primordial theory of comet origin is based on the assumption, supported by several theoretical and numerical studies, that the Jovian planets could not have reached their present masses without the ejection of a large amount of residual matter (e.g. Safronov 1969, 1972). Furthermore, some recent numerical studies have shown that bodies starting out in the outer planetary region can evolve under the combined action of planetary and stellar perturbations into Oort-cloud-type orbits (Fernández 1980, Fernández and Ip 1981, Weissman 1982).

Napier and Staniucha (1982) and Clube and Napier (1984) have criticized the theory of a primordial comet origin on the basis that an extended comet cloud would not have survived over the age of the solar system. We shall return to discuss further some aspects of this theory in Section 7.

b) In-situ formation. Some authors have proposed that comets formed in the outer regions of the collapsing solar nebula (Cameron 1973, Biermann and Michel 1978, Hills 1982). Such comets might have filled the circumsolar space up to distances of 10^3-10^4AU. Perturbations from giant molecular clouds and passing stars could remove comets from the outer portions of this more tightly bound cloud giving rise to transient loosely bound comet clouds (Bailey 1983a).

The main criticism to the in-situ theory is the seemingly lack of physical conditions for the accumulation of grains into comet-sized bodies at large heliocentric distances (Öpik 1973). Hills (1982) argues that radiation pressure from the protosun and protostars might have forced nebular grains to coagulate into comets, although his result depends on some simplifying assumptions. Little more can be added to the discussion of a more tightly bound comet cloud because it is beyond our current possibilities of detection. Bailey (1983b) and Bailey et al. (1984) have discussed some observational procedures, such as a far-infrared survey, to try to detect or at least to put upper limits to its mass.

c) Capture of interstellar comets. Laplace already visualized comets as interstellar bodies captured by the solar system mainly through Jupiter's perturbations (see modern discussions by, e.g.,Radzievskii and Tomanov 1977, Valtonen and Innanen 1982). The lack of observed comets with original strongly hyperbolic orbits has been a severe objection against the capture hypothesis (Sekanina 1976). Furthermore, Valtonen and Innanen (1982) show that the probability of capture of interstellar comets into elliptic orbits by Jupiter's perturbations is extremely low so that the capture of a sizable comet cloud could only occur in the very improbable event of encounter velocities of about 0.5 kms^{-1}. Another dynamical difficulty noted by Fernández (1981a) is that capture of interstellar comets would lead to a comet population strongly concentrated toward small inclinations, which is against the observations.

Realizing the difficulties inherent to the capture mechanism by the Sun-Jupiter system, Clube and Napier (1984) argue that comet capture might occur during transits of the solar system through giant molecular clouds. Thus, they regard giant molecular clouds as hierarchic structures each comprising about 25 medium-size molecular clouds (MMC) of mass ~ 2x$10^4M_\odot$ with a fraction of it being under the form of cometary bodies. The authors conclude that encounters of the solar system with discrete MMCs would have a twofold effect: 1) disruption of already existent loosely bound comet clouds and 2) capture of transient comet clouds from MMCs as the Sun recedes from them. It is too early to judge this mechanism on a quantitative basis given our poor knowledge of the structure, size and frequency of encounters with giant molecular clouds. Valtonen (1983) has found Clube and Napier's mechanism to be inadequate to supply a sizable comet cloud unless an undetected solar companion is postulated.

From our previous discussion of comet birthplaces a key question emerges, namely: is the dynamical lifetime of the comet cloud too short as compared to the solar system age so as to preclude a primordial comet origin?. We shall next discuss this point.

3. DYNAMICAL STABILITY OF THE COMET CLOUD

Passing stars, giant molecular clouds (GMCs) and the Galaxy itself all perturb the orbits of cloud comets. Those cloud comets reaching the planetary region are also perturbed by the planets and actually many of them are removed from the comet cloud. We shall leave the analysis of planetary perturbations for the next section and focus now on the perturbing sources outside the solar system.

The action of stellar perturbations on solar system bodies has already been analyzed in detail, starting with Öpik's (1932) pioneer work, complemented later by Oort (1950), Sekanina (1968) and Rickman (1976) among others. Because the star's relative velocity V_* is much larger than the velocity of the comet when it moves far from the Sun, the comet can be assumed to be at rest in a heliocentric frame of reference. As a result of a stellar passage, Sun and comet will experience velocity changes given by

$$\vec{\Delta v}_\odot = \frac{2\mu}{v_* D_\odot} \frac{\vec{D}_\odot}{D_\odot} \, , \qquad (1a)$$

$$\vec{\Delta v}_c = \frac{2\mu}{V_* D_c} \frac{\vec{D}_c}{D_c} \, , \qquad (1b)$$

where $\mu = GM$, G being the gravitational constant and M the stellar mass, $D_\odot$ and D_c are the minimum distances to the Sun and the comet. The change in the comet's velocity with respect to the Sun will be: $\vec{\Delta v} = \vec{\Delta v}_c - \vec{\Delta v}_\odot$

The perturbing effect of a single star will generally be very small. Yet the effect of many stellar encounters will cumulate quadratically so as to produce significant orbital changes over long time scales. As the comet is much more strongly perturbed when it is close to its aphelion, its perihelion distance q and inclination i will be the orbital elements that undergo the greatest changes. Fernández (1980, 1981b) has derived the changes Δq and Δi experienced by a cloud comet at a heliocentric distance r perturbed by a passing star. They are

$$\Delta q = \frac{q}{v_t^2} \ (\Delta v_t^2 + 2 \ \Delta v_t \ x \ v_t \ \cos \beta) \qquad (2)$$

$$\Delta i = \frac{\Delta v_t \ \sin \beta \ \cos \alpha}{(\Delta v_t^2 + v_t^2 + 2v_t \ x \ \Delta v_t \ \cos \beta)^{1/2}} \, , \qquad (3)$$

where $\vec{\Delta v}_t = \vec{\Delta v} \cos \theta$ is the transverse component of the velocity change $\vec{\Delta v}$, θ being the angle between $\vec{\Delta v}_t$ and $\vec{\Delta v}$. The transverse velocity is $v_t^2 \sim 2GM_\odot q/r^2$. β is the angle between $\vec{\Delta v}_t$ and $\vec{v}_t$ and α the angular distance of the comet to the ascending node.

Given a stellar flux n_o, the cumulative changes of q and i after a time T will be

$$\Delta q_T^2 = \int_{D_L}^{D_U} (\overline{\Delta q^2}) n_o T D_\odot dD_\odot \, , \qquad (4)$$

$$\overline{\Delta i^2}_T = \int_{D_L}^{D_U} \overline{(\Delta i^2)} n_o T D_\Theta dD_\Theta \ , \qquad (5)$$

where $\overline{\Delta q^2}$, $\overline{\Delta i^2}$ are averages over Θ, β and α. D_L, D_U are the lower and upper limits for the stellar distances of closest approach to the Sun during T. We take $D_L = (2\pi n_o T)^{-1/2}$, i.e. the probability for a star to pass from the Sun a distance $D_\Theta < D_L$ is 0.5, and $D_U = 2.5r$ meaning that we limit ourselves to the close stellar encounters. It can be shown that the more distant encounters do not change drastically the results obtained from eqs. (4) and (5). By adopting $n_o = 10$ stars $pc^{-2}Myr^{-1}$, an average stellar mass of 0.7 M_Θ, and average star's relative velocity $V_* = 30$ km s^{-1} (Rickman 1976) and T = 4.5 x 10^9yr, we get the results shown in Figs. 4 and 5.

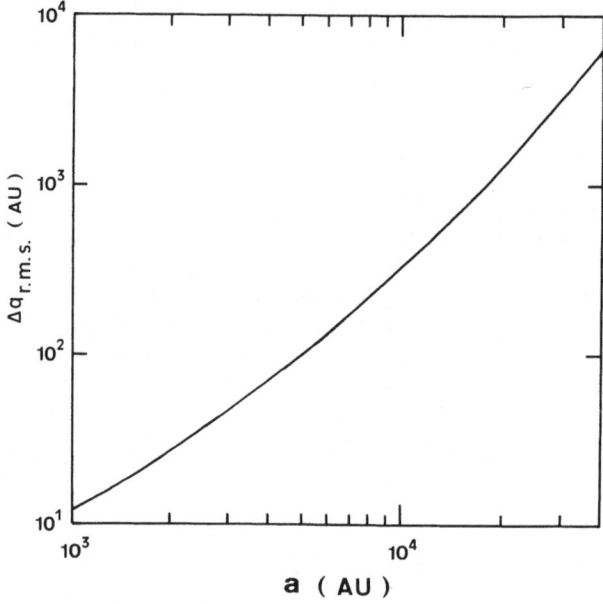

a (AU)

Figure 4. The expected change of the perihelion distance of a near-parabolic comet, caused by perturbing stars throughout the solar system lifetime, as a function of the comet's semimajor axis. The comet is assumed to have an initial q = 20 AU.

A sharp increase in the r.m.s. change Δq is found for increasing a (Fig.4). Comets with a $\sim 10^3$AU and perihelia within the planetary region (q<30 AU) are not expected to have their perihelia removed from the planetary region by stellar perturbations over the age of the solar system. The r.m.s. change of Δq attains several 10 AU for a of a few 10^3AU. Comets with a $\sim 10^4$AU have already an expected change of ~ 300 AU so that we should expect their perihelia were long since removed from the planetary region. In realistic terms, the removal of the comet's perihelion from the planetary region should depend on whether the rate of change of q by stellar perturbations is greater or smaller than the diffusion speed of

the orbital energy caused by planetary perturbations. For comets with a
of a few 10^4AU will suffice a single revolution to cause a change of
their perihelion distance of several 10^2AU. Summing up, comets with
a $\sim 10^4$AU will be perturbed by passing stars fast enough to be removed
from the planetary region in the course of a few revolutions. Such a dis
tance defines the lower limit of the Oort region.

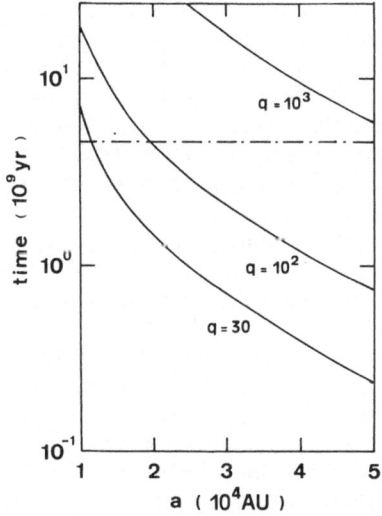

Figure 5. Time scales for randomization by stellar perturbations of the
orbital planes of comets with q=30 AU and q=10^2AU , q=10^3AU as a function of their semimajor axes. The horizontal line indicates the age of
the solar system.

The orbital planes of comets with q≲100 AU and a≳2x10^4AU should have got randomized by stellar perturbations over the age of the solar sys
tem (Fig. 5). The criterion for randomization is that the r.m.s. change
Δi_T as given by eq. (5) reaches the value π. Should comets form close to
the ecliptic plane, a certain concentration might still be present for
comets with a ≲ 2x10^4AU. However comets driven into the inner planetary
region will drastically change their inclinations so as to produce a
randomly oriented influx of cloud comets.

The cumulative change of the comet's energy per unit mass can be
roughly estimated by considering only close encounters, i.e. $\Delta v^2 =$
($\Delta \vec{v}_c - \Delta \vec{v}_\odot$)$^2 \sim \Delta v_\odot^2$. Thus we have

$$2 \Delta \varepsilon_T = \int_{D_L}^{D_U} \Delta v_\odot^2 n_o T D_\odot dD_\odot$$

$$= (\frac{2\mu}{V_*})^2 n_o T \, Ln(\frac{D_U}{D_L}) . \qquad (6)$$

Introducing the numerical values quoted above with T = 4.5x10^9yr, we ob-
tain an energy change $2\Delta\varepsilon_T = 1.15x10^8$ cm^{2}s^{-2} or a r.m.s. change in the

comet's velocity of 1.07×10^4 cm s^{-1} which is in fairly good agreement
with Weissman's (1980) result and somewhat smaller than Bailey's (1983a).
We should expect our result to be somewhat underestimated because of the
neglect of distant encounters in eq. (6). By setting the condition
$2 \Delta \epsilon_T = v_{esc}^2 = 2 GM_\odot / r \sim GM_\odot / a$ we find that comets with a $\gtrsim 7.6 \times 10^4$ AU have
got enough energy to escape from the solar system.

No analytical or numerical studies have so far been carried out on
the change of the perihelion direction of comets subject to stellar perturbations. The perihelion points of comets formed in the outer planetary region should have shown a preference for the ecliptic plane. No such
a preference is currently observed in LP comets which in principle can
be attributed to stellar perturbations. Quantitative studies on this problem would be highly desirable so as to ascertain time scales for randomization of perihelion directions of cloud comets as a function of their
semimajor axes and perihelion distances.

Encounters with GMCs are much less frequent although their effects
on the stability of the comet cloud may be much more drastic because of
their large masses. GMCs have typical masses of $\sim 5 \times 10^5 M_\odot$ and radii of
about 20 pc (Solomon and Sanders 1980, Gordon and Burton 1980). Sanders
et al. (1984) find the surface density of H_2 – main component of the molecular clouds – to peak at ~ 6 kpc from the Galactic center. It may be
however a factor of five smaller at the Sun's distance (~ 10 kpc). The
perturbing effects of molecular clouds on cloud comets was considered by
Biermann (1978). Shortly after Napier and Clube (1979) presented a theory according to which quasiperiodic encounters of the solar system with
GMCs during its passage through the spiral arms of the Galaxy would result in the capture of temporary "Oort clouds" and in a heavy comet bombardment of the planets. From numerical simulations, Napier and Staniucha
(1982) have concluded that a primordial comet cloud would have been lost
at present as a consequence of perturbations from GMCs. However, their
results are highly sensitive to the adopted numerical values. For example, Napier and Staniucha consider rather low encounter velocities with
GMCs (5 and 10 km s^{-1}) which favors their disruptive influence. This problem has been reviewed in detail by Bailey (1983a) who comes to the conclusion that an inner core of the comet cloud of radius $\sim 10^4$ AU would have withstood encounters with GMCs over the age of the solar system.

We can compare the perturbation exerted by a GMC on the solar system during a penetrating encounter to that caused by a very close stellar passage. We assume the GMC to be a uniform sphere of radius $R_{GMC} =$
20 pc and mass $M_{GMC} = 5 \times 10^5 M_\odot$. Biermann (1978) and Bailey (1983a) have
developed the mathematical expression for the mean perturbation caused
by a penetrating encounter with a GMC, which can be written as

$$2 \, \overline{\Delta \epsilon}_{GMC} = \frac{2}{3} \left(\frac{2 GM_{GMC}}{V} \right)^2 \frac{r^2}{b^4} \left\{ 1 - \left(1 - \frac{b^2}{R_{GMC}^2} \right)^{3/2} \right\}^2 , \qquad (7)$$

where b is the impact parameter, V is the encounter velocity and r is an
average heliocentric distance for cloud comets. We shall adopt: $r \approx 1.7 a \approx$
4×10^4 AU. The energy change experienced by a cloud comet due to a close
stellar passage is

$$2 \, \Delta \varepsilon_* \approx \left(\frac{2GM_*}{D_L V_*}\right)^2 , \qquad\qquad (8)$$

where D_L is the closest approach of a star to the Sun expected during the age of the solar system T. We have: $D_L = (\pi n_0 T)^{-1/2} \sim 550$ AU.

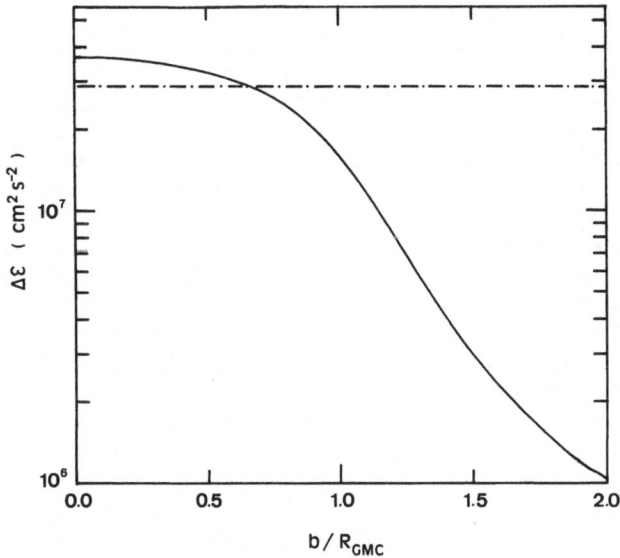

Figure 6. The mean energy change of a typical cloud comet at a heliocentric distance $r = 4 \times 10^4$ AU, due to an encounter with a GMC, as a function of the impact parameter b (in units of the radius of the GMC). The condition $b/R_{GMC} < 1$ corresponds to penetrating encounters. The horizontal line indicates the energy change caused by the closest stellar approach to the Sun expected during its lifetime.

As shown in Fig.6, the energy imparted to a comet by a star passing at the closest solar distance expected during the solar system lifetime can be comparable to that imparted by a GMC in a penetrating encounter. Furthermore, penetrating encounters with GMCs do not seem to be as frequent as Napier and Clube claim. According to Bailey (1983a), the number of such penetrating encounters during the solar system lifetime can be placed in the range 1 - 10. Encounters with smaller molecular clouds are probably much more frequent. Rampino and Sothers (1984) estimate that one of such encounters takes place during every transit of the solar system through the galactic disk (~ 30 Myr). However the perturbation caused by a typical molecular cloud of radius ~ 5 pc and mass $\sim 10^4 M_\odot$ is much smaller than that obtained from eq. (7) for a GMC. Of course, our assumption of a homogeneous cloud is an oversimplification. The computation of the energy transfer rate will depend somewhat on the GMC's substructure (Bailey 1983a). On the other hand, the consideration by Napier and Staniucha (1982) and Clube and Napier (1984) of GMCs as divi-

ded in 25 discrete clouds, each of mass $2 \times 10^4 M_\odot$ and radius 2 pc, is also
very likely an oversimplification. For instance, Solomon and Sanders
(1980) conclude that only a small fraction of a GMC mass is in condensed
cores. Summing up, even though the action of GMCs and molecular clouds
should be taken into consideration as a perturbing source of cloud comets, it does not seem to change drastically the picture obtained before
with the exclusive consideration of stellar perturbations (e. g. Oort
1950, Weissman 1980, Fernández 1980).

    Due to galactic perturbations, stable motion is only possible for
comets in eccentric orbits up to distances of $\sim 8 \times 10^4$ AU over long periods
of time (Chebotarev 1966). This is beyond the limits imposed by stellar
and GMC perturbations, so that galactic perturbations are not of primary
concern for the stability of the comet cloud over cosmogonic time scales.
Nevertheless, comets of smaller a, say $\sim 2.5 \times 10^4$ AU, will be perturbed by
the Galaxy so that galactic effects might be present in the observed
"new" comets. From the numerical integration of orbits in the restricted
three-body problem: Sun-galactic nucleus-comet, Byl (1983) finds that
changes in q of cloud comets are smallest for aphelia near the galactic
equator and poles. Therefore, it may be possible that galactic perturbations introduce some correlation between the galactic structure and the
aphelion distribution of LP comets.

    We are still left with the possibility of an as yet undetected so-
lar companion as a perturbing source of the comet cloud. Indeed, the dis
covery that most stars have stellar companions (Abt and Levy 1976) gives
some support to the idea that the Sun also has or had a companion. Kirk
(1978) has shown that such a hypothetical companion cannot be close to
the Sun, say at $\sim 10^3$ AU, otherwise we should not observe the clustering
of comet energies in the interval $0 > \varepsilon > -10^{-4}$ AU^{-1}. A more distant solar
companion with a = 8.8×10^4 AU has recently been proposed by Whitmire and
Jackson (1984) and Davis et al. (1984) to explain a possible 26-Myr periodicity in biological mass extinctions. The authors suggest that the
unseen companion stirs comets of the Oort cloud during its perihelion
passages every 26 Myr giving rise to comet showers. The problem with a
distant companion is to explain its dynamical stability over periods of
more than 10^8 yr (see Weissman's contribution to this book).

4. PLANETARY PERTURBATIONS

A fraction of the Oort cloud population will be deflected to the planetary region where is perturbed by the planets. Indeed, all cloud comets
should have been subject to planetary perturbations at early times if
they formed in the planetary region. Planetary perturbations will greatly
change the orbital energy ε of long-period orbits but very little the
other orbital elements (Yabushita 1972). From numerical experiments,
Kerr (1961) found that the distribution of energy changes per perihelion
passage, $\Delta\varepsilon$, could be fitted to a Gaussian distribution. Everhart (1968)
further noted a departure from the Gaussian distribution in the form of
long $\Delta\varepsilon$-tails accounting for drastic energy changes produced in close
encounters. Fernández (1981a) derived as an acceptable approximation
for the $\Delta\varepsilon$-distribution up to energy changes $\Delta\varepsilon \sim 2.5\ \Delta\varepsilon_T$, the expression

$$f(\Delta\varepsilon) \; \alpha \; \exp(-3/4 \; \Delta\varepsilon^2/\Delta\varepsilon_T^2), \hspace{2cm} (9)$$

where $\Delta\varepsilon_T$ is the typical energy change for comets with perihelion distances and inclinations in the ranges (q, q+Δq) and (i, i+Δi) computed as the r.m.s. of their energy changes per perihelion passage. As shown in Fig. 7, $\Delta\varepsilon_T$ is strongly dependent on the orbital elements q, i.

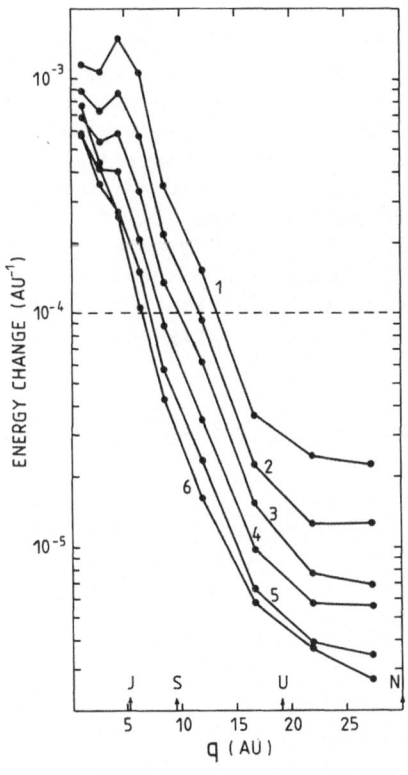

Figure 7. The typical energy change per perihelion passage as a function of the comets' perihelion distance and for six inclination ranges: $0<i<30^o$ (curve 1)...... $150^o<i<180^o$ (curve 6) (Fernández 1981a).

LP comets will random-walk in the energy space until they are finally lost to the interstellar space or to the inner planetary region. For a comet of orbital parameters q, i coming from the Oort region ($\varepsilon \sim 0$), its expected energy ε after n perihelion passages will be

$$\varepsilon^2 = n \; \Delta\varepsilon_T^2(q,i). \hspace{2cm} (10)$$

As mentioned before, during the dynamical evolution the other orbital elements, such as q and i, are not expected to change significantly as long as the comet moves on a long-period orbit.

An initial population of N_0 near-parabolic comets ($\varepsilon \sim 0$) will be dynamically depleted due to diffusion through the boundary $\varepsilon = 0$ to positive energies (hyperbolic orbits). Everhart (1976) showed from numerical experiments that the number N of comets that still remain bound to the solar system after n perihelion passages is

$$N(n) = \frac{1}{2} N_0 \; n^{-1/2}. \hspace{2cm} (11)$$

As mentioned, a Gaussian distribution is valid as long as we neglect close encounters with the Jovian planets which cause drastic chan-
ges in ε. To make allowance for close encounters, $\Delta\varepsilon_T$ should be multiplied by a factor of $\sim$3 (Fernández 1981a).

5.   HOW MANY COMETS ARE REALLY "NEW"?

This question is of great significance for our understanding of the dynamical properties of the comet cloud. Numerical experiments carried out
by Fernández (1982) have shown that $\sim$85% of the so-called new comets
coming into the inner planetary region have passed before by the planetary region beyond Jupiter's orbit. We can roughly estimate the fraction
F of cloud comets with orbital energies $0 > \varepsilon_c > \varepsilon_L (\sim -10^{-4} AU^{-1})$, passing by
the planetary region with perihelion distances q, that will return to
the Oort region (namely with energies $0 > \varepsilon > \varepsilon_L$). For this to happen, the
energy change the comet undergoes after a perihelion passage has to fall
in the energy range $\varepsilon_L - \varepsilon_c < \Delta\varepsilon < -\varepsilon_c$. By using the distribution function
of energy changes $f(\Delta\varepsilon)$ given by eq. (9), we can readily obtain

$$F = \int_{\varepsilon_c - \varepsilon_L}^{\varepsilon_c} f(\Delta\varepsilon)\, d\Delta\varepsilon = \frac{1}{2}\left\{ \mathrm{erf}\left[\sqrt{3/4}\left(\frac{\varepsilon_c}{\Delta\varepsilon_T}\right)\right] + \mathrm{erf}\left[\sqrt{3/4}\left(\frac{\varepsilon_L - \varepsilon_c}{\Delta\varepsilon_T}\right)\right] \right\}, \quad (12)$$

where erf is the error function.

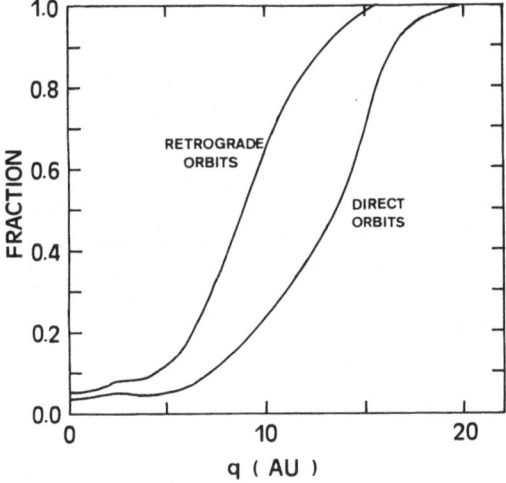

Figure 8.   The fraction of cloud comets returning to the Oort region
($a > 10^4$ AU) after a perihelion passage as a function of the comet's perihelion distance. Results are for two groups of comets: direct and retrograde orbits, both assumed to be randomized.

The fraction of cloud comets returning to the Oort region has been computed separately for comets in: a) direct and b) retrograde orbits. The computed values of F are presented in Fig. 8 as a function of the comet perihelion distance. We see that a large fraction of the incoming cloud comets with perihelia beyond Jupiter will return to the Oort region and that F will be greater for comets in retrograde orbits. Even for cloud comets coming into the inner planetary region we should expect that ∿5% of them will return to the Oort region, while the fraction will be close to unity for the Uranus-Neptune region.

6. THE DISTRIBUTIONS OF PERIHELION DISTANCES AND INCLINATIONS OF LONG-PERIOD COMETS

Let us start considering cloud comets driven into the planetary region by stellar perturbations for the first time. It is easy to show that such comets will have an uniform q-distribution. For this, let v_c be the velocity of a cloud comet. For comets deflected into the planetary region, v_c forms a very small angle θ with the radius vector Sun-comet. We have for the transverse comet's velocity

$$v_T^2 \simeq v_c^2 \theta^2 \simeq \frac{2GM_\odot q}{r^2}$$

hence

$$v_c^2 \theta \, d\theta \simeq \frac{GM_\odot}{r^2} \, dq \quad . \tag{13}$$

Under the assumption of randomization of the vector v_c by stellar perturbations we have

$$f_\theta(\theta) \, d\theta = \frac{1}{2} \sin \theta \, d\theta \sim \frac{\theta}{2} \, d\theta \quad . \tag{14}$$

Finally, by combining eqs. (13) and (14) we get the q-distribution

$$f_q(q) \, dq = \frac{GM_\odot}{2v_c^2 r^2} \, dq \quad . \tag{15}$$

As v_c and r are independent of q, the distribution f_q turns out to be uniform. This is in agreement with the q-distribution observed for dynamically young LP comets (cf. Fig.3).

    Let us now compute the q-distribution for all the LP comets with different dynamical ages. For this, let us consider an initial population of N_0 cloud comets passing through the planetary region with perihelia in the range (q, q+dq). The number of comets surviving dynamical ejection will decrease with the number of perihelion passages n following eq. (11). Through the whole dynamical evolution, the initial population of N_0 comets will perform a number of n_T perihelion passages before being ejected or transferred to periodic orbits (T < 200 yr). From

Fernández (1981a) we find

$$n_T = \sum_{n=1}^{n_M} \frac{1}{2} N_o n^{-1/2} \simeq N_o n_M^{1/2} \quad , \tag{16}$$

where n_M is the maximum number of perihelion passages a comet avoiding ejection can perform that brings it from a near-parabolic orbit to a periodic one. For comets random-walking in the energy space ε, the number of passages n_M required to pass from an energy $\varepsilon_o \sim 0$ to an energy $\varepsilon_p \sim -0.03$ AU^{-1} (corresponding to an orbital period $P \simeq 200$ yr) is

$$n_M = \frac{(\varepsilon_p - \varepsilon_o)^2}{\Delta\varepsilon_T^2} \sim \frac{\varepsilon_p^2}{\Delta\varepsilon_T^2} \quad . \tag{17}$$

By substituting eq. (17) into eq. (16) we finally obtain

$$n_T \sim N_o \frac{\varepsilon_p}{\Delta\varepsilon_T} \quad . \tag{18}$$

Since a population of LP comets is made up of comets with different dynamical ages, the longer their dynamical lifetime n_T, the larger their number (thus, the influx rate $\dot{N}_{LP}$) considering those LP comets with inclinations in the range (i, i+di) passing perihelion in the range (q, q+dq). Thus we have

$$\dot{N}_{LP} \propto n_T \propto \Delta\varepsilon_T^{-1} \quad . \tag{19}$$

Since $\Delta\varepsilon_T$ decreases with increasing q, the influx rate $\dot{N}_{LP}$ will increase as we go farther away in the planetary region (Fig.9). The results are very impressive: we should expect the influx rate of LP comets to be about two orders of magnitude greater in the outer planetary region than in the region of the terrestrial planets. A previous numerical study by Fernández (1982) led to a similar result. As Fig. 9 shows, for Saturn's zone the influx rate of LP comets should be about one order of magnitude greater than for the inner planetary region. Analyses of the cratering rate for outer solar system bodies should take into consideration such an increase in the flux of LP comets. In this respect, a recent study by Zimbelman (1984) incorporating this effect leads to rather similar crater production rates from LP comets throughout the planetary region.

We should note that physical decay will limit the lifetime of LP comets in the inner planetary region well below their dynamical lifetime given by n_T. Obviously, the smaller q, the smaller the number of revolutions a comet can perform before physical decay which results in a decreasing number of observed LP comets as we approach the Sun (Fig. 3).

The i-distribution of LP comets should also be affected by planetary perturbations. A greater fraction of cloud comets in retrograde orbits passing through the outer planetary region will return to the Oort region as compared to those in direct orbits. This effect will be

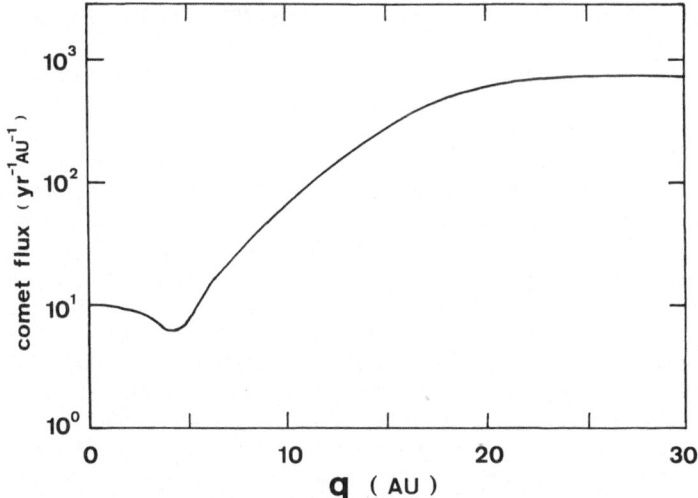

Figure 9. The computed influx rate of LP comets per unit of q as a function of the comet's perihelion distance. The ordinate values have been normalized to an influx rate of 10 LP comets per year per AU in the inner planetary region.

more pronounced in the Jupiter-Uranus region (see Fig. 8). Therefore, more cloud comets in retrograde orbits will have a chance to pass again through the planetary region, and some of them will cross Jupiter's orbit because of the diffusion of their perihelia by stellar perturbations (Fernández 1981b). We should bear in mind that a typical cloud comet with a=2.5x10^{4}AU will experience an average change of $\Delta q \sim 5$ AU in a revolution (Fernández 1980). Presumably, comets passing beyond Jupiter are not strongly depleted of their volatiles so we might expect them to show up very active when passing through the inner planetary region. The observed strong excess of retrograde orbits for dynamically young comets with q>2 AU (Fig. 2) can thus be explained in terms of the greater survival rate of cloud comets in retrograde orbits passing by the outer planetary region.

The strong excess of retrograde orbits observed for q>2 AU will tend to disappear for smaller perihelion distances following the greater variation of i that accompanies a drastic reduction of q by stellar perturbations. In addition to this dynamical effect, the author (Fernández 1981b) has shown that selection effects may favor the discovery of LP comets in prograde orbits for the range 1<q$\lesssim$2 AU. This may explain the apparent conflict with Delsemme's conclusion that a predominance of prograde orbits is present among "new" comets (see his presentation in this book). Thus, Delsemme has drawn his conclusion from the sample of observed "new" comets with well determined original orbits, the overwhelming majority of which have q<2 AU (cf. Marsden et al. (1978) catalogue).

7. COMET ORIGIN AS A BYPRODUCT OF THE FORMATION OF THE JOVIAN PLANETS

Our review of the action of stellar and planetary perturbations on cloud
comets will allow us to analyze further the primordial theory of comet
formation. Oort (1950) proposed that comets formed in the asteroid belt.
However, by noting the icy nature of comets and because of dynamical
considerations, Kuiper (1951) suggested that the region close to Neptune's orbit was a more suitable place for comet formation. From the study
of the accretion of the Jovian planets, Safronov (1969, 1972) concluded
that they could have reached their present masses only at the expense of
ejecting large amounts of residual matter. This is specially applicable
to the cases of Uranus and Neptune for which the accretion of solid
bodies played a much greater role than in the hydrogen-dominated Jupiter
and Saturn. Numerical experiments carried out by Fernández and Ip (1981)
confirm that a large amount of matter of perhaps several tens of $M_\oplus$ is
ejected during the late stage of formation of Uranus and Neptune. We
can give a simple explanation of why this should happen: as proto-Uranus
and proto-Neptune grew they started to stir up planetesimals of their
accretion zones until their encounter velocities reached values of $U >$
$\sqrt{2}-1$ (with respect to the circular velocity of the protoplanet). At this
point, collision (accretion) with the protoplanet virtually becomes to a
halt because ejection becomes a much more probable event (Weidenschilling
1975).

Comets random-walking in the energy space will finally be ejected
unless stellar perturbations can remove their perihelia from the planetary region. For this to happen it is required that such comets pass by
the Oort region (energies $\varepsilon_L < \varepsilon < 0$). Since the typical step in the random-
walk is $\Delta\varepsilon_T$, we can see from Fig. 7 that bodies under the gravitational
control of Neptune will have a very smooth orbital diffusion with a
large probability of passing by the Oort region since $\Delta\varepsilon_T << |\varepsilon_L|$. By con-
trast, bodies under the gravitational control of Jupiter and Saturn are
subjected to much stronger perturbations with $\Delta\varepsilon_T >> |\varepsilon_L|$, so that in the
diffusion process such bodies will probably overshoot the Oort region.
The ratios of comets placed into the Oort region to those ejected on
hyperbolic orbits, as derived by Fernández and Ip (1981) from numerical
experiments, are shown in TABLE I. As expected, the ratio turns out to
be greater for bodies under the gravitational control of Neptune and
lower for Jupiter's.

TABLE I

Ratio of comets placed into the Oort region to those ejected on hyperbolic orbits	
Jupiter	0.03 ± 0.01
Saturn	0.16 ± 0.04
Uranus	1.30 ± 0.50
Neptune	2.60 ± 0.70

Uranus' contribution to the comet cloud was partially hindered because a large part of the residual material of its accretion zone fell
under the gravitational control of Jupiter which finally ejected the bo
dies. This can be understood in terms of probabilities of ejection and
transfer to the influence zone of an inner Jovian planet. A body of
Uranus' region requires a relative velocity $U = 0.35$ to be able to reach
Jupiter's influence zone. This means that before the body gets a veloci
ty $U > \sqrt{2}-1$ for ejection to be possible, it will probably fall under the
gravitational control of Jupiter (Fernández and Ip 1984). However, for
a body of Neptune's region the relative velocity required for transfer
to Jupiter's region is $U = 0.46$ which is larger than the velocity requi
red for getting a near-parabolic orbit. Therefore, before falling under
Jupiter's control the body can reach the Oort region.

For bodies scattered by a Jovian planet we can roughly estimate
their timescale for ejection or for reaching the Oort region. In the
diffusion process a body will move throughout the energy range: $\varepsilon_L - \varepsilon_0$,
where ε_0 is its starting energy. ε_0 will be of the order of $- 0.1 \, AU^{-1}$
for Jupiter's controlled bodies in starting low eccentricity orbits and
$\sim -3 \times 10^{-2} AU^{-1}$ for Neptune's. The probability that the body falls in a
certain energy range $(\varepsilon, \varepsilon+d\varepsilon)$ during its orbital diffusion will be

$$f_\varepsilon(\varepsilon) \, d\varepsilon = \frac{d\varepsilon}{\varepsilon_L - \varepsilon_0} \sim \frac{d\varepsilon}{|\varepsilon_0|} \quad . \tag{20}$$

The time t_{oc} that will take the comet to reach the Oort region (or to be
ejected) is given by

$$t_{oc} = \sum_{n=1}^{n_M} T_n \quad ,$$

where $n_M = \dfrac{\varepsilon_0^2}{\Delta \varepsilon_T^2}$ is the average number of steps to cover the energy range
$\varepsilon_L - \varepsilon_0$ and T_n is the comet's revolution period during the revolution n. The average revolution period $\overline{T}$ is defined as

$$\overline{T} = \frac{\displaystyle\int_{T_m}^{T_M} T \, f_T(T) \, dT}{\displaystyle\int_{T_m}^{T_M} f_T(T) \, dT} \quad , \tag{21}$$

where $T = |\varepsilon|^{-3/2}$ and f_T is the probability distribution of T. The maxi-
mum and minimum periods T_M and T_m correspond to the orbital energies ε_L
and ε_0, respectively.

From eq. (20) we obtain: $f_T(T) = C \, T^{-5/3}$, C being a normalizing
factor. By introducing this relation into eq. (21) we get

$$\overline{T} = 2 \, T_M^{1/3} \, T_m^{2/3} = 2 \, |\varepsilon_L|^{-1/2} \, |\varepsilon_0|^{-1} \quad . \tag{22}$$

The time t_{oc} can then be approximated by

$$t_{oc} \simeq n_M \overline{T} = \frac{2\,|\varepsilon_o|\,|\varepsilon_L|^{-1/2}}{\Delta\varepsilon_T^2} \quad \text{years,} \qquad (23)$$

which is appropriate for bodies of Uranus' and Neptune's region. Given the large steps $\Delta\varepsilon_T$ of bodies scattered by Jupiter and Saturn, rather than $T_M = |\varepsilon_L|^{-3/2}$ as defining the maximum revolution period, we should use $T_M = \Delta\varepsilon_T^{-3/2}$ as a more adequate value for the comet's orbital period previous to ejection. With this modification, eq. (23) becomes

$$t_{oc} \simeq \frac{2\,|\varepsilon_o|}{\Delta\varepsilon_T^{5/2}} \qquad (23')$$

suitable for bodies of Jupiter and Saturn regions.

TABLE II

Time scale for reaching the Oort region (yr)	
Jupiter	2.3×10^5
Saturn	1.0×10^7
Uranus	6.0×10^8
Neptune	1.3×10^9

Computed time scales t_{oc} are shown in TABLE II for the four giant planets. The values of $\Delta\varepsilon_T$ used in eqs. (23) and (23') have been multiplied by a factor of three (see Section 4) to make allowance for close encounters that very probably will occur during the multiple perihelion passages of the bodies. The computed times of TABLE II are in rather good agreement with those found by Ip (1977) using a Monte Carlo procedure. As seen, for bodies of Neptune's region t_{oc} is very long which suggests a slow buildup of the comet cloud (Fernández and Ip 1981). Therefore, *the formation of planets and comets would have been coeval although the buildup of the comet cloud would have occurred later on in the history of the solar system*. Furthermore, the long dynamical time scale t_{oc} for bodies under the gravitational control of Neptune suggests that a tail of more dynamically stable bodies remains bound to its influence zone forming a flat structure we can call the "cometary belt" (Fernández and Ip 1983).

The previous discussion suggests that two comet populations with different dynamical histories might reach the inner planetary region: 1) The so-called "new" comets on near-parabolic orbits. They are comets that got diffused to the Oort region where they were perturbed by passing stars and GMCs. 2) The "belt" comets coming from the above-discussed tail of bodies with long dynamical time scales. Comets with a smaller than a few 10^3AU would have kept their primordial concentration towards

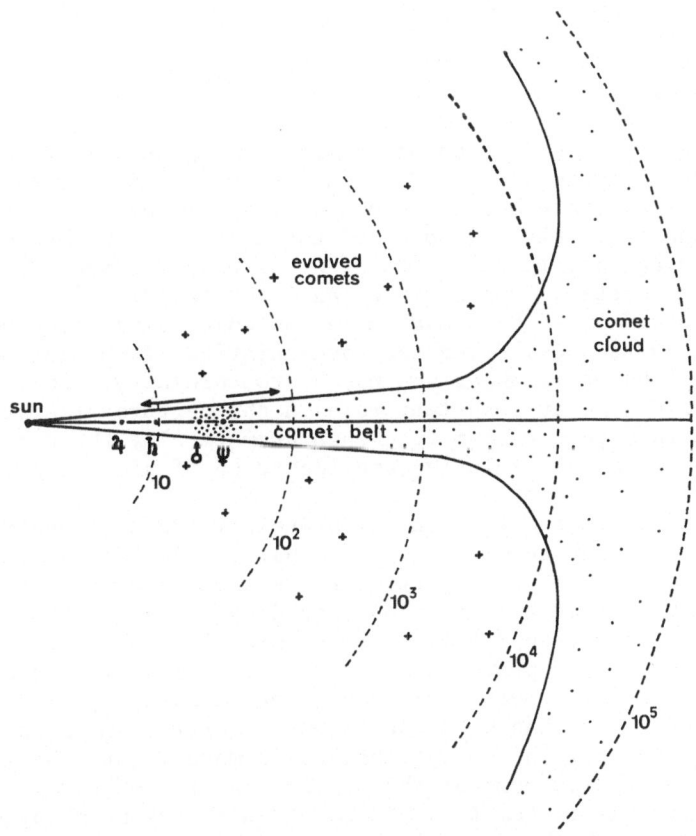

Figure 10. Hypothetical space distribution of comets bound to the solar system. A comet origin in the Uranus–Neptune region is assumed. The arrows indicate the sense of scattering of the residual bodies. "Evolved comets" would correspond to those derived from the Oort cloud whose semi major axes have been shortened by planetary perturbations.

the ecliptic plane.

According to what was discussed before, we may speculate that the i-distribution of near-parabolic comets reaching the outer planetary region presents an excess of retrograde orbits following the discussion of Section 6. In addition, there might be an excess of small-i comets from the contribution of belt comets. We might presume that the large number of observed short-period comets – not well understood as a capture process of near-parabolic comets (Joss 1973, Fernández and Ip 1983) – might be explained as an inward diffusion of low-i belt comets by perturbations of the Jovian planets. At the beginnings of the solar system, such population of belt comets might have been several orders of magnitude greater and contributed to the heavy bombardment of the terrestrial planets (Wetherill 1975, Fernández and Ip 1983). Figure 10 sums up our view of the current comet structure under the assumption of comet formation in the region of Uranus and Neptune.

8.   CONCLUDING REMARKS

Several pieces of evidence point to the presence of a large comet reservoir surrounding the solar system up to distances of several 10^4AU. Members of this reservoir evolve under the perturbing action of passing
stars, sporadic encounters with GMCs and perturbations from the Galaxy
itself. The ultimate fate of most cloud comets will be ejection to the
interstellar space. A critical point of the discussion is how stable is
the comet cloud against the disrupting forces of stars and GMCs over the
age of the solar system. Supporters of an interstellar origin argue that
a loose comet cloud would be dynamically unstable mainly due to catastro
phic encounters with GMCs. Given the uncertainties of the parameters
characterizing GMCs, this assertion may be too premature. As discussed,
GMCs might have played an important role in the perturbation of the comet
cloud although they do not seem to change dramatically the conclusions
already reached with the exclusive consideration of stellar perturbations.

The idea of a comet origin in the planetary region is attractive
because it fits current cosmogonic ideas that the Jovian planets would
have formed at the expense of ejecting large amounts of residual material. The objection that a comet cloud would have been lost over the
solar system age is weakened further by considering the long time scales
of bodies scattered by Neptune to reach the Oort region. We should also
note that a primordial comet origin would not necessarily lead to a
loose comet cloud which has been the target of criticisms by supporters
of the interstellar origin theory. Instead, a more tightly bound comet
cloud may well be the outcome of the combined action of planetary and
stellar perturbations on comets scattered from the Uranus-Neptune region.
The energy range in which comets were stored should mainly depend on the
field of extra solar system objects during the buildup of the comet
cloud.

Finally, we shall mention several lines of research that may help
to put further constraints on theories of formation of the comet cloud:

1) A better knowledge of the structure, mass and frequency of
encounters with GMCs is required to assess their impact on the survival
of the comet cloud.

2) A better estimate of the average comet mass and influx rate of
cloud comets will allow us to set more precise limits on the total mass
of the comet cloud. Hence an estimate of the mass removed by the outer
planets will be possible under the assumption of a comet origin in the
planetary region. The derived amount of material removed from the outer
planetary region can then be compared to that expected on cosmogonic
grounds.

3) Better statistics of comet orbits with large q will permit to
check several theoretical forecasts, such as the excess of retrograde
orbits among LP comets and the overall increase in the number of LP
comets with increasing q beyond Jupiter.

4) The distribution of the perihelion points of LP comets do not
show a preference for the ecliptic plane. However, comets formed in the
planetary region should have had low-inclination orbits, whereby their
perihelia should have been concentrated toward the ecliptic plane. We

may presume that the cometary perihelia "lost memory" of their primordial locations through the combined perturbing action of passing stars, GMCs and the planets. This still needs to be proved by numerical or analytical studies.

5) The deviation from randomness of the directions of the perihelia of LP comets needs to be further analyzed once a better comet sample in the southern hemisphere is available, in particular, to prove or disprove the alledged relationship with the solar apex.

REFERENCES

Abt, H.A., and Levy, S.G. 1976. 'Multiplicity among solar-type stars'. *Astrophys. J. Suppl.* 30, 273-306.

Bailey, M.E. 1983a. 'The structure and evolution of the Solar System comet cloud'. *Mon. Not. Roy. Astron. Soc.* 204, 603-633.

Bailey, M.E. 1983b. 'Theories of cometary origin and the brightness of the infrared sky'. *Mon. Not. Roy. Astron. Soc.* 205, 47p-52p.

Bailey, M.E., McBreen, B., and Ray, T.P. 1984. 'Constraints on cometary origins from isotropy of the microwave background and other measurements'. *Mon. Not. Roy. Astron. Soc.* 209, 881-888.

Biermann, L. 1978. 'Dense interstellar clouds and comets'. Symp. on *Important Advances in 20th Century Astronomy*, Copenhagen.

Biermann, L., and Michel, K.W. 1978. 'On the origin of cometary nuclei in the presolar nebula'. *Moon Planets* 18, 447-464.

Bogart, R.S., and Noerdlinger, P.D. 1982. 'On the distribution of orbits among long-period comets'. *Astron. J.* 87, 911-917.

Byl, J. 1983. 'Galactic perturbations on near-parabolic cometary orbits'. *Moon Planets* 29, 121-137.

Cameron, A.G.W. 1973. 'Accumulation processes in the primitive solar nebula'. *Icarus* 18, 407-450.

Chebotarev, G.A. 1966. 'Cometary motion in the outer solar system'. *Soviet Astron. - AJ* 10, 341-344.

Clube, S.V.M., and Napier, W.M. 1984. 'Comet capture from molecular clouds: a dynamical constraint on star and planet formation'. *Mon. Not. Roy. Astron. Soc.* 208, 575-588.

Davis, M., Hut, P., and Muller, R.A. 1984. 'Extinction of species by periodic comet showers'. *Nature* 308, 715-717.

Everhart, E. 1968. 'Change in total energy of comets passing through the solar system'. *Astron. J.* 73, 1039-1052.

Everhart, E. 1976. 'The evolution of comet orbits'. In *The Study of Comets*. IAU Coll. No. 25 (B. Donn, M. Mumma, W. Jackson, M. A'Hearn, and R. Harrington, eds.) pp. 445-464, NASA SP-393.

Everhart, E., and Marsden, B.G. 1983. 'New original and future cometary orbits'. *Astron. J.* 88, 135-137.

Fernández, J.A. 1980. 'Evolution of comet orbits under the perturbing influence of the giant planets and nearby stars'. *Icarus* 42, 406-421.

Fernández, J.A. 1981a. 'New and evolved comets in the solar system'. *Astron. Astrphys.* 96, 26-35.

Fernández, J.A. 1981b. 'On the observed excess of retrograde orbits among long-period comets'. *Mon. Not. Roy. Astron. Soc.* 197, 265–273.

Fernández, J.A. 1982. 'Dynamical aspects of the origin of comets'. *Astron. J.* 87, 1318–1332.

Fernández, J.A., and Ip, W.-H. 1981. 'Dynamical evolution of a cometary swarm in the outer planetary region'. *Icarus* 47, 470–479.

Fernández, J.A., and Ip, W.-H. 1983. 'On the time evolution of the cometary influx in the region of the terrestrial planets'. *Icarus* 54, 377–387.

Fernández, J.A., and Ip, W.-H. 1984.'Some dynamical aspects of the accretion of Uranus and Neptune: The exchange of orbital angular momentum with planetesimals'. *Icarus* 58, 109–120.

Fernández, J.A., and Jockers, K. 1983. 'Nature and origin of comets'. *Rep. Prog. Phys.* 46, 665–772.

Gordon, M.A., and Burton, W.B. 1980. 'The distribution and size of giant molecular clouds in the galaxy' In *Giant Molecular Clouds in the Galaxy*. Third Gregynog Astrophys. Workshop (P.M. Solomon and M.G. Edmuns, eds.) pp.25–39, Pergamon Press, Oxford.

Hills, J.G. 1982. 'The formation of comets by radiation pressure in the outer protosun'. *Astron. J.* 87, 906–910.

Hurnik, H. 1959. 'The distribution of the directions of perihelia and the orbital poles of non-periodic comets'. *Acta Astron.* 9, 207–221.

Ip, W.-W. 1977. 'On the early scattering processes of the outer planets'. In *Comets – Asteroids – Meteorites: Interrelations, Evolution and Origin* (A.H. Delsemme, ed.) pp.485–490, Univ. of Toledo Press, Toledo, Ohio.

Joss, P.C. 1973. 'On the origin of short-period comets'. *Astron. Astrophys.* 25, 271–273.

Kerr, R.H. 1961. 'Perturbations of cometary orbits'. *Proc. 4th Berkeley Symp. of Mathematical Statistics and Probability*. Vol.3, pp.149–164, Univ. of California Press, Berkeley.

Kirk, J. 1978. 'On companions and comets'. *Nature* 274, 667–669.

Kuiper, G.P. 1951. 'On the origin of the solar system'. In *Astrophysics* (J.A. Hynek, ed.) pp.357–427, McGraw-Hill.

Lyttleton, R.A. 1974. 'The non-existence of the Oort cometary shell'. *Astrophys. Space Sci.* 31, 385–401.

Marsden, B.G. 1982. *Catalogue of Cometary Orbits*. Fourth Ed.

Marsden, B.G., and Sekanina, Z. 1973. 'On the distribution of "original" orbits of comets of large perihelion distance'. *Astron. J.* 78, 1118–1124.

Marsden, B.G., Sekanina, Z., and Everhart, E. 1978. 'New osculating orbits for 110 comets and analysis of original orbits'. *Astron. J.* 83. 64–71.

Napier, W.M., and Clube, S.V.M. 1979. 'A theory of terrestrial catastrophism'. *Nature* 282, 455–459.

Napier, W.M., and Staniucha, M. 1982. 'Interstellar planetesimals – I. Dissipation of a primordial cloud of comets by tidal encounters with massive nebulae'. *Mon. Not. Roy. Astron. Soc.* 198, 723–735.

Oja, H. 1975. 'Perihelion distribution of near-parabolic comets'. *Astron. Astrophys.* 43, 317–319.

Oort, J.H. 1950. 'The structure of the cloud of comets surrounding the solar system and a hypothesis concerning its origin'. *Bull. Astron.*

Inst. Neth. 11, 91–110.

Öpik, E.J. 1932. 'Note on stellar perturbations on nearly parabolic orbits'. *Proc. Am. Acad. Arts Sci.* 67, 169–183.

Öpik, E.J. 1973. 'Comets and the formation of planets'. *Astrophys. Space Sci.* 21, 307–398.

Radzievskii, V.V., and Tomanov, V.P. 1977. 'On the capture of comets by the Laplace scheme'. *Soviet Astron. - AJ* 21, 218–223.

Rampino, M.R., and Stothers, R.B. 1984. 'Terrestrial mass extinctions, cometary impacts and the Sun's motion perpendicular to the galactic plane'. *Nature* 308, 709–712.

Rickman, H. 1976. 'Stellar perturbations of orbits of long-period comets and their significance for cometary capture'. *Bull. Astron. Inst. Czech.* 27, 92–105.

Safronov, V.S. 1969. *Evolution of the Protoplanetary Cloud and Formation of the Earth and the Planets* (Translated from Russian (1972) by the Israel Program for Scientific Translations, Jerusalem).

Safronov, V.S. 1972. 'Ejection of bodies from the solar system in the course of the accumulation of the giant planets and the cometary cloud'. In *The Motion, Evolution of Orbits, and Origin of Comets* (G.A. Chebotarev, E.I. Kazimirchak-Polonskaya, and B.G. Marsden, eds.) pp.329–334, IAU Symp. No. 45.

Sanders, D.B., Solomon, P.M., and Scoville, N.Z. 1984. 'Giant molecular clouds in the Galaxy. I. The axisymmetric distribution of H_2'. *Astrophys. J.* 276, 182–203.

Sekanina, Z. 1968. 'On the perturbation of comets by near-by stars: Encounters of comets with fast moving stars'. *Bull. Astron. Inst. Czech.* 19, 291–301.

Solomon, P.M., and Sanders, D.B. 1980. 'Giant molecular clouds as the dominant component of interstellar matter in the Galaxy'. In *Giant Molecular Clouds in the Galaxy*. Third Gregynog Astrophys. Workshop (P.M. Solomon, and M.G. Edmuns, eds.) pp.47–73, Pergamon, Oxford.

Tyror, J.G. 1957. 'The distribution of the directions of perihelia of long-period comets'. *Mon. Not. Roy. Astron. Soc.* 117, 369–379.

Valtonen, M.J. 1983. 'On the capture of comets into the solar system'. *Observatory* 103, 1–4.

Valtonen, M.J., and Innanen, K.A. 1982. 'The capture of interstellar comets'. *Astrophys. J.* 255, 307–315.

Weidenschilling, S.J. 1975. 'Close encounters of small bodies and planets'. *Astron. J.* 80, 145–153.

Weissman, P.R. 1980. 'Stellar perturbations of the cometary cloud'. *Nature* 288, 242–243.

Weissman, P.R. 1982. 'Dynamical history of the Oort cloud'. In *Comets* (L.L. Wilkening, ed.) pp.637–658, Univ. of Arizona Press, Tucson.

Weissman, P.R. 1983. 'The mass of the Oort cloud'. *Astron. Astrophys.* 118, 90–94.

Wetherill, G.W. 1975. 'Late heavy bombardment of the Moon and terrestrial planets'. *Proc. Sixth Lunar Sci. Conf.* 2, 1539–1561.

Whitmire, D.P., and Jackson, A.A. 1984. 'Are periodic mass extinctions driven by a distant solar companion?'. *Nature* 308, 713–715.

Yabushita, S. 1972. 'Planetary perturbation of orbits of long-period comets with large perihelion distances'. *Astron. Astrophys.* 16,

471-477.

Yabushita, S. 1979. 'A statistical study of the evolution of the orbits of long-period comets'. *Mon. Not. Roy. Astron. Soc.* <u>187</u>, 445-462.

Zimbelman, J.R. 1984. 'Planetary impact probabilities for long-period comets'. *Icarus* <u>57</u>, 48-54.

DISCUSSION

<u>S. Yabushita</u>: You say the excess of retrograde orbits is real. But by stellar encounters, direct orbits can be converted into retrograde orbits and viceversa.

<u>J. A. Fernández</u>: That will only occur after a long dynamical time scale provided q is not very small. A typical Oort cloud comet of, say q $\sim$ 10 AU and a $\sim$ 2.5x10^{4}AU, will experience an average inclination change of only $\sim$15? after an orbital revolution (see Fernández 1981b).

<u>R. Lüst</u>: In a sample of 89 comets with semimajor axes > 10^{4}AU (original values from the Marsden-Sekanina-Everhart catalogue) we counted 46 comets with direct and 43 with retrograde orbits. Could you comment on your sample?.

<u>J. A. Fernández</u>: Your sample contains a large fraction of comets with small perihelion distances. Comets getting very small q ($\lesssim$1 AU) will very probably be randomized by stellar perturbations. For 1$\lesssim$q$\lesssim$2 AU selection effects may favor the discovery of comets in prograde orbits. The excess of retrograde orbits should show up for LP comets with rather large perihelion distances, say q > 2 AU. It is just for this sample that I have found about 60% of retrograde orbits.

<u>P. Farinella</u>: Dr. Greenberg's study on the origin of comets was based on the assumption of a 10% efficiency of Uranus and Neptune in ejecting proto-comets into the Oort cloud. Do you agree with this estimate?.

<u>J. A. Fernández</u>: I think this gives the correct order of magnitude, even though we should bear in mind that Neptune's efficiency is much greater than Uranus'.

EMPIRICAL DATA FROM OORT'S CLOUD

A. H. Delsemme
Department of Physics and Astronomy
The University of Toledo
Toledo, OH 43606 USA

ABSTRACT. *Empirical evidence about the size and the origin of the Oort's cloud of comets is confronted with theories about its origin. The slow diffusion of the orbits of the "new" comets into the inner solar system implies a redefinition of the concept of "new" comet. A gradual transfer of orbital angular momentum occurs from the planets to the comets as the comets grow older on shorter period orbits. The observed retrograde to prograde ratio of the new comets is difficult to explain. Either it comes from a poorly understood observational bias, or from a neglected secular action of the Galaxy, or it implies a recent asymmetrical perturbation of the Oort's cloud (less than 10-20 million years ago). The grazing incidence of a giant molecular cloud or an exceptionally close stellar passage would introduce such an asymmetry; this would also be true for the unseen hypothetical stellar companion of the Sun recently invoked to explain the periodicity of the geological extinction of species through violent cometary showers.*

1. INTRODUCTION

A renewed interest in the elucidation of the origin of comets is apparent in the literature of the 1980's. Besides the classical approach which links the origin of the Oort Cloud with that of the solar system, alternate theories have multiplied recently, as is also clear from a number of other papers in this Colloquium.

This is a healthy situation. However, it is important not only to remember the empirical data that are related to the problem, but also to take them into account inasmuch as it is possible, even if they are scanty, because they may be used as bounds to limit the possibilities, or as criteria to compare models.

I have discussed in the past several sets of empirical data that are relevant to the "new" comets (Delsemme 1977, 1979, 1983). I will first mention them shortly; I will then explore a new set of data, that are relevant to the possible orbital angular momentum of the Oort's cloud.

71

A. Carusi and G. B. Valsecchi (eds.), Dynamics of Comets: Their Origin and Evolution, 71–85.
© *1985 by D. Reidel Publishing Company.*

2. THE DIFFUSION OF NEW COMETS INTO THE PLANETARY SYSTEM

The first set of observational data comes from the statistics in the
binding energy (per unit mass) of the very long period comets. We have
now a list of two hundred and twenty long-period comets whose original
orbits are known with accuracy. The most complete list is given in
particular by Marsden and Roemer (1982) and is based mainly on the
work of Marsden and Sekanina (1973) and Marsden, Sekanina and Everhart
(1978). They have also shown that whatever the reason, the effective
distance of the outer margin of the Oort's cloud is best established by
the extrapolation of the binding energies of "new" comets with larger
perihelion distances, that is, with smaller non-gravitational forces
(NGF). They find that the extrapolated mean binding energy is $1/a = 46$
(in 10^{-6} AU^{-1}) and the scatter is only ± 10. This corresponds to an
effective margin extending nominally between 35,700 AU and 55,600 AU,
with a mean distance of $2a = 43,500$ AU. This does not mean that Oort's
mechanism does not work any more at 200,000 AU, but only that aphelia
in that zone do not make a significant contribution. Oort (1950) had
found 220,000 AU for the same outer margin, but this was from poor
statistics based on ten orbits with $1/a < 50 \times 10^{-6}$ AU^{-1}, also ignoring
the variable influence of NGF.
    The fact that the Oort's cloud is five times as small as believed
before has important consequences that have not escaped the attention
of recent authors (Bailey 1977, Weissman 1978). The major consequence
is that the mean velocity perturbation ΔV_{RMS} introduced by stellar
passages is smaller than Oort believed. Using Oort's (1950) approach
with proper numerical changes, I find a formula which is almost the
same as that agreed with by most authors (Faintich 1971; Weissman 1982):

$$\Delta V_{RMS} = 1.8 \ T^{1/2} \ (\text{in m s}^{-1}) \tag{1}$$

where T is the duration of the perturbation in millions of years. With
a mean $2a = 43,500$ AU, the mean period of a new comet is 3.2 M yrs, with
no more than 3M years spent in the outer margin of the Oort's cloud.
Therefore the ΔV_{RMS} introduced by stellar perturbations is 3.0 m s^{-1}
per revolution. The transverse component Δv of that velocity change is
only 2.5 m s^{-1} according to:

$$\Delta v = \Delta V_{RMS} \sqrt{\frac{2}{3}} = 0.82 \Delta V_{RMS} \tag{2}$$

The transverse velocity v (at distance r near aphelion) is linked to
the perihelion distance q by the formula:

$$v = r^{-1} (2GMq)^{1/2}. \tag{3}$$

    Any assumed "new" comet observed with a perihelion q of the order
of 1 or 2 AU (aphelion v between 1 and 1.4 m s^{-1}) is therefore unlikely
to be a new comet (in Oort's meaning) because on the average, it came
from a previous orbit that had been modified only by $v = 2.5$ m s^{-1}; the
perturbation is random in the velocity space, but the largest possible
value of the previous mean transverse velocity v_0 is:

$$v_o = v + \Delta v \tag{4}$$

which puts its previous mean perihelion at most between 10 and 15 AU.
Taking into account its random nature, the velocity perturbation Δv is
described by a Gaussian distribution whose mean will be added on the
average at a right angle to its original v in the velocity space. For
this reason, it is easy to check numerically that an assumed "new"
comet with perihelion q = 1 AU has already passed an average of five
times through the outer solar system; all these passages took place
outside of Jupiter's orbit, the innermost one (before q = 2 AU) being
on the average near 8 AU. How then is it possible that most of these
new comets still have an unchanged orbital energy, corresponding to the
mean distance of the margin of the Oort's cloud?

The answer is that they have really changed. However, the perturbations of Uranus and Neptune are almost negligible; their $\Delta(1/a)$ is of
the order of 10 x 10^{-6} AU^{-1} in most cases, only slightly widening the
peak in 1/a of the orbits' distribution (Everhart and Raghavan 1970).
Second, most of the recent statistics that have refined our knowledge
of the mean distance of the outer fringe of the Oort's cloud, have been
derived by giving more weight to those new comets with large perihelia
(from 2 to 7 AU); many more of these have indeed come from unperturbed
orbits with perihelia beyond Saturn's distance. The original orbits
with the smaller perihelia are much more scattered in 1/a as shown by
Marsden and Sekanina's 1973 study.

We will therefore accept to change slightly the traditional definition of a "new" comet. From now on, we will call "new" those comets
whose binding energies are smaller than 850 x 10^{-6} AU^{-1}: this is the
place in the distribution of all known orbits, where the wing of the
peak discovered by Oort becomes conspicuous. Most of these "new" comets
are new for us only, since they are usually at their fourth or fifth
passage (with larger perihelion distances) through the system of the
major planets. However, all share a major property: before their
present passage, they had not yet exchanged much energy or momentum
with the planetary system. Those that have, are already beyond our
binding energy limit and have therefore been rejected out of the new
definition.

Is there any empirical evidence of this very slow diffusion of the
orbits of "new" comets towards the inner solar system? Table I and
Fig. 1 represent the ratio N/L of "new" comets N to all long-period
comets L as a function of their perihelion distance. The very fact that
the ratio N/L climbs from 11% (within 0.8 AU) up to 81% (from 3 to 7 AU)
is a strong empirical evidence that the previous discussion is right.
The major jump on Fig. 1 takes place from 1.8 to 3.0 AU; it can be
interpreted by the fact that previous perihelia of those comets (now
near 3 AU) were beyond 10 AU and were therefore undisturbed by the major
perturbations of Jupiter and Saturn. There is no observational bias
that could be large enough to explain away the large statistical
difference (a factor of eight) that is apparent in the N/L ratio, from
0.8 AU to more than 3 AU.

Another empirical evidence comes from the fact that the vaporization dependence on distance, observed for new comets, is the same as

that for older comets: they are all controlled by the sublimation of
water ice (Delsemme 1983); the nine "new" comets studied so far all have
perihelion distances smaller than 1.5 AU; therefore, they are likely to
have all had previous perihelion passages at a distance where any material more volatile than water would have already been lost.

TABLE I

Ratio N/L of the New Comets to all Long-Period Comets,
Versus their Perihelion Distance q

q in AU	L	N	N/L	
0 – 0.8	271	30	11%	
0.8 – 1.2	158	24	15%	
1.2 – 1.6	60	13	22%	
1.6 – 2.0	30	14	47%	
2.0 – 3.0	39	20	51%	
3.0 – 7.0	31	25	81%	
0 – 7.0	589	126	21%	TOTAL

N means the number of NEW comets within each q interval.
L is the number of all LONG-PERIOD comets, including new
comets, within the same interval.

3. A BIMODAL DISTRIBUTION FOR "NEW" COMETS

The second set of data is about the brightness distributions of the
"new" comets (Delsemme 1979); it is bimodal, presumably separating
pristine comets from the other ones: 82% have a brightness distribution
peak near absolute magnitude 5.5 (presumably, the true "new" comets),
whereas 18% show a peak near absolute magnitude 10 (presumably, long
period comets that have lost their original brightness by fragmentation).
Incidentally, in my Table I page 267 (Delsemme 1979) a copying mistake
has put Comet 1932 VI in the 9.1 to 10 line; it should be moved to the
3.1 to 4 line; this brings the number of fragmented comets from 19%
to 18%.

  The shape of the distribution of the true "new" comets implies
that their formation mechanism has not been influenced by fragmentation:
their brightnesses imply a nuclear radius from 5 to 1 km, with an average
of 3 km. They match the size predicted by Goldreich and Ward (1973)
from gravitational instabilities in the protosolar nebula for those
planetesimals accreting near the terrestrial planets, but not in the
zone of Uranus and Neptune where it is usually assumed they were born.
However, the settling of dust was not necessarily homologous in the
solar nebula, and the numerical mismatch cannot be construed as an argument against this type of origin. The essential fact of this discussion
is that the observed size distribution suggests a straightforward
accretion without later fragmentation.

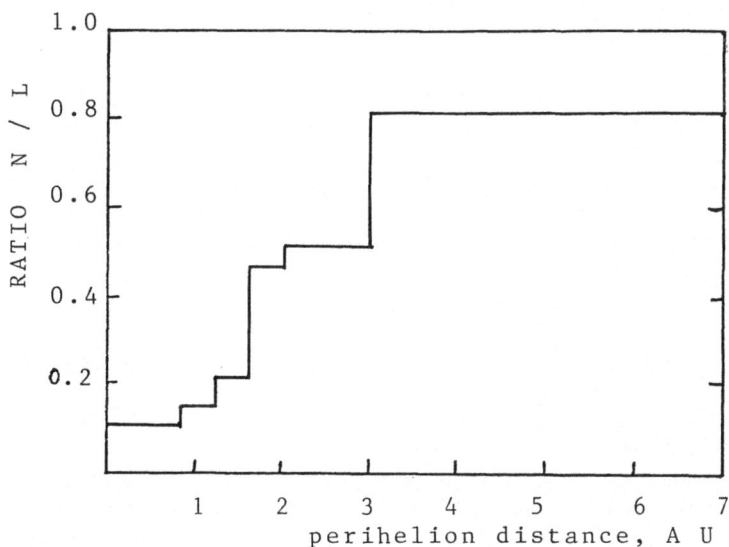

Fig.1.- Ratio of the New (N) to the long-period (L) comets.

4. RETROGRADE VERSUS PROGRADE ORBITS

The third set of data, relevant to the rotation of the Oort's cloud,
must be introduced first by showing the observed regularities in cometary
orbits. We all know that the short-period comets turn in the prograde
direction; it is also conventional wisdom to state that the long-period
comet orbits have a more or less isotropic distribution in space. In
Table II we have refined somewhat this description, by using Marsden's
(1983) Catalog of Cometary Orbits to classify all comets into subsets,
mostly according to their periods.
    According to our new definition, line (7) lists as "new" those
long-period comets whose orbits were known accurately enough to compute
original orbits with an $1/a \leq 850$ (in 10^{-6} AU^{-1}); they include the residue
of (weakly) hyperbolic comets and they probably contain a very large
proportion of really new comets. A few very-long-period comets present
in line 6 are duplicated in lines 4 and 5, since we did not use the
original orbits for those two lines. However, the general trend is
clear: the ratio R/P (retrograde over prograde number of orbits) varies
monotonically as a function of the mean period of each subset (see Fig.
2). The interpretation of this is also obvious: statistically speaking
the more the comets have exchanged energy with the planets, the more
they have exchanged angular momentum.
    In Table II, the group of poorly defined parabolic comets is not
very useful because it is ambiguous: they are listed as having parabolic orbits, either because no deviation from a parabola can be detected, or because the appropriate calculations have not been made.
One of the major reasons for this situation is that the observed arc
may not be long enough. For parabolic comets passing close to the sun,
there is therefore a differential Holetschek (1891) effect. The

TABLE II

COMETS	PERIOD, YEARS	TOTAL	PROGRADE	RETROGRADE	RATIO R/P
(1) Short p.	$3<P<20$	104	104	0	0.00
(2) Interm. p.	$20<P<200$	17	13	4	0.31
(3) Long p. I	$200<P<1000$	37	24	13	0.54
(4) Long p. II	$10^3<P<10^4$	101	$56\frac{1}{2}$	$44\frac{1}{2}$	0.79
(5) Long p. III	$10^4<P<10^5$	43	23	20	0.87
(6) Long p. IV	$10^5<P$, & hyperbolic	139	74	65	0.88
(7) "New"	----	126	65	61	0.94
(8) Poorly defined "parabolic" orbits		312	170	142	1.20
(9) General totals:		879	$529\frac{1}{2}$	$349\frac{1}{2}$	0.66

(3) Five long-period I orbits have almost identical elements; this well-known sungrazing group has been assumed to come from one comet that has recently split, and counted for one in the statistics.

(4) Comet 1970 II, with an inclination of 90.0°, was counted for $\frac{1}{2}$ in the prograde and $\frac{1}{2}$ in the retrograde columns.

(5) The long-period I, II and III are taken straight from Marsden's Catalog. Group II and mainly Group III contain already a few "new" comets.

(6) Group IV contains many "new" comets.

(7) The "new" comet list is from Marsden and Roemer (1982) with a cutoff at $1/a=850\times10^{-6}$ AU^{-1} (see text for discussion).

(8) The 5 identical sungrazing comets have only been counted for one. The differential Holetschek effect probably explains the large R/P ratio of this group.

(9) The general totals are not very meaningful, because the duplication of group (7) has not been removed from the totals.

differential Holetschek effect between prograde and retrograde parabolic comets comes from the Earth's motion on its orbit: since prograde comets turn in the same direction as the Earth, on the average they are hidden by the sun's glare longer. The effect is maximal for q/cos i = 0.7 AU because the perihelion apparent angular velocity component along the ecliptic matches the Earth's velocity on its orbit; therefore, more comets of this type are in the category of the poorly defined orbits, which fall into the parabolic bin. The observed excess of retrograde comets classified as parabolic is present but not very important except for short heliocentric distances, where it reaches R/P = 2 between 0.4 and 0.8 AU; this consistency with the qualitative prediction confirms that the retrograde excess probably comes from observational selection. A correction of this observational selection effect could be undertaken (à la Everhart, 1967) but it was judged that it was beyond the scope of this paper.

The previous discussion clarifies the apparent discrepancy that the reader may detect between our results and those of Fernandez in another

chapter of this book: the R/P ratio is inverted when ambiguous comets
are eliminated from the statistics, and I have shown that we probably
understand why.

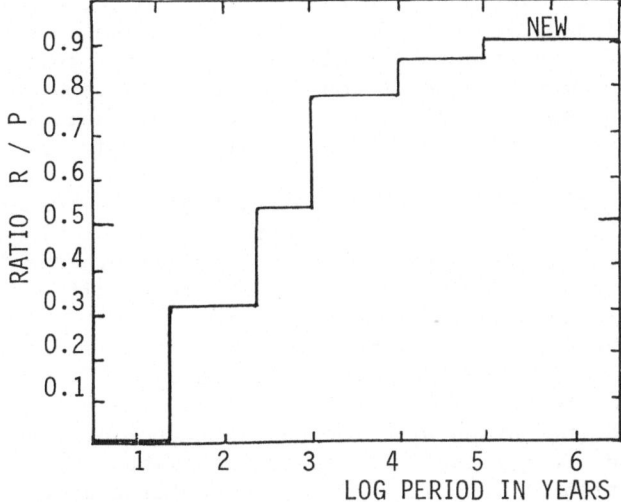

FIG.2.- The ratio R/P of the number of retrograde to prograde
comets grows steadily from zero up to 0.94 as a function of
their period.

5. THE ROTATION OF THE OORT'S CLOUD

If we accept the conventional wisdom that short-period comets derive
from the orbital diffusion of "new", then of long-period comets, the
smooth but steady orbital angular momentum transfer from the planets to
the comets shown in Table II and Fig. 2 comes in step with the random
transfer in binding energy, with a difference: the change in binding
energy is a random walk, whereas the momentum transfer, although random
in magnitude, is always prograde since it remains on the average in the
direction of motion of the interacting planets.

     However, the fact that the R/P ratio remains asymptotically near
94%, and does not reach 100% even for very large periods, is puzzling.
It cannot be dismissed easily as an observational bias. As a matter of
fact, there are more retrograde comets in the small number of "new"
comets observed with perihelia beyond 2 AU; this difference is likely
to be introduced by the slow diffusion of the orbits described before:
retrograde comets are less perturbed than prograde during their passage
through the solar system; for this reason, they are kept longer within
the 1/a limits that we have arbitrarily accepted to define a "new"
comet. However, since this bias is in favor of more retrograde comets,
the fact that R/P remains much lower than unity is even more puzzling.
Since it seems to remain in the purest possible sample of "new" comets,
that have obviously neither exchanged much energy nor rotational
momentum with the solar system, could this dissymetry be the telltale
of a primeval rotational momentum of the Oort's cloud?

6. MODELS OF THE SOLAR NEBULA

The classical theories relate the origin of the Oort's cloud to that of
the solar system; they vary only in the details about the place of
cometary accretion and the mechanism of their ejection into Oort's
cloud. Oort's early hypothesis (origin within the asteroid belt) has
generally been ruled out because it would yield early temperatures incompatible with the survival of an icy conglomerate nucleus (Whipple
1950); all other distances, starting at Jupiter's orbit, up to a few
10^4 AU have been proposed.

A circular ring of protocometary bodies implies a primordial rotational momentum presumably prograde in the equatorial plane of the
protosolar nebula, that is in the ecliptic (for all practical purposes).
According to Kepler's third law, its magnitude (per unit mass) would
increase with the square root of the ring's radius r.

Of course, the transfer of cometary aphelia to the outer fringe of
the Oort's cloud, followed by the growing diffusion of their aphelion
velocities (due to the stellar perturbations) is likely to have introduced a Gaussian distribution of velocites with an r.m.s. velocity
orders of magnitude larger than the initial aphelion velocity. But this
Gaussian distribution of velocities would be centered on a prograde
primeval velocity. The question is: could this prograde velocity be
detected by the dissymmetry R/P? As can be seen in Fig. 3, the R and
the P comets that come back from the Oort's cloud would be from two
different heights in the slope of the Gaussian curve: their ratio R/P
would be lower than unity, and it would become smaller with larger q's.

To clarify the ideas, we have compared the small-mass models of
the nebula, from the Russian School (Otto Schmidt 1949, Safronov 1972-
1977) and the large-mass models of the American School (Kuiper 1951 -
Cameron 1978-82).

In the small-mass models, Safronov (1972) shows that the comets
are planetesimals ejected by the giant planets during their accretion;
100 Earth masses are ejected by Jupiter, 80 by Saturn, 50 by Uranus
and 60 by Neptune; but because of the varying efficiency of the process,
the mass eventually bound into Oort's cloud is 0.2 Earth masses from
Jupiter's contribution, 0.4 from Saturn's, 0.6 from Uranus' and 1.3
from Neptune's. The average weighted distance from which comets were
ejected is therefore rather in the vicinity of 30 AU. For a 1 solar
mass nebula, the Keplerian velocity at this distance is 5.5 km/s. When
ejected at the nominal distance of 50,000 AU, conservation of angular
momentum gives a transverse velocity at aphelion of 3.3 m/s.

In the large-mass model, Cameron (1978) has submitted a novel
theory on the origin of comets. Using a typical 3 solar mass nebula,
it spreads a massive disk at distances going to 300-600 AU. The disk's
mass varies with time; after a nominal maximum mass of 2 solar masses,
the accretion disk loses half of its mass in $t = 2 \times 10^4$ years, by a
photospheric loss mechanism. For those planetesimals whose period is
short compared to t, the mass loss inside their orbit enlarges their
radius only. However, planetesimals (that is, pristine comets) further
away than 500 AU have periods longer than 2×10^4 years. Those for
which the mass loss is slightly smaller than a factor of 2 see their

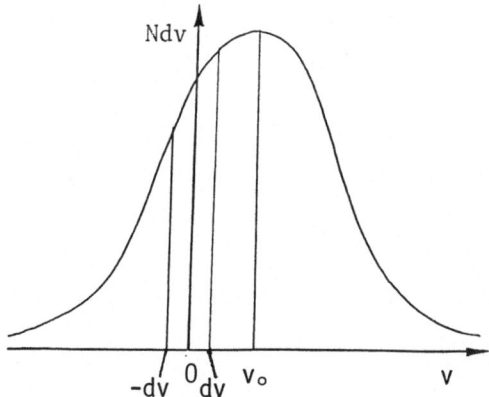

Fig.3. The Gaussian distribution of the transverse velocities
v is centered on the velocity v_o of the primordial ring of
comets. The slope of the distribution near velocity zero
gives an R / P ratio different from unity.

orbits impulsively elongated to those large distances which are now the
outer margin of the Oort's cloud, with a perihelion remaining the same
(500 AU, typically). If we assume a 3 solar mass nebula, the 500 AU
ring of comets turns with a Keplerian circular velocity of 2.3 km/s,
yielding 23 m/s when at the nominal aphelion in the Oort's cloud.

The two scenarios yield two Oort's clouds with a different primeval
momentum, that we have expressed by 2 different transverse velocities
at the nominal (and arbitrary) distance of 50,000 AU. (See Table III).

In Cameron's case, the primeval cloud remains a ring. In Safronov's
case, planetesimals are ejected by their grazing passages near the
giant planets on hyperbolic orbits (related to the planet's center of
mass), going to all directions. At an average of 30 AU, the orbital
momentum transfer per unit mass goes from 0 to 2V (V= 5.5 km/s). The
mean orbital momentum transferred by planetary interaction gives an
average extra 1 m/s in the prograde direction at 50,000 AU.

In both cases, during 4.5 billion years, stellar random perturbations scatter the velocities according to a Gaussian distribution
following formula (1). This yields an RMS velocity of:

$$\Delta V = 121 \text{ m/s}. \tag{5}$$

The RMS transverse component in the plane of the ecliptic is

$$\sigma = \Delta V \sqrt{\frac{1}{3}} = 0.58 \, \Delta V = 70 \text{ m/s}. \tag{6}$$

In RMS units of the present-day distribution, the primeval velocity is
0.33σ (Cameron's model) or 0.06σ (Safronov's model). Translated in
terms of R and P for the retrograde and prograde comets that come back
from the Oort's cloud within 0<q<7 AU, we can write R/P=0.97 (Cameron's)
or 0.995 (Safronov's). These results concern only the velocity components projected onto the ecliptic. In order to compare with the

TABLE III

The Orbital Angular Momentum of the Oort's Cloud

	Large–Mass Theory	Small–Mass Theory
Major reference	Cameron 1978	Safronov 1972
Mass of disk + sun	3 $m_\odot$	1.04 $m_\odot$
Mean radius of protocomets' ring	500 AU	30 AU
Keplerian velocity of ring	2.3 km/s	5.5 km/s
Transverse velocity at 50,000 AU	+23 m/s	+3.3 m/s
Extra velocity due to ejection	0	+1.0 m/s
Primeval velocity in Oort's cloud	+23 m/s	+4.3 m/s
Δv (RMS) in ecliptic (from stars)	±70 m/s	±70 m/s
Primeval velocity in RMS units	0.33σ	0.06σ
Predicted R/P for observed "new" comets	0.97	0.995

observations and reduce somewhat the noise coming from the transverse velocity component which is not in the ecliptic, I have also projected the transverse velocity onto the plane of the ecliptic for each of the 126 new comets, according to the formula:

$$v_e = v(1-\cos^2\omega\,\sin^2 i)^{1/2}. \tag{7}$$

v_e is the transverse velocity projected on the plane of the ecliptic, v is the transverse velocity of the comet, ω is the argument of its perihelion and i the inclination of its orbit.

The 126 transverse velocities projected onto the ecliptic v_e have been classified in the histogram of Fig. 4. To understand the significance of the histogram, a transverse velocity of 2.24 m/s brings the comet's perihelion at the distance of Jupiter's orbit. The well documented depletion for short heliocentric distances is quite apparent near $v_e = 0$. Assuming that the asymmetry of the data comes entirely from the slope present in the Gaussian distribution of the velocities (see Fig. 3) it is concluded that this Gaussian distribution is superimposed on a mean orbital momentum of 152 m/s at the nominal distance of 43,500 AU. The simpleminded use of dN/N = 6% excess of retrograde comets already gives a good approximation of this orbital momentum, very close to the same value. Expressing the number of comets of velocity v by the Gaussian formula:

$$N\,dv = \exp(-\tfrac{1}{2}\,\frac{v^2}{\sigma^2})\,dv \tag{8}$$

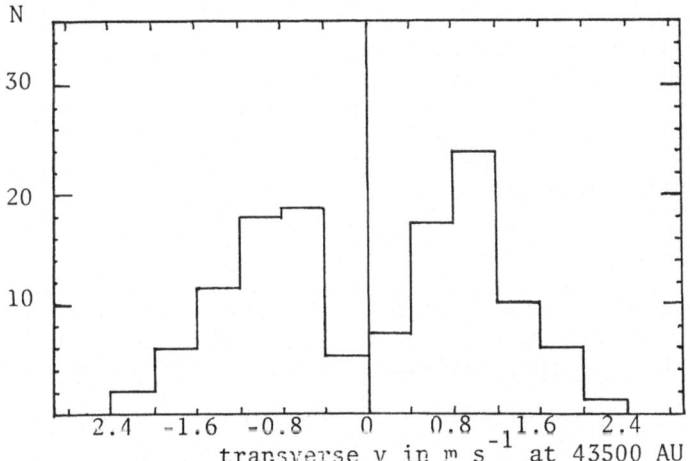

Fig.4.- Number N of new comets classified by transverse
momentum intervals.

the logarithmic differentiation gives:

$$-\frac{v}{\sigma} = \frac{dN/N}{dv/\sigma} \tag{9}$$

with dv = 1.9 m/s between the mean prograde and mean retrograde velocity
of our sample, and σ = 69.4 m/s for the transverse r.m.s. component
over the age of the solar system, dv/σ = 0.027 and dN/N = 0.060,
yielding v = 2.22 σ or 154 m/s.

DISCUSSION

First, it seems prudent to remove from the sample all comets whose
perihelion is close to the orbit of Jupiter. In a computer simulation,
Fernandez (1981) has shown that at Jupiter's distance, the perturbation
on prograde orbits is (statistically speaking) six times larger than on
retrograde orbits. The net result is that prograde orbits have been
removed from the narrow range in binding energy that defines "new"
comets, six times more effectively than retrograde orbits, introducing
an unwanted bias.

        Since the transverse velocity corresponding to Jupiter's orbit is
2.24 m/s, if we use a cutoff of 2.00 m/s for v, we lose only three
comets (one prograde and two retrograde, as would be expected from
Fernandez' results). We are left with 123 "new" comets with R/P = 0.92,
that is an 8% asymmetry in favor of the prograde direction for new
comets. Table IV shows the final results. The oddity of the result
is reinforced by the fact that Fernandez' (1981) mechanism, applied to
all outer planets, predicts a larger depletion of prograde comets than
of retrograde comets. We conclude that the original R/P is even

smaller than 0.92.
      Are the results statistically significant? Because of the small
numbers involved, the chances that the asymmetry is in favor of the
prograde direction are only about 2 to 1. However, when the probabilities are computed for each speed interval, then multiplied, the total
probability in favor of the prograde direction grows to 75%.

TABLE IV

Retrograde/Prograde ratio of the "new" comets, as a function
of their transverse velocity component v_e projected onto
the plane of the ecliptic.

$v_e ms^{-1}$	Prograde	Retrograde	R/P
0 −0.4	7	5	0.71
0.4–0.8	17	19	1.12
0.8–1.2	24	18	0.75
1.2–1.6	10	11	1.10
1.6–2.0	6	6	1.00
2.0–2.4	1	2	2.00
Total	65	61	0.94

Can the results be biased by an observational selection? We have
mentioned a differential Holetchek effect that could explain the excess
of retrograde comets. This observed excess is serious only for small
perihelion distances, as predicted by its interpretation. An inverse
differential effect is therefore predicted for the "good" orbits (those
not classified as parabolic). But at short perihelion distances, the
number of "new" comets (retrograde or not) is severely depleted, as are
all comets (Everhart 1967). For this reason, the inverse differential
Holetchek effect should be practically negligible for the "new" comets.
It is concluded that the spurious excess of quasi-parabolic comets on
retrograde orbits introduces an insignificant bias for new comets.
      Let's assume therefore that the observed asymmetry of at least 8%
is real. It is incompatible with a primordial rotational momentum due
to a ring of protocomets at 30 AU (R/P = 0.995) or at 500 AU (R/P = 0.97).
Because of the uncertainties due to small number statistics, it could
be more easily consistent with Cameron's model than with Safronov's.
However, a primordial ratio R/P = 0.94, taken at its face value, would
rather be suggestive of a very massive protosolar nebula or of a nebular
satellite of the protostar system. In such a dense nebular satellite,
the formation of comets could have taken place on a more reasonable time
scale than at the low density of the interstellar medium. Another
poooibility is Hill's (1982) mechanism: in presence of turbulence
the infall is no more strictly radial (Cameron 1984).
      Another possible interpretation is that the detected rotational
momentum is real but not primordial, it would be connected to a recent

non-radial impulse upon the cloud of comets. The observed "new" comets have had their perihelion lowered into the inner solar system during their last 3 to 6 orbits only ($\Delta v = 2.5$ ms^{-1} per orbit), that is during the last 10-20 million years. This gives the time scale for a "recent" impulse. Three distinct possibilities fall into this category: a giant molecular cloud, the close passage of a foreign star, or an unknown stellar companion of the Sun.

First possibility: each passing star produces a small non-radial impulse; it is only because of the large number of stars (more than 25,000) that have randomly influenced the Oort's cloud in the last 4.5 billion years, that we speak about a symmetrically Gaussian distribution of velocities within the cloud. However, during the last 10-20 million years, we deal with a population of 50-100 perturbing stars only, and the very large influence of one single close stellar encounter becomes a serious possibility. Biermann, Huebner and Lust (1983) believe that indeed they have identified one star track by aphelion clustering in the Oort's cloud.

Second possibility: following the confirmation of the existence of a system of giant molecular clouds in the Galaxy (Solomon and Edmunds 1980), their passage close to the Solar System has been shown to have an action competitive with that of the nearby stars. The impact of such a molecular cloud at grazing incidence on the Oort's cloud could have introduced all the rotational momentum needed to explain the R/P ratio observed on "new" comets some 10-20 million years later; however, since the number of passages of the Solar System through giant molecular clouds during its lifetime is estimated to be in the range of 1 to 10 (Bailey 1983), the mean duration between passages would be ½ to 2 billion years, and the possibility of a passage during the last 10-20 million years would therefore be of the order of 1%, which makes the hypothesis rather unattractive. For this reason, Clube and Napier (1984) go one step further, opening up vistas on new mechanisms modifying the traditional Oort's cloud concept by cometary captures during passages through "grainy" molecular clouds with a mass structure that plays a fundamental role; a complete discussion of these original ideas is beyond the scope of this paper.

Third possibility: an unknown stellar companion of the Sun, whose 26-28 million year period, on a rather elongated orbit, has been proposed to explain the quasi-periodicity of the mass extinctions of living species on Earth, reported by Raup and Sepkoski (1984). The impact that has produced the major extinction separating the Cretaceous from the Tertiary has now been well substantiated (Bohor et al. 1984) by the existence of shock-metamorphic features of high pressure in quartz grains (duplicated by laboratory experiments at 90 kilobars) and excluding all other hypotheses trying to circumvent an extraterrestrial impact. Since other geologic layers contain the same extraterrestrial concentration (30 to 300 times) of iridium and of heavy metals of the platinum group (Alvarez et al. 1984) sometimes in multiple horizons, there is not much doubt left that there was a multiple impact of several large bodies, concentrated in 10 thousand to 1 million years, and possibly repeated at quasi-periodic intervals of 26-28 million years. Comets are clearly good candidates for this purpose, and the hypothetical stellar companion of the Sun is one of the two mechanisms proposed so far (the frequency of molecular clouds collisions, if modulated by oscillations of the sun around the mid-plane of the Galaxy, is another possibility). Future will tell whether scenarios can be found to off-set the inherent orbital instability of the stellar companion of the Sun, over the age of the Solar System (even a circular orbit near 50,000 AU has a survival probability of the order of 1-2%, Shoemaker 1984).

Such a stellar companion would clearly be a good explanation for the asymmetry R/P, since it would have transmitted its last impulse 11 or 12 million years ago, that is about four "new" comets' periods. A constraint introduced by the wing of the Oort's peak, in the distribution in (1/a), implies that the mean perihelion distance of the companion could not be much less than 10,000 AU. This would imply a quite acceptable aphelion at 170,000 AU to reach the proper period, as well as a transit time of the order of one million years to go across the Oort's cloud.

The scarcity of the data does not allow to choose any further among the different hypotheses that we have discussed here.

On the other hand, the influence of the forces exerted by the Galaxy on the Oort's cloud have been completely neglected in the present paper. We intend to compute soon the other spatial components of the orbital angular momentum of the set of our 126 "new" comets, in order to verify if it could be explained by tidal or epicycle effects in the Galaxy.

Orlando Naranjo and Boon Soonthornthum verified most of my computations. Grants from NSF: AST-82-07435 and from NASA: NSG-7301 (planetary atmospheres) are gratefully acknowledged.

NOTE ADDED IN PRESS

Important developments have taken place in the six months following the Rome meeting: The orbital angular momenta of the 126 new comets show a much larger anisotropy in a plane perpendicular to the ecliptic; it is 54% larger in the retrograde than in the prograde direction. This is too large to come from primordial or galactic effects; it would be dissipated by orbital diffusion in 20-30 million years, hence it must be due to a recent impulsive event. Fast moving bodies (stars or molecular clouds) are ruled out: only a slow body like Nemesis is acceptable. The presumed orbit of Nemesis has been deduced; larger anomalies are found (at the 2σ to 3σ level) along the predicted orbit, over a strip of the sky of 180°, suggesting the place of the perihelion of an eccentric orbit for this massive object. Details will be published at the Tucson colloquium "The Galaxy and the Solar System" (January 1985) and in a letter to "Nature" (Delsemme 1985).

REFERENCES

Alvarez W., Kauffman E.G., Surlyk F., Alvarez L.W., Asaro F.,

Michel H.V. (1984). Science 223, 1135.

Bailey M.E. (1977). Astrophys. Space Sci. 50, 3.

Bailey M.E. (1983). Mon. Not. R. Astron. Soc. 204, 603.

Biermann L., Huebner W. and Lüst Rh. (1983). Proc. Natl. Acad. Sci. USA 80, 5151.

Bohor B.F., Foord E.E., Modreski P.J., Triplehorn D.M. (1984). Science 224, 867.

Cameron A.C.W. (1978) p. 49 in "Origin of Solar System", ed. S.F. Dermott; Wiley NY.

Cameron A.C.W. (1979). Moon Planets, 21, 173.

Cameron A.C.W. (1984). Formation and Evolution of Solar Nebula (Harvard-Smithsonian preprint).

Cube S.V.M. and Napier W.M. (1984). Mon. Not. R. Astron. Soc., (in press).

Delsemme A.H. (1977) p. 453 in "Comets, Asteroids, Meteorites", ed. A.H. Delsemme, Univ. of Toledo Bookstore.

Delsemme A.H. (1979) p. 265 in "Dynamics of the Solar System" ed. R.L. Duncombe, publ. IAU, Reidel, Holland.

Delsemme A.H. (1983). Proc. "Ices in the Solar System", NATO, Nice, Jan. 1983 (in press).

Delsemme A.H. (1985). Letter to "NATURE" (submitted, preprint available).

Everhart E. (1967). Astron. J. 72, 716.

Fernandez J.A. (1981). Astron. Astrophys. 96, 26.

Goldreich P. and Ward W. (1973). Astrophys. J., 183, 1051.

Hills, J.G. (1982). Astronom. J. 87, 906.

Holetscheck J. (1981). Astron. Nachr., 126, 75.

Marsden B. (1983). Catalog of Cometary Orbits, Enslow Publ., Hillside, NJ.

Marsden B. and Roemer E. (1982). p. 720 in "Comets" ed. L. Wilkening, Univ. of Ariz. Press.

Marsden B., Sekanina Z. and Everhart E. (1978). Astronom. J. 83, 64.

Marsden B. and Sekanina Z. (1973). Astronom. J. 78, 1118.

Oort J.H. (1950). Bull. Astron. Inst. Netherlands, 11, 91.

Raup D.M. and Sepkoski J.J. (1984). Proc. Nat. Acad. Sci. USA, 81, 801.

Safronov V.S. (1972). p. 329 in Proc. IAU Symposium No 45, ed. Chebotarev et al.; Reidel Dordrecht, Holland.

Shoemaker E. (1984). Division for Planetary Science Meeting, AAS, Hawaii, October 1984.

Solomon P.M. and Edmunds M.G. (1980). Editors, Proc. Third Gregynog Workshop on Giant Molecular Clouds.

Weissman P.R. (1982) p. 637 in "Comets" ed. L. Wilkening, U. of Arizona Press, Tucson, Arizona.

DYNAMICAL EVOLUTION OF THE OORT CLOUD

Paul R. Weissman
Earth and Space Sciences Division
Jet Propulsion Laboratory
Pasadena, CA 91109   USA

ABSTRACT. New studies of the dynamical evolution of cometary orbits in the Oort cloud are made using a revised version of Weissman's (1982) Monte Carlo simulation model, which more accurately mimics the perturbation of comets by the giant planets. It is shown that perturbations by Saturn provide a substantial barrier to the diffusion of cometary perihelia into the inner solar system; Jupiter also. Perturbations by Uranus and Neptune are rarely great enough to remove comets from the Oort cloud, but do serve to scatter the comets in the cloud in $1/a$. The new model gives a population of 1.8 to 2.1 x 10^{12} comets for the present-day Oort cloud, and a mass of 7 to 8 earth masses. Perturbation of the Oort cloud by giant molecular clouds in the galaxy is discussed, as is evidence for a massive "inner Oort cloud" internal to the observed one. The possibility of an unseen solar companion orbiting in the Oort cloud and causing periodic comet showers is shown to be dynamically plausible but unlikely based on the observed cratering rate on the earth and moon.

1.   INTRODUCTION

Weissman (1982, hereafter referred to as Paper I) developed a new Monte Carlo simulation model for studying the dynamical evolution of comets in the Oort cloud under the influence of random stellar perturbations. By running a large number of hypothetical comets through the model, it was possible to test the dynamical plausibility of various theories of cometary origin. One of the more important results of Paper I was that the population of the Oort cloud had been depleted by between 30 and 84% depending on where in the primordial solar nebula the comets actually formed. Also, using the simulation model it was possible to calculate the population of the cloud and the distribution of orbital elements of comets in the Oort cloud, and to show that stellar perturbations had so randomized the cometary orbits over the history of the solar system that it was impossible to choose among several proposed formation sites for the Oort cloud comets.

The simulation model developed in Paper I assumed a relatively simple mechanism for perturbation of comets passing through the planetary region: comets approaching within 1.4 times the semimajor axis of Sat-

87

urn's orbit had a 65% chance of being removed from the Oort cloud either by hyperbolic ejection or capture to a short-period orbit due to Saturn perturbations, and those approaching Jupiter's orbit had a 94% probability of removal. Comets which returned to the Oort cloud were assumed to have no change in their orbital energies, thus ignoring the small, but non-zero, perturbations by Jupiter and Saturn, as well as those by Uranus and Neptune. Perturbations by the two outer jovian planets are rarely great enough to remove comets from the Oort cloud, but can yield a significant change in semimajor axis, orbital period, and aphelion distance.

In the revised model presented here, the perturbations on the orbital energy of comets entering the planetary region, q < 40 AU, is chosen randomly from a gaussian distribution with rms value equal to the sum of the rms perturbation from each of the planets whose orbits were crossed. For Jupiter the rms perturbation is 620. x 10^{-6} AU^{-1}, and for Saturn, Uranus, and Neptune the values are 101., 7.7, and 5.9 x 10^{-6} AU^{-1}, respectively. The new semimajor axis is then tested for hyperbolic ejection, an aphelion distance beyond the sun's sphere of influence (2 x 10^5 AU), or capture to a shorter period orbit (P < 10^6 years, Q < 2 x 10^4 AU) where the comet is relatively unaffected by stellar perturbations. Comets which survive these end-states and are returned to the Oort cloud do so with the new, perturbed semimajor axis and continue to evolve until they fall into one of the possible end-states, or the total time followed exceeds the age of the solar system.

2. DYNAMICAL EVOLUTION OF COMETS

As in Paper I, the revised dynamical model was used to test different hypotheses of cometary origin. Initial perihelion distances of 20, 200, and 10^4 AU were chosen to represent cometary formation as icy planetesimals in the Uranus-Neptune zone, on the edge of the primordial solar nebula accretion disk, or in distant subfragments of the primordial nebula. The initial aphelion distance for all the orbits was chosen as 4 x 10^4 AU, and the total rms velocity perturbation on the Oort cloud over the history of the solar system was chosen to be 0.120 km s^{-1}. Cases were typically run for between 10^4 and 5 x 10^4 hypothetical comets.

The relative fraction of comets falling into each of four possible end-states: direct hyperbolic ejection, stellar loss (aphelion beyond 2 x 10^5 AU), planetary loss (ejection or capture due to planetary perturbations), and survivor is given in Table 1. These results are almost identical with those for similar cases described in Paper I, indicating that the previous, approximate treatment of planetary perturbations was acceptable. Thus, the major results of Paper I are validated. The lack of direct ejections by stellar perturbations is a result of a weakness in the dynamical model which emphasizes the sum of distant random perturbations, rather than the more violent though less frequent perturbations by stars passing directly through the Oort cloud.

A new result of the revised model is the ability to more accurately study the diffusion of cometary perihelia into the planetary region. The perihelion distribution of all new comets passing within 200 AU of the sun over the history of the solar system is shown in Figure 1. Pre-

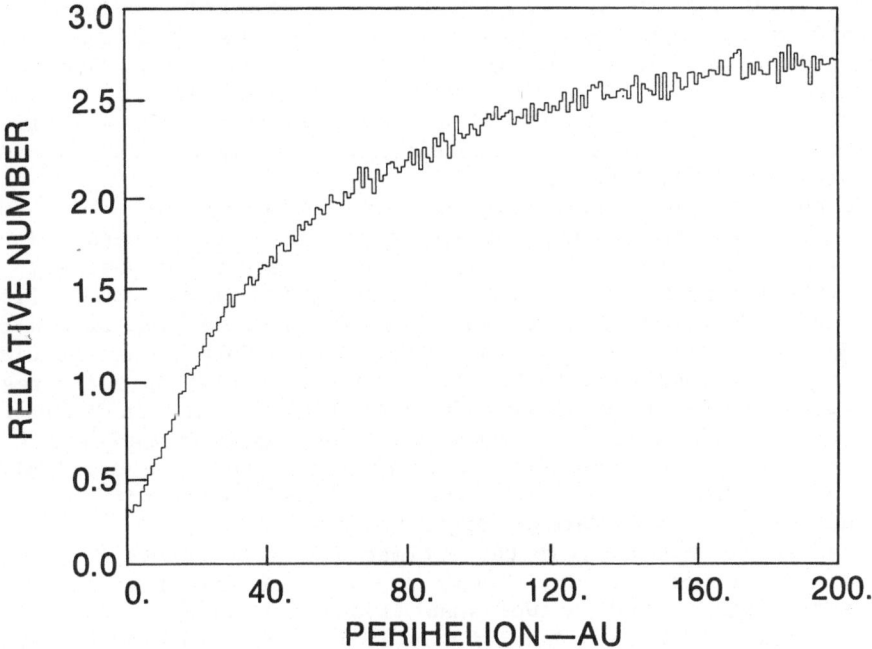

Figure 1. Perihelion distribution for new comets from the Oort cloud passing through the planetary region over the history of the solar system. The number of comets increases rapidly with increasing perihelion distance.

TABLE I. Oort Cloud End-States

	Initial perihelion	- AU	
End-state	20	200	10,000
Planetary loss	0.773	0.477	0.076
Ejected	0.0	0.0	0.0
Stellar loss	0.068	0.068	0.214
Survivor	0.159	0.455	0.710

vious estimates of the perihelion distribution by Oort (1950) and Weissman (1980) predicted that it would be flat with increasing perihelion distance, while Kresak and Pittich (1978) believed it could be fit by $N(q) \propto q^{1/2}$. In fact, the number of comets per interval in perihelion distance increases only slowly in the observable region (q < 5 AU), then rises rapidly out to about 20 AU before beginning to level off and asymptoticly approach a maximum value about ten times that at 1 AU. The number of comets passing through perihelion is 1.5, 2.1, 3.6, and 5.0 times the value at 1 AU at the orbits of Jupiter, Saturn, Uranus, and Neptune, respectively. At 200 AU the flux of comets is 9.3 times the flux at 1 AU and still slowly increasing. This result was also demonstrated by Fernandez (1982).

To enter the observable region of the solar system, q < 5 AU, the comets must "leap frog" over the orbits of the outer planets, with some probability that they will be lost at each perihelion passage. Thus, the inner solar system is, in fact, undersupplied in new comets from the Oort cloud. Conversely, to account for the observed flux of new comets from the Oort cloud requires a larger total Oort cloud population. Following the method used in Paper I, the new population estimates from the revised dynamical model are shown in Table 2. As in Paper I, the different possible formation sites for the cometary nuclei all yield approximately the same current population estimate. The original population of the Oort cloud varies, however, based on the different depletion factors for the different origin hypotheses.

Also shown in Table 2 is the estimated mass of comets in the original and current Oort clouds. These estimates are based on the cometary mass distribution found by Weissman (1983) but using a revised value for the albedo of cometary nuclei surfaces. Weissman (1983) assumed an albedo of 0.6 based on observational results by Delsemme and Rud (1973). But laboratory studies of the albedo of dirty ice mixtures by Clarke and Lucey (1984) have shown that albedos between 0.1 and 0.3 are more likely to be expected. Using an assumed albedo of 0.3 raises the previous estimate of cometary masses in Weissman's paper by a factor of 2.8. These revised values are shown in Table 2.

The number of comets passing within 1.4 times the semimajor axis of each of the Jovian planets and their dynamical fate is shown in Table 3. Note, that although this particular case (initial q = 20 AU) followed 5×10^4 hypothetical comets, there are over 2.4×10^5 passages within the planetary region. Thus the average Oort cloud comet has possibly made between four and five passes through the planetary region, though fully 80% of them are through the Uranus-Neptune zone only. Note that Saturn actually removes slightly more comets from the Oort cloud than does Jupiter, because its lower removal efficiency is compensated for by a higher flux of comets crossing its orbit.

Lastly, the distribution of orbital elements for the hypothetical comets surviving in the present-day Oort cloud is shown in Figure 2. This, again, is for the case with initial q = 20 AU. Although the orbits all had the same initial energy, $1/a = 50 \times 10^{-6}$ AU^{-1}, and eccentricty, e = 0.9990, they have now diffused widely through the diagram to fill most of the possible orbits. The sharp cut-off of orbits along the left edge of the scatter diagram is the result of the limit on the sun's sphere of influence of 2×10^5 AU.

The especially dense band of comets extending down from the middle-top of Figure 2 towards the left are comets with aphelia of about 4.5×10^4 AU. Though the orbits in the cloud have been largely randomized, there appears to be some concentration close to the original aphelion

TABLE II. Oort Cloud Population and Mass

	Initial perihelion - AU		
	20	200	10,000
Current poulation - 10^{12}	1.8	2.1	1.9
Original population - 10^{12}	11.7	6.5	3.7
Current mass - earth masses	7.1	8.2	7.4
Original mass - earth masses	45.5	18.4	10.5

TABLE III. Comets Passing through the Planetary Region

	Planet				
End-state	Jupiter	Saturn	Uranus	Neptune	Total
Ejected	8899	8594	28	5	17526
Stellar loss	122	1021	202	86	1431
Captured	8749	9280	1104	556	19689
Returned	1056	9837	87449	108646	206988
Total	18826	28732	88783	109293	

distance, 4×10^4 AU. In reality the expected aphelion distribution of initial Oort cloud orbits would be some range of values, rather than the single one used in this test case, so the concentration shown in Figure 2 would not exist.

In general, the cometary aphelia tend to diffuse outward (towards the left in Figure 2) rather than inward. This results from the pumping up of orbital energy and angular momentum by the stellar perturbations, but is also a reuslt, in part, of limitations in the dynamical simulation model. Because the comets are typically in such long period, highly eccentric ellipses it is assumed that stellar perturbations occur at aphelion only. Future models will attempt to correct this deficiency.

3. PERTURBATIONS BY GIANT MOLECULAR CLOUDS

Clube and Napier (1982) suggested that the Oort cloud would be severely perturbed by random encounters with giant molecular clouds in the galaxy. They estimated that the solar system would have about 10 to

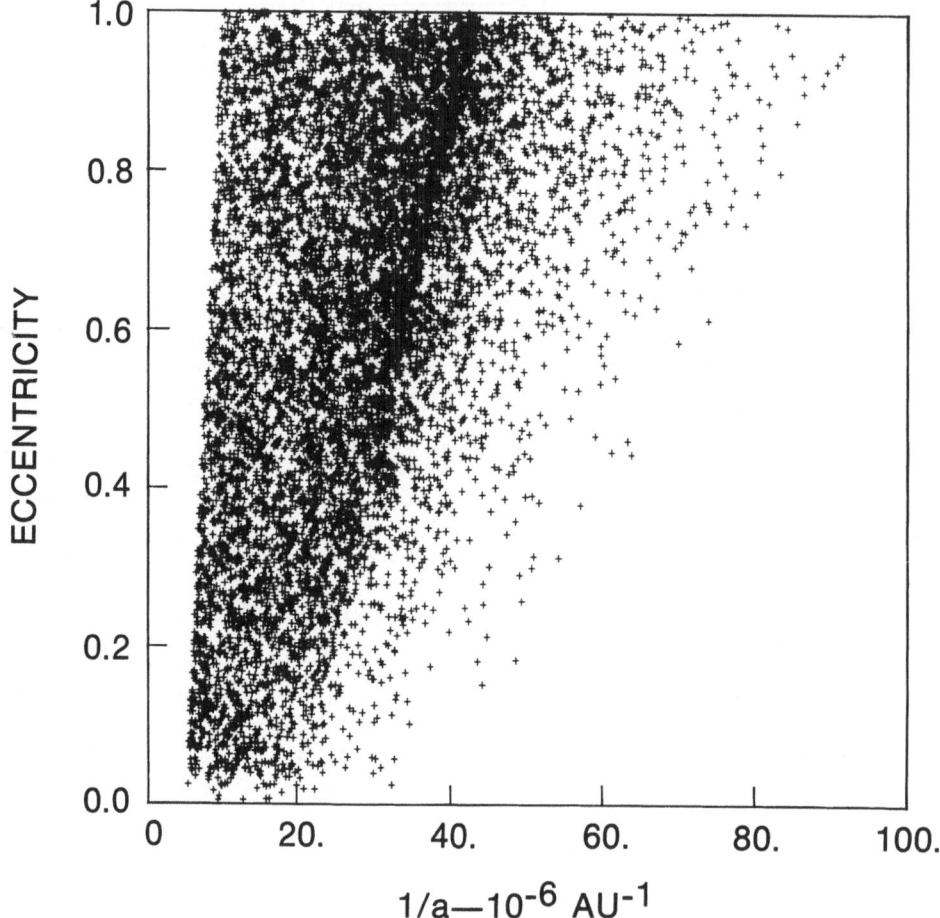

Figure 2. Scatter diagram in energy and eccentricity for the orbits of 7928 hypothetical comets in the current Oort cloud. The comet orbits have been randomized and have diffused widely away from their initial values of 1/a and eccentrictiy of 50×10^{-6} AU^{-1} and 0.9990, respectively. The sharp cut-off at the left is caused by the aphelion limit of 2×10^5 AU.

20 encounters with GMC's in its lifetime, and that the tremendous mass of the GMC's would strip away a major fraction of the comets in the cloud. They also suggested that the cloud was replenished by capturing large numbers of comets which were originally a part of the perturbing GMC.

An examination of Clube and Napier's ideas by Bailey (1983a) has shown that they overestimated the perturbations on the Oort cloud by a factor of between two and ten. The range in uncertainty is caused by our lack of knowledge of the sun's past dynamical history in the galaxy. The sun's current velocity relative to the Local Standard of Rest is about 16 km s^{-1}, as compared with typical values of about 60 km s^{-1} for G-type stars in the galaxy. If the sun has always had this relatively low velocity then the perturbations by the GMC's are reduced by only a factor of two, and are still a major problem. But if the current low velocity of the sun is only a statistical fluke, and its past velocity has been closer to the observed value for G-type stars, then the GMC's represent only a modest addition to the total perturbation from random passing stars.

Since the acceleration of random motions of stars during their life-times generally results from encounters with GMC's, the latter of the two alternatives described above is more likely the correct one. It would also be more likely from a statistical point of view.

Bailey also found that the distribution of observed cometary orbits was consistent with a "centrally condensed cloud" in which there was a signifcant number of comets interior to those actively being perturbed by passing stars. Clube and Napier (1984) found that such an "inner Oort cloud" could serve as a replenishment source for the outer, observable cloud after a catastrophic encounter with a GMC. This eliminated the difficulty of capturing comets from a GMC, typically encountered at a relatively high velocity, ~20 km s^{-1} or more.

The existence of a massive inner Oort cloud has been suggested for a number of other reasons. Shoemaker and Wolfe (1984) believe that planetesimals scattered out of the Uranus-Neptune zone would form such an inner cloud and account for the late heavy bombardment of the terrestrial planets and Jovian satellite systems. Cameron (1978) has suggested that such an inner cloud would come from icy planetesimals formed in a primordial solar nebula accretion disk extending out to 10^3 AU or more. Fernandez (1980) has shown that such a cloud would be a more efficient source for the short-period comets, and both Whipple (1964) and Bailey (1983b) have suggested that a comet belt or cloud beyond Neptune could explain the perturbations on Uranus's orbit.

An inner Oort cloud would be difficult to sense dynamically because the aphelia of the comet orbits would be too small, $< 10^4$ AU, for them to be significantly perturbed by random passing stars, and the perihelia would be beyond the orbit of Neptune. Hills (1981) suggested that stars passing particularly close to the sun would perturb the inner cloud and send showers of comets into the planetary region perhaps once every 5 x 10^8 years.

Bailey (1983c) calculated that the thermal radiation of comets in the inner cloud might be detectable using the Infrared Astronomical Satellite (IRAS), and Low et al. (1984) have reported that IRAS observed unexplained emission which they interpret as clouds of cold material in the outer solar system. Weissman (1984) interpreted the IRAS observations of clouds of fine particulate material around Vega and some 30 other stars in the solar neighborhood as the dust from inner Oort clouds around each of these stars.

4.   UNSEEN SOLAR COMPANIONS IN THE OORT CLOUD

Whitmire and Jackson (1984) and Davis et al. (1984) suggested that a small, unseen companion star to the sun is in a distant orbit which periodically takes it through the denser regions of the Oort cloud. Although this companion star must be of low mass and luminosity to have escaped detection, its low orbital velocity makes it a very significant perturber. According to the authors above, the companion star perturbs the inner cloud and causes periodic comet showers which eventually result in catastrophic biological extinctions on the earth. They suggest that the star has a period of 2.8×10^7 years to match the estimated periodicity in the extinctions found from paleontological records.

The stability of a star in such a distant orbit can be studied using the revised Oort cloud simulation program described in this paper. According to the authors above, the perihelion of the companion star is about 3×10^4 AU, giving an aphelion distance of 1.55×10^5 AU to yield the correct orbital period. Five thousand hypothetical stars in such an orbit, randomly oriented on the celestial sphere, were integrated through the revised model.

It was found that 23% of the stars failed to survive for ten orbits, the estimated number of observed cycles of extinction in the fossil record. Most of those lost diffused to aphelia beyond the sun's sphere of influence, a likely result considering the large initial aphelion. Continuing the integration further, 86% of the stars failed to survive for more than 10^9 years. Additionally, the orbital period of some stars which remained bound tended to oscillate considerably, the average period change per orbit being on the order of 10%.

In conclusion, one can not exclude the possibility of such a star remaining bound to the sun in a reasonably constant period orbit during the recent past, but the dynamical simulation results suggest that it is far from a certainty. The probability is small that the star should be in just such an orbit at this point in the solar system's history if it formed concurrently with the planetary system.

A more conclusive argument against the existence of an unseen companion star involves the observed cratering record on the earth and moon. Current estimates of the mean cratering rate in the terrestrial region based on craters counted on dated surfaces is in rough agreement with that expected from the observed flux of earth-crossing comets and asteroids. Repeated comet showers would result in far more craters, between 5 and 18 times the current number (depending on the assumed mass distribution for cometary nuclei) if the companion star had been in the same orbit over the history of the solar system.

Assuming that the companion star has evolved outward from a more tightly bound initial orbit does not alleviate this problem; it in fact makes it worse since the shorter initial orbital period would result in more frequent comet showers. The only apparent solution is to assume that the companion star was captured relatively recently, about $3 - 5 \times 10^8$ years ago. The probability of a capture event is very low, on the order of 10^{-13} per star passage. Still, comet showers over only that recent interval may be enough to double the total cratering on the earth and moon in the past 4×10^9 years. That result would be incompatible with the presently estimated cratering rate from dated craters.

Acknowledgments. The author thanks A. L. Lane for providing computing facilities used in a part of these studies. This work was supported by the NASA Planetary Geosciences Program and was performed at the Jet Propulsion Laboratory under contract with the National Aeronautics and Space Administration.

References:

Bailey, M. E. 1983a. The structure and evolution of the solar system comet cloud. Mon. Not. Roy. Astron. Soc., 204, 603-633.

Bailey, M. E. 1983b. Comets, planet X, and the orbit of Neptune. Nature, 302, 399-400.

Bailey, M. E. 1983c. Theories of cometary origin and the brightness of the infrared sky. Mon. Not. Roy. Astron. Soc., 205, 47P-52P.

Cameron, A. G. W. 1978. The primitive solar accretion disc and the formation of the planets. In The Origin of the Solar System, ed. S. F. Dermott, John Wiley & Sons, New York, pp. 49-75.

Clarke, R. N., and Lucey, P. G. 1984. Spectral properties of ice-particulate mixtures: Implications for remote sensing I: Intimate mixtures. J. Geophys. Res., 89, 6341-6348.

Clube, S. V. M., and Napier, W. M. 1982. Spiral arms, comets, and terrestrial catastrophism. Quart. J. Roy. Astron. Soc., 23, 45-66.

Clube, S. V. M., and Napier, W. M. 1984. Comet capture from molecular clouds: A dynamical constraint on star and planet formation. Mon. Not. Roy. Astron. Soc., 208, 575-588.

Davis, M., Hut, P., and Muller, R. A. 1984. Extinction of species by periodic comet showers. Nature, 308, 715-717.

Delsemme, A. H., and Rud, D. A. 1973. Albedos and cross-sections for the nuclei of comets 1969IX, 1970II, and 1971I. Astron. & Astrophys., 28, 1-6.

Fernandez, J. A. 1980. On the existence of a comet belt beyond Neptune. Mon. Not. Roy. Astron. Soc., 192, 481-491.

Fernandez, J. A. 1982. Dynamical aspects of the origin of comets. Astron. J., 87, 1318-1332.

Hills, J. G. 1981. Comet showers and the steady state infall of comets from the Oort cloud. Astron. J., 86, 1730-1740.

Kresak, L., and Pittch, E. M. 1978. The intrinsic number density of active long-period comets in the inner solar system. Bull. Astron. Inst. Czech. 29, 299-309.

Low, F. J., et al. 1984. Infrared cirrus: New components of the extended IR emission. Astrophys. J., 278, L19-L22.

Oort, J. H. 1950. The structure of the cometary cloud surrounding the solar system and a hypothesis concerning its origin. Bull. Astron. Inst. Neth., 11, 91-110.

Shoemaker, E. M., and Wolfe, R. F. 1984. Evolution of the Uranus-Neptune planetesimal swarm (abstract). In Lunar and Planetary Science XV, 780-781.

Weissman, P. R. 1980. Stellar peturbations of the cometary cloud. Nature, 288, 242-243.

Weissman, P. R. 1982. Dynamical history of the Oort cloud. In Comets, ed. L. L. Wilkening, Univ. Arizona Press, Tucson, pp. 637-658.

Weissman, P. R. 1983. The mass of the Oort cloud. Astron. & Astrophys., 118, 90-94.

Weissman, P. R. 1984. The Vega particulate shell: Comets or asteroids? Science, 224, 987-989.

Whipple, F. L. 1964. Evidence for a comet belt beyond Neptune. Proc. Natl. Acad. Sci., 51, 711-718.

Whitmire, D. P., and Jackson, A. A. IV. 1984. Are periodic mass extinctions driven by a distant solar companion? Nature, 308, 713-715.

DISCUSSION

Fernandez: Have you computed the perturbations caused by the hypothetical solar companion on the Oort cloud?

Weissman: I have only done some very preliminary runs with one of my Monte Carlo simulation programs and it seems that an unseen solar companion would be very damaging to the Oort cloud, removing perhaps 10% of the comets on each orbit. It is difficult to estimate this accurately because the impulse approximation used in the program is not valid when the star's velocity is so low. To do it correctly requires a detailed integration of the orbits.

Lissauer: Will the magnitude of Nemesis' perturbations of the Oort cloud comets vary greatly from orbit to orbit?

Weissman: Yes. The perihelion of the star's orbit random walks up and down which would greatly vary the number of comets showering from the inner Oort cloud into the planetary region. This fact has been used by Hut to explain why some extinctions are more catastrophic than others.

Bailey: Can you comment on the increase in the rate of perihelion passages versus perihelion distance found by yourself (factor ~ 10) and the last speaker, Fernandez (factor ~ 100)?

Weissman: Fernandez follows the comets throughout their lifetimes while I terminate my runs sooner, when the comets can no longer be considered part of the Oort cloud. Comets with perihelia in the Uranus-Neptune zone continue to survive for many additional returns because of the relatively small perturbations in 1/a caused by those planets.

STELLAR PERTURBATIONS ON COMETS

F. REMY and F. MIGNARD
CERGA
Avenue Copernic
06130, Grasse (France)

ABSTRACT. The stochastic processes involved in the evolution of a hypothetical cloud of comets are investigated. The cloud is assumed to be randomly perturbed by passing stars approaching the Sun at distance less than 1 pc. Within the frame of the impulse approximation we derive analytical expressions for the probability distribution of the impulse imparted to comets for both close and distant approaches. An application to the rate of ejection of comets is presented.

I. INTRODUCTION

The existence of a reservoir of comets lying in the outskirts of the solar system is now largely admitted since its introduction by Oort in 1950. The dutch astronomer also thought of a possible mechanism to send comets from the cloud to the inner regions of the solar system where they become eventually active and visible. According to his views, close approaches with passing stars change the comet's orbital elements up to the point at which its perihelion distance may become small enough for the comet to be strongly influenced by planetary perturbations and turned into a short period comet. Within the framework of this model the role of the gravitational effect of stars is limited to the injection of comets to the sphere of influence of the planets, primarily that of Saturn and Jupiter. These comets are currently referred to as new comets in contrast with those comets repeating passages after they were perturbed by the planets.

A convenient way to test the efficiency of the model is to simulate the effect of cumulative perturbations due to passing stars on the orbital elements of comets within a cloud. Such a work has been in particular carried out by Weissman (1980, 1982, 1983). Weissman models the stellar perturbation on comets as a single impulse imparted to the comet at aphelion. This single impulse every orbital period is assumed to be the synthesis of the three or four perturbations by passing stars experienced by a comet during one revolution about the Sun. By selecting a cloud of initially very eccentric comets, Weissman follows the long term evolution of aphelion and perihelion distance. He shows that perihelion diffuses into the planetary region and fraction of the initial population may

97

A. Carusi and G. B. Valsecchi (eds.), Dynamics of Comets: Their Origin and Evolution, 97–104.

become visible. He demonstrates as well that a small fraction of the cloud is lost in the interstellar medium, mainly because of a diffusion of aphelia.

More recently, Remy and Mignard (1985) have refined the study by Weissman by allowing the perturbations to occur at random, whatever the location of the comet on its orbit. In addition the magnitude of the perturbation imparted to the comets during the passage of a star is evaluated through a modelling of the arrival of stars in the vicinity of the Oort cloud both in term of the distance to the Sun and to the comet and also by drawing at random the direction of the star's velocity vector.

Certainly the only way to follow properly the evolution of the Oort cloud during the past 4.5 billion years is to rely on numerical simulations. However since the passage of stars happens at random the impulse imparted to the comets by these stars retains also a certain randomness. As a consequence the evolution of the size of the comets orbits must random walk with the time. Thus it is possible to start from the probability distribution of the parameters related to the perturbing stars to derive intermediate results connected to the underlying stochastic process that ultimately rules the evolution of the Oort cloud.

It is the aim of this paper to present some properties of the probability distribution of the impulse generated by passing stars. In the limited space of this report we will restrict ourselves to the presentation of the results referring for the detailed computation to a paper in preparation (Mignard and Remy, 1985).

2.   THE IMPULSE DISTRIBUTION

When a star crosses the region surrounding the Sun and the comet it passes at a certain time at a minimum distance from the Sun and later or before at a minimum distance from the comet. Let $\vec{R}_S$ and $\vec{R}_C$ the radius vectors Sun-star and comet-star at these closest approaches. It can be shown (Rickman, 1976) that the comet undergoes during the star passage an impulse,

$$\vec{I} = \frac{2GM}{V} \left[\frac{\vec{R}_c}{R_c^2} - \frac{\vec{R}_s}{R_s^2} \right] \tag{1}$$

where G is the gravitational constant, M and V respectively the mass and the speed of the star. Typical values of I are of the order of some tens of centimeters per second.

The probability that a star has a closest approach to the Sun in the range (R, R+dR) and a velocity vector directed in a solid angle $d\Omega$ is,

$$dP = \frac{2}{R_M^2 - R_m^2} \; R \; dR \; \frac{d\Omega}{4\pi} \tag{2}$$

where R_M and R_m are respectively the maximum and the minimum distance of

passage of stars to the Sun allowed. The latter boundary is introduced
to prevent a cloud from being disrupted by a single, but unlikely passage of star very close to the Sun. A close approach to the Sun would make
R_S small in Eq.(1) and generate a large impulse for all the comets in
the cloud. As for R_M it is introduced facilitate comparisons with numerical simulations in which we want to generate a substantial fraction of
the passing stars likely to perturb the comets in the cloud. If we increase R_M we must also increase the rate of passing stars as R_M^2 and results as the probability of ejection of comets given in Eq. (8) is obviously independent of R_M. The probability distribution (e.g. Eq. 4) depends strongly of R_M but not the number of times a given perturbation
occurs during a certain timespan, since the number of passing stars is
proportional to R_M^2. So the introduction of R_M is a matter of mathematical conveniency for intermediate computations while it has no effect on
the physical results. In numerical calculations we have used,

$$R_M = 2 \quad 10^5 \text{ a.u.}$$

$$R_m = 2 \quad 10^3 \text{ a.u.}$$

$$V = 20 \text{ km.s}^{-1}$$

From Eq.(2) it is in principle possible to derive the probability
distribution for the impulse $\vec{I}$. Such an approach proved to be untracta-
ble. However we have derived simple expressions for two limiting cases :
i) the central region of the distribution for impulses in the range of
-1 to $+1$ m.s^{-1} ii) the tails of the distribution valid for larger impulses.
These two regions are connected, the first, to distant passage of
stars, when the Sun-comet distance is much smaller than the Sun-star
distance and the second to close approaches when the star comes very
close to the Sun or to the comet in comparison with their mutual distance.

2.1. Close approaches

In this case Eq.(1) reduces to,

$$\vec{I} = \frac{2GM}{V} \frac{\vec{R}}{R^2} \tag{3}$$

where R is distributed according to Eq.(2). With the three components of
$\vec{I}$, hereafter referred to as α , β , γ , it can be demonstrated that the
probability distribution of I is isotropic and given by (Mignard and Re-
my, 1985) ,

$$P(I > I_o) = I_M^2/I_o^2 \tag{4}$$

where,

$$I_m = 2GM/VR_M \sim 40 \text{ cm.s}^{-1}$$

and the probability law for the components is such that the probability that α is in the range dα is $f(\alpha)$ dα with,

$$f(\alpha) = 1/3I_m \qquad \text{for } |\alpha| < I_m$$
$$f(\alpha) = I_m^2/3|\alpha|^3 \qquad \text{for } |\alpha| > I_m$$

(5)

The above expression was obtained with the reduced impulse Eq.(3) and is likely to represent the tails of the impulse distribution. The distribution decreases as $1/\alpha^3$ instead of an exponential decrease for the normal distribution. The tails of the velocity distribution are then more pronounced that it would be for a gaussian law, as anticipated by Weissman (1982).

We have carried out a numerical simulation by using the reduced impulse to model the interaction between comets and passing stars. The result is shown in Fig.(1) and is in excellent agreement with the above theory.

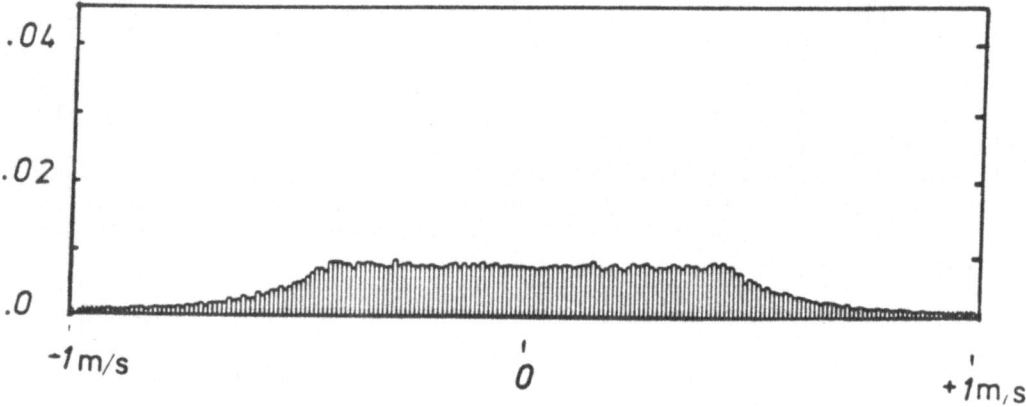

Figure 1. Simulated distribution for the impulses imparted to a comet by close encounters with a passing star.

Finally it must be pointed out that the distribution does not possess second order moment because of a slow decrease toward the large impulses. This fact prevents one from invoking the central limit theorem to synthesize several successive perturbations by a single impulse distributed according to the Maxwell law.

2.2. Distant approaches

In this case we have $R_c \sim R_s$ a fact which enhances the role of the differential effect expressed in the impulse equation. It is then reasonable to expand Eq.(1) up to its dipole term in,

$$\vec{\delta} = \vec{R}_c - \vec{R}_s$$

By taking account of the fact the closest approach to the comet and to the Sun are not simultaneous we obtain with $\vec{R} = \vec{R}_s$,

$$\vec{I} = \frac{2GM}{V} \left[-\frac{\vec{r}}{R^2} + \frac{\vec{r}\,\vec{V}}{V^2 R^2}\,\vec{V} + 2\,\frac{\vec{r}\,\vec{R}}{R}\,\vec{R} \right]$$

where $\vec{r}$ is the Sun-comet vector and is considered as a small parameter with respect to R (see Mignard and Remy, 1985 for a detailed derivation). The modulus of the impulse has a very compact expression,

$$I = \frac{2GM}{V}\,\frac{r}{R^2}\,\sin\theta$$

where θ is the angle between the comet's radius vector and the star velocity vector.

With the probability distribution given in Eq.(2) and by assuming a uniform distribution of points of closest approaches on the sphere of radius R we obtained the probability distribution for the modulus of the impulse and for the three components α, β, γ, in the central region of the distribution.

With $\bar{I}_m = I_m r r_m$ or R_M we have,

$$\bar{I} < \bar{I}_m \qquad f(I) = \bar{I}_m \left[\frac{1}{3}\frac{I}{\bar{I}_m^3} + \frac{1}{10}\frac{I^3}{\bar{I}_m^5} + (\frac{\pi}{4}\frac{1}{3}\frac{1}{10})\frac{I^5}{\bar{I}_m^7} \right]$$

$$I > \bar{I}_m \qquad f(I) = \frac{\pi}{4}\frac{\bar{I}_m}{I^2} \tag{6}$$

Typical values of $\bar{I}_m$ range between 5 to 15 cm.s^{-1} for a comet's distance to the Sun between $2\,10^4$ to $6\,10^4$ a.u. For a very large R_M we must consider the number distribution nf(I) rather than the probability distribution f(I), where n is the number of star passages during a certain lapse of time. With $n \propto R_M^2$, only is the second branch of the frequency distribution meaningful since I_m goes to 0 while nI_m remains bounded.

The distribution of the components follows from the distribution of the modulus after an integration of the joint distribution for the three components over β and γ. The computation is straightforward and yields,

$$|\alpha| < \bar{I}_m \qquad f(\alpha) = \frac{1}{\bar{I}_m} \left(A + B\,\frac{|\alpha|}{\bar{I}_m} + C\,\frac{|\alpha|^3}{\bar{I}_m^3} + D\,\frac{|\alpha|^5}{\bar{I}_m^5} \right)$$

$$|\alpha| > \bar{I}_m \qquad f(\alpha) = \frac{\pi}{16}\frac{\bar{I}_m}{\alpha^2} \tag{7}$$

with A = 0.415   B = -1/6   C = -1/60   D = -0.0352

With the variable $z = \alpha/\bar{I}_m$ the above distribution becomes dimensionless and well adapted for simulations.

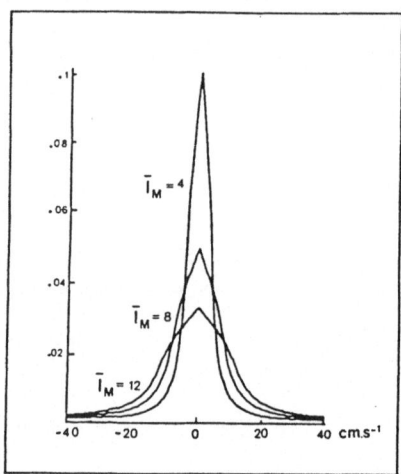

Figure 2. Computed impulse distribution for distant encounters with the passing stars.

The expansion of the impulse formula up to its dipole term makes the above results valid for a limit ranges of impulse, namely those generated by distant approaches. For a typical comet whose distance from the Sun is 3 10^4 a.u. we will restrict distant approaches to passing stars not approaching the Sun closer than twice the Sun-comet distance. Then the impulses to be considered are in the range -1 to +1 ms^{-1}. Then within this range we have the following moments for the frequency distribution,

$$E(\alpha) = 0, \ E(\alpha^2) = \sigma^2(\alpha) \sim 4 \ \bar{I}^2_m$$

The distribution differs from a normal law mainly because of its slow decrease reflected by the distribution function,

$$P(\ \alpha < \bar{I}_m) \ = 0.62$$

$$P(\ \alpha < 3\bar{I}_m) \ = 0.85$$

$$P(\ \alpha < 5\bar{I}_m) \ = 0.92$$

However in its central part the distribution can be matched with a gaussian of standard deviation $\sigma = \bar{I}_m$. But the tails become prominent as soon as $\alpha > \bar{I}_m$.

   The distribution $f(\alpha)$ is plotted in Fig.(2) for three sample values of $\bar{I}_m$, corresponding to comets distance from the Sun, respectively 18, 36 and 54 10^4 a.u. We have also simulated the exact impulse distribution by using the complete equation (1) for the impulse to the comet. In the simulation comets have been maintained at fixed distance from the Sun

but in different locations. Twenty thousand stars were drawn at random
in distance and in orientation of the velocity vector. Results are plotted in Fig. (3) for $\bar{I}_m$ = 4 and 8 cm.s^{-1}. A comparison with figure (2)
shows that the fit to the theory is very good. A similar simulation was
conducted for the modulus of the impulsion ans showed a similar agreement with Eq.(6).

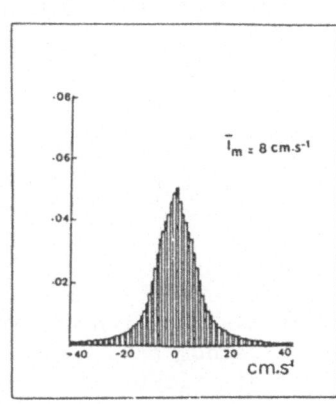
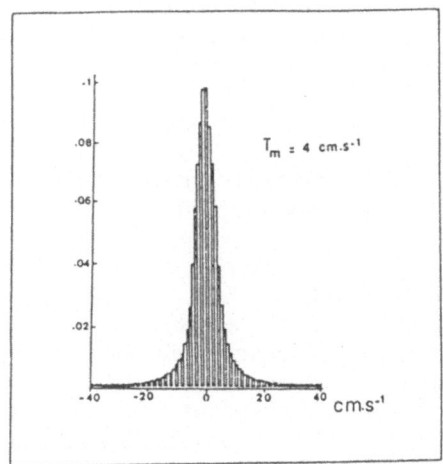

Figure 3. Simulated distribution for the impulses imparted to the comet
during distant encounters with passing stars. The Sun-comet distance is
36 10^4 a.u. ($\bar{I}_m$ = 8 cm.s^{-1}) and 18 10^4 a.u. ($\bar{I}_m$ = 4 cm.s^{-1}).

3. APPLICATION : EJECTION OF COMETS

A comet is ejected from the solar system whenever its orbital velocity
is larger than the escape velocity V_e = $(2GM/r)^{1/2}$ where r is the Sun-
comet distance. For a typical comet we have V_e ~ 200 m.s^{-1} much larger
than the comet's velocity in the vicinity of aphelion. Therefore an ejection would result from a single close encounter with a star, such that
the impulse imparted to the comet is larger than V_e.

      From Eq.(4) the probability for the impulse to be larger than V_e du-
ring one encounter is,

$$P_1 \text{ (ejection)} = I_m^2/V_e^2 \sim 6 \ 10^{-6}$$

      After N passages of stars the probability for a comet to be ejected
is,

$$P_N \text{ (ejection)} \sim 1 - \exp (- NI_m^2 / V_e^2) \tag{8}$$

      Then if ρ denotes the number density of stars in the solar neighbor-
hood we have for the probability of ejection during the time T,

$$P_N = 1 - \exp(-T/\tau)$$

where   is the characteristic time,

$$\tau = V/(2\pi\rho\ GMr)$$

The waiting time for ejection follows an exponential distribution, hence the number of comets ejected per unit time follows a Poisson distribution. Over the age of the solar system the relative depletion of the Oort cloud by ejection is then,

$$P_N = 0.09 \quad \text{with} \quad \rho = 0.08\ \text{star/pc}^3, \quad M = 1M_\odot$$

for T = 4.5 billions years. This number is similar to the result first given by Weissman (1980).

4. CONCLUSION

In this paper we have obtained the probability distribution of the impulse imparted to comets within the Oort cloud by random passing stars. The main result is the fact that the tails of this distribution are much more extended than it would be for a gaussian distribution. In principle this distribution should be sufficient to allow the derivation of results more closely related to the dynamical properties of the Oort cloud, such as the frequency distribution of the orbital parameters after the passage of a certain number of stars. In practice an efficient method to achieve such results needs to be devised and it is our intent to advance in this direction.

REFERENCES

Mignard, F., Remy, F., 1985,'Dynamical evolution of the Oort cloud. II. A theoretical approach', submitted to *Icarus*.
Oort, J.H., 1950,'The structure of the cloud of comets surrounding the solar system and an hypothesis concerning its origin', *Bull. Astron. Neth.*, 11, 91-110.
Remy, F., Mignard, F., 1985, 'Dynamical evolution of the Oort cloud. I. A monte-Carlo simulation', submitted to *Icarus*.
Rickman, H., 1976, 'Stellar perturbations of orbits of long period comets and their significance for cometary capture', *Bull. of the Astron. Inst. of Czechoslovakia*, 27, 2, 92-105.
Weissman, P.R., 1980, 'Stellar perturbations of the cometary cloud', *Nature*, 288, 242-243.
Weissman, P.R., 1982, 'Dynamical history of the Oort cloud', in *Comets*, L.L. Wilkening Ed., Univ. of Arizona Press.

SOME REMARKS ABOUT THE APHELION DISTRIBUTION OF LONG PERIOD COMETS ON THE SKY

Rhea Lüst
Max-Planck-Institut für Physik und Astrophysik
Institut für Astrophysik
Karl-Schwarzschild-Str. 1
D-8046 Garching b. München, FRG

ABSTRACT.   The possibility of an observational biasing in the distribution of cometary aphelia is confirmed by statistics. Furthermore, the tendency of aphelia to form clusters and their distribution with respect to different coordinate systems is investigated.

1. INTRODUCTION

The evolution of Oort's cloud has repeatedly been discussed in the past by several authors (e.g. Hasegawa 1976, Oja 1975, Tyror 1957). The dynamical influence of stellar passages or dense interstellar clouds may also provide clues for the origin of the comets (Biermann, 1978). The obvious fact that the aphelion distribution of distant comets on the sky is not really isotropic, but shows a tendency to form clusters, raised the question whether such clusters might still reflect dynamical perturbations in the past.

Arguments that one of these dense clusters might result from a stellar passage some million years ago have recently been given by Biermann, Huebner and myself (1983). The present contribution shall present the statistical basis for the former investigation. Moreover, some interesting results came out which shall be outlined here.

2. CATALOGUES

For this kind of investigation it is necessary to separate those comets coming from Oort's cloud for the first time into the inner planetary system from the rest. It was possible to do this by using a catalogue of 200 long period comets compiled from high precision orbital elements by Marsden, Sekanina and Everhart in 1978. This catalogue contains the original reciprocal semimajor axes (corrected for the perturbations of all planets), together with their osculating and future values. With the data of this catalogue it became for the first time possible to classify the long period comets reliably according to their "age", which means according to the number of their passages through the inner solar system. There

105

A. Carusi and G. B. Valsecchi (eds.), Dynamics of Comets: Their Origin and Evolution, 105–111.
© *1985 by D. Reidel Publishing Company.*

exist earlier calculations by various authors (e. g. Hasegawa 1976, Oja 1975, Tyror 1957) similar to this one. However, a reliable classification was not yet possible at that time. Therefore the results of the different investigations do not agree in their details with ours or with each other, though some general trends are common.

The Marsden-Sekanina-Everhart catalogue has last year been extended by adding 25 further comets (Everhart and Marsden 1983). Since we have counted three comets of the Kreutz group – which belong to the "old" comets – as one, our sample comprises 223 comets, which we divided into 3 subgroups according to their reciprocal original semimajor axes, Tab. I).

TABLE I

	N	a[a. u.]	$1/a[10^6$ a. u.$^{-1}]$	P[y]	N_P
New comets	89	$>10^4$	<100	10^6-10^7	1
Interm. c.	69	$10^4>a>400$	$100<1/a< 2500$	$8000-10^6$	≤ 20
Old comets	65	$400>a>40$	$2500<1/a<25000$	$250-8000$	multiple

The criteria for a comet to be regarded as "new", "intermediate" or "old" are stated in the Marsden-Sekanina-Everhart catalogue. They depend somewhat on the perihelion distances. The periods of the "intermediate" comets range between about a few thousand and one million years, while the periods of the "new" comets are in the range of a few million years. This is also the time scale for stellar passages.

For these three groups the aphelion directions in the galactic, the equitorial and the solar apex coordinate systems were calculated by using the orbital elements given in Marsden, 1982. By this we wished to find out whether the distributions might show symmetries with respect to one of these reference systems. The computerized equal area plots give a first impression for this, though carful statistical work is necessary for a meaningful interpretation (Fig. 1).

All former investigators have stated a larger number of aphelia on the southern celestial hemisphere in all three coordinate systems, and it has been proposed that this might at least partly be caused by an observational biasing due to better observing facilities on the northern hemisphere of our globe. Since conclusions about the dynamics of Oort's cloud should of course rest on an unbiased sample, the question of an observational influence was one aim of this work. Secondly we tried to verify the tendency for clustering. Furthermore, the distribution in the galactic system which shows pronounced symmetries towards the galactic plane was discussed. In the following, these questions are discussed in more detail.

3. EQUATORIAL COORDINATES, OBSERVATIONAL BIASING

The equatorial system is appropriate for discovering a north-south biasing. The declination of the aphelion or – which is equivalent – the heliocentric perihelion influences among other orbital parameters the

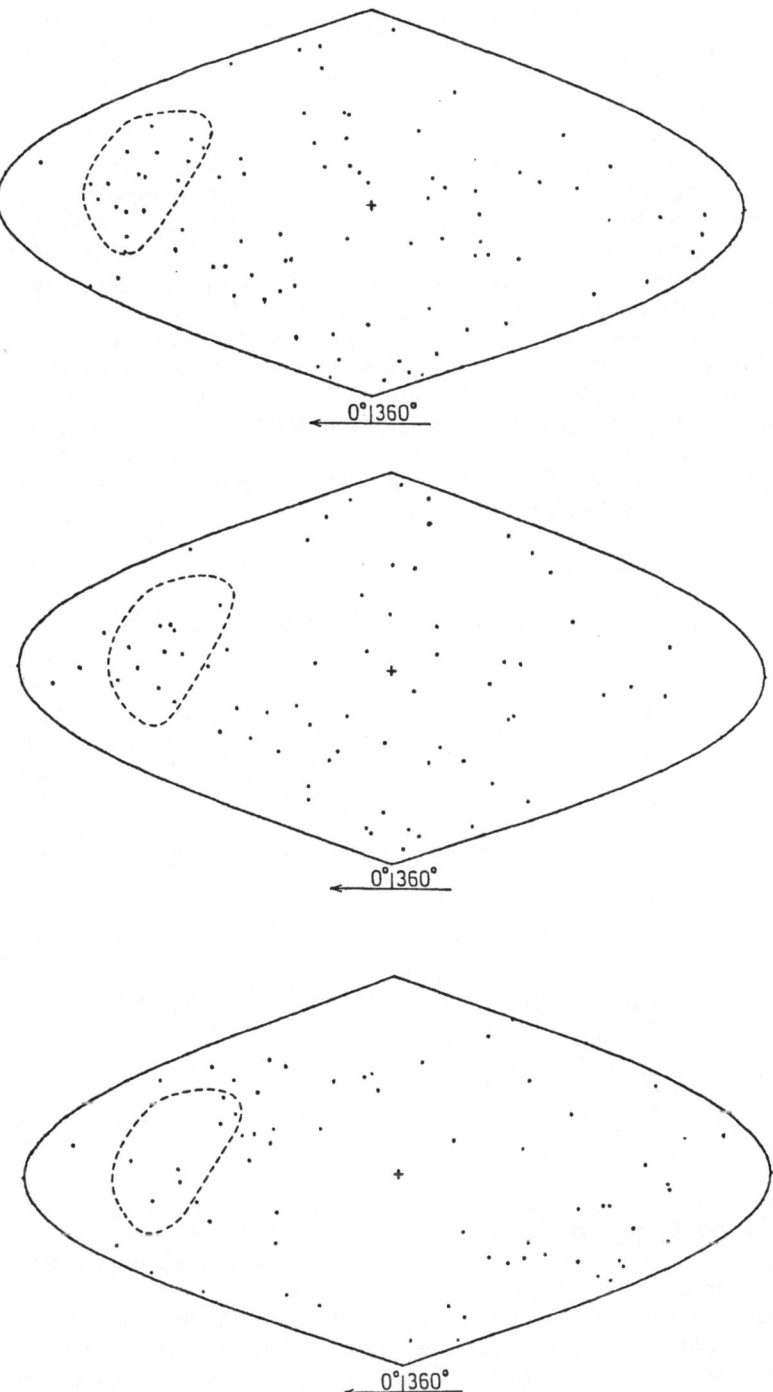

Fig. 1.   Distribution of 89 "new", 69 "intermediate" and 65 "old" comets in galactic coordinates.

detectability of a comet. The relation is, however, rather complex. So the intrinsic brightness determines together with the perihelion distance the time, or more specifically the numerical anomaly at detection. Furthermore, the position of the earth in its orbit is of major importance (see e.g. Everhart, 1967). But on the average, the best period of observation will roughly be around the date of the perihelion passage, and therefore the declination of the perihelion is an important parameter. However, the geocentric and heliocentric perihelia can differ widely from each other, while the aphelia are identical. For the visibility the geocentric positions are relevant. The maximum angle between the geocentric and the heliocentric perihelion is $90°$ for $q = 1$ a.u. and decreases rapidly with increasing perihelion distances. It is already less than $20°$ for $q > 3$ a.u.. For comets inside the earth's orbit, the two positions are practically not correlated. This weakens the correlation between the aphelion direction and detectability, especially for comets which come close to the sun. On the other hand, this fact offers a possibility to detect an observational biasing by separating the comets according to their perihelion distances.

TABLE II

Declin. Sky area %	$\delta < -60°$ 6.7	$-60° < \delta < 0°$ 43.3	$0° < \delta < 60°$ 43.3	$\delta > 60°$ 6.7	100
Total	24	94	93	12	223
q < 1	8	32	40	8	88
q > 1	16	62	53	4	135
New	10	37	37	5	89
Intermed.	9	29	26	5	69
Old	5	28	30	2	65

TABLE III

Declin. Sky area %	$\delta > 40°$ 18	$40° > \delta > 0°$ 32	$0° > \delta > -40°$ 32	$\delta < -40°$ 18
Observed	16	31	38	4
Isotropic distrib.	16	28.5	28.5	16

Tab. II contains comet numbers with aphelia in the two polar caps $|\delta| > 60°$ and in the northern and southern hemispheres for the two groups with small and large perihelion distances. We found 4 times as many aphelia around the south pole than around the north pole among the comets with perihelia outside the earth's orbit, and the equal numbers for the q < 1 group. The same trend is demonstrated by the numbers in the remaining two regions south and north of the equator. For the complete sample we got equal numbers (94:93) in these two regions. With other words: The asymmetry is exclusively caused by the regions around the poles. The same behavior is qualitatively found in the separated groups of the new, the intermediate and the old comets. As a further test we have calculated the geocentric perihelia for the 89 new comets, Tab. III. As was expected, the biasing is even more pronounced here. These results point very strongly to a north–south observational biasing.

4. GALACTIC COORDINATES, CLUSTERING

In galactic coordinates, the distribution of the aphelia indicates some interesting trends depending on the "age" of the comets. The trends are most pronounced among the primordial comets, they are still there among the "intermediate" comets, while the "old" comets show a different distribution.

The histograms of Fig. 2 demonstrate these findings in more detail.
a) A belt around the galactic equator 20° wide is very poorly populated (see also Fig. 1, dotted lines). This is even more evident if we omit the hatched areas which belong to a dense cluster of comets in a limited longitude space between 180° and 230°. There are only 4 aphelia out of 158 "new" and "intermediate" comets in a belt with an area of 1/7 of the total sky area. Statistical calculations give a binomial probability frequency for this to happen of about $3 \ 10^{-6}$ and a deviation from the mean which is about 4 times the standard deviation. These numbers are definitely meaningful. They are also in line with a recent investigation of Byl (1983) on the influence of the galactic gravity field. The sample of the "old" comets does not show any deviation from a random distribution.
b) On the other hand, the apparent deficiency around the two poles has not been proven meaningful by statistics. The total number of aphelia in both polar caps is about 1/2 of the statistical mean. Furthermore, in this case the "old" comets do not differ generally in their behavior from the more primordial ones.
c) This higher population in the mid–latitudes among the "new" comets, especially in the south, is mainly caused by the distribution of several clusters. Though a classification into clusters is somewhat arbitrary, one can locate about 5 denser regions, four of which happen to lie on the southern galactic hemisphere. The densest cluster which has recently been discussed in detail by Biermann et al. (1983) is encircled by a dashed line in Fig. 1 and lies north of the galactic plane.

5. APEX COORDINATES

The distribution in the apex system shows an accumulation in the hemi-sphere of the antapex. This is mainly the result of the afore-mentioned dense cluster which is located between -60° and -20° latitude. The antapex region shows only a slight increase of aphelia which is, however, statistically not significant.

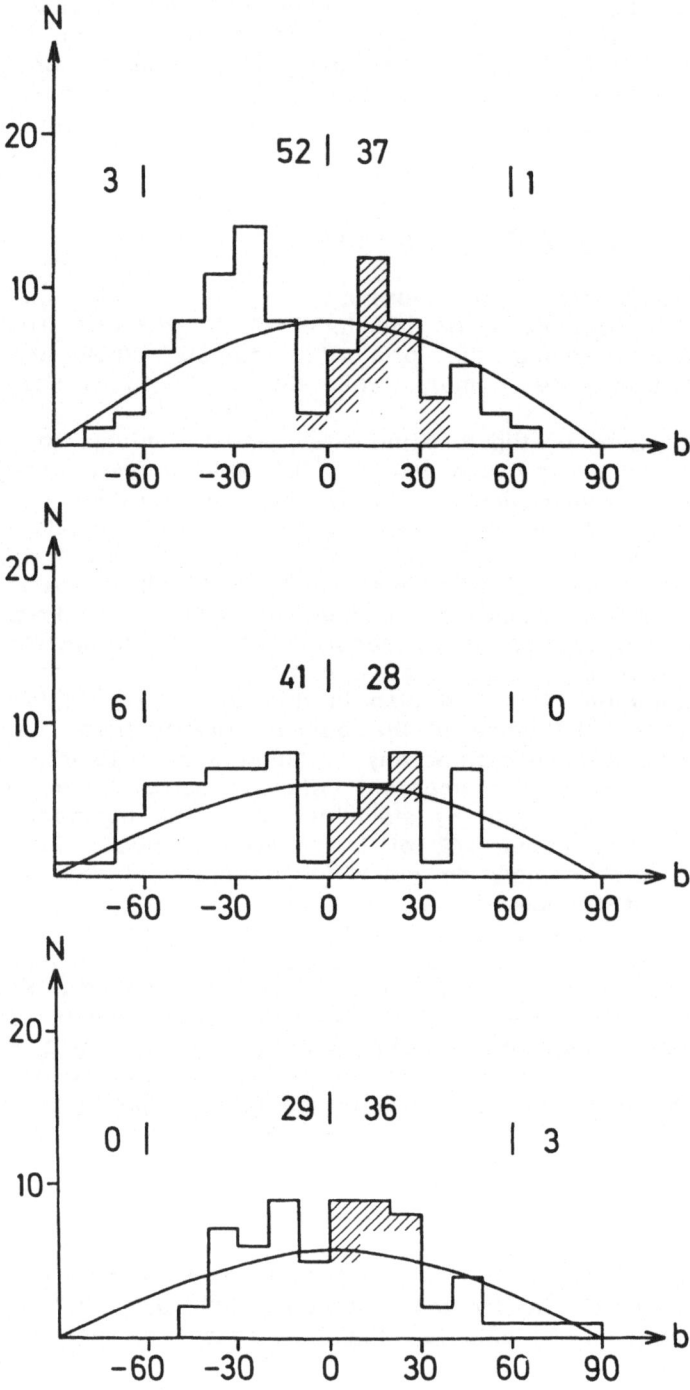

Fig. 2. Aphelion numbers of 89 "new", 69 "intermediate" and 65 "old" comets versus galactic latitude.

6. SUMMARY

The investigation has shown that the possibility of observational biasing has to be taken into account in any conclusion about an anisotropy in the distribution of long period comet aphelia. The numbers point very strongly to the fact that comets with aphelia in high southern declinations are more easily detected that those coming from the region near the north pole. Such an effect should gradually decrease in the future with the better observing facilities on the southern hemisphere of the earth. Therefore, future investigations of this problem are desirable to verify the reality of such a bias.

Furthermore the statistics confirm the tendency of the aphelia to form clusters. The clustering is especially pronounced among the "new" comets coming from the outer regions of Oort's cloud. This is compatible with the conception that these comets have been injected into the inner solar system by the influence of stars which have encountered the solar system some million years ago. Moreover the aphelia seem to avoid a narrow zone around the galactic equator. Though the numbers point to a real effect, it is not yet clear which mechanism might be responsible for it.

REFERENCES

Biermann, L., 1978 in Astronomical Papers dedicated to Bengt Strömgren, ed. A. Reiz and T. Andersen, Copenhagen Univ. Obs., p.327.
Biermann, L., Huebner, W.F., Lüst, Rh., 1983, Proc. Natl. Acad. Sci. USA 80, 5151.
Byl, J., 1983, The Moon and the Planets 29, 121.
Everhart, E., 1967, Astron. J. 72, 716.
Everhart, E. and Marsden, B.G., 1983, Astron. J. 88, 135.
Hasegawa, I., 1976, Publ. Astr. Soc. Japan 28, 259.
Marsden, B.G., Sekanina, Z., Everhart, E., 1978, Astron. J. 83, 64.
Marsden, B.G., 1982, Cat. of Com. Orbits, 4th Ed., Smiths. Astroph. Obs. Cambridge, Mass.
Oja, H., 1975, Astron. Astrophys. 43, 317.
Tyror, J.G., 1957, Mon. Not. Roy. Astr. Soc. 117, 370.

DISCUSSION

Weissman: Do you still believe that these concentrations of aphelia are caused by recent passages of stars through the Oort cloud in that direction, as stated in your paper?

Lüst: Yes, I think that good evidence has been given in various papers for this possibility, (e.g. Biermann et al., 1983).

SECTION III

METEOR STREAMS AND INTERRELATIONS WITH MINOR PLANETS

THE FORMATION AND EVOLUTION OF METEOR STREAMS

I P Williams
Theoretical Astronomy Unit,School of Mathematics
Queen Mary College, Mile End Road, London E1   4NS

ABSTRACT. The physical processes which may affect the evolution of
meteor streams are discussed and a review is then given of the work
carried out to date on the evolution of meteor streams. It is clear that
they evolve principally due to the effect of planetary perturbations and
radiation pressure. The formation of streams from the breakup of a comet
is also discussed. All the evidence, including the recent discovery of
1983TB points to the correctness of this hypothesis.

1. INTRODUCTION

The existence of meteor streams is inferred from the observations of
meteor showers, where a number of meteors, significantly in excess of
the sporadic (or random) is observed, the radiant (or apparent point of
origin) of these meteors being situated within a small area of the sky.
This observation indicates that the original orbits of all the meteors
were parallel to each other. Since most showers are annual, that is they
are seen regularly at approximately the same time each year, the meteor
shower phenomenon is simply explained in terms of a stream of
meteoroids, all moving on similar orbits about the Sun, with these
orbits intersecting the orbit of the Earth close to a fixed point in
space. This point will, by definition, be either the ascending node or
the descending node of the meteoroid orbit.

As well as being detectable as a visible trail of light in the sky,
some smaller meteoroids can be detected by radio when they enter the
atmosphere of the Earth.

The overwhelming majority of meteors detected in the atmosphere lie
in the size range 0.02 to 0.5cm, with corresponding masses in the range
3×10^{-5} to 1.5×10^{-1}g, (Hughes,1978). The number of meteors seen per hour
can also be recorded and from this can be inferred the rate which would
have been observed had the atmospheric conditions been perfect and had
the radiant of the stream been at the zenith of the observer. This
inferred rate is called the Zenith Hour Rate, Z.H.R.. In many streams
the Z.H.R. is less than 20 and so the actual observed rate is comparable
to the sporadic rate. For these streams, there is still considerable
debate as to whether they are real as opposed to being a fluctuation in
the sporadic background. Most of the well known streams have a Z.H.R. in

115

A. Carusi and G. B. Valsecchi (eds.), Dynamics of Comets: Their Origin and Evolution, 115–127.
© *1985 by D. Reidel Publishing Company.*

the range 20-100 and some of the relevant data on these showers is given in Table 1. The criterion for inclusion in the Table is a Z.H.R. in excess of 20 either visually or by radar, the data being taken from Cook(1973). In a number of showers, while the normal Z.H.R. is minimal, a very spectacular display has been observed on a few occasions, where the Z.H.R. has reached many thousands. Good examples of this phenomenon are the Andromedids with a Z.H.R. of 13000 in 1885 and the Pegasids with 14000 in 1833. The Leonids and Draconids also display this phenomenon and both these streams are mentioned again later.

In discussing the age of a stream, it is important to realize that there are two concepts involved, the time that has elapsed since the formation of the stream, which can correctly be termed the age of the stream, and also the period of time for which the stream has been intersecting the orbit of the Earth and so has been seen on Earth as a meteor shower. This latter time should correctly be termed the age of the shower. The two times are not generally related, though of course, the shower age cannot exceed the stream age. The shower age is principally governed by the rate at which perturbations and radiation forces cause the meteoroid orbits as a family to evolve, while the stream age is governed by destruction and dispersal mechanisms. A non-quantitative guide to the age of a stream may be gained from the homogeneity of the meteoroid distribution around the stream orbit and the stream density. In an old stream, the meteoroids will be distributed fairly uniformly and at a low density, while a young stream will be expected to show high density, probably in clumps. The picture in reality may of course be more complicated, especially if the source of the meteoroids has been active over a long period of time, so that the stream has no unique age. A further complicating factor is the fact that it is the number of orbits completed, rather than the number of years that has elapsed, that is the dominating factor in determining the amount of evolution in a stream. Thus, a stream with a five year period like the Quadrantid may show signs of old age, like uniformity, in a much shorter interval than a stream like the Orionid, with a period in excess of half a century.

Some indication of the age of a shower may be obtained from the study of historical records. In the mid nineteenth century, when interest in meteors and meteor streams had just started, Biot(1848) produced a catalogue of observed "fireballs" through antiquity, though in reality these were more concentrated in the eleventh century. Astapovic and Terenteva(1968) used this catalogue in an attempt to determine the changes in the present day meteor streams. Unfortunately, they took no account of planetary perturbations, which of course, moves both the time of observation (that is, the position of the node of the orbit) and also the position of the radiant. In cases where no continuous record of a shower exists, identification of a particular shower with a set of fireballs ten centuries earlier can therfore be doubtful. They found a scatter of observations around January 9th with a Right Ascension and Declination roughly comparable with those of the Quadrantid shower. The change in the day of observation between then and the present represents a nodal retrogression rate of 0.7 degrees per century. This value is consistent with, but somewhat higher than, both the current observed rate and the rate calculated by considering the effects of planetary perturbations (Hughes *et al*, 1979). They also found a set of fireballs in the period December 6 to 18 with the radiant being some 10 degrees away from that of the current Geminid shower. If the identification is correct, it would imply that the node of the Geminid stream had not changed over ten centuries while the inclination has changed considerably. This behaviour is in contradiction with the predictions of most computer models of the Geminid stream. There are strong suggestions that a secondary radiant exists in the neighbourhood of that of the main Geminid shower (Webster *et al*,1966, Hindley,1969). Kresakova(1974) also suggested that there may be two streams crossing the ecliptic close to the same point and that the more minor of the two may be associated with comet Mellish. It may be that the eleventh century observations refer to the minor stream rather than to the main Geminid stream. The real problem with both the Geminid and Quadrantid observation from the eleventh century is that no other mention seems to have been made of either of them until the 1830's when recorded observations become prolific. Other showers notably the Eta Aquarids, Orionids and Perseids have more continuous records stretching back over 1500 years and so a far more reliable guide to their evolution is obtainable from a study of historical records.

In this review, we discuss the formation and subsequent evolution of a family of small particles, concentrating on the stream aspects though some attention will also be given to the observed characteristics of a shower. We will discuss the evolution first and proceed to draw conclusions about the formation process. We proceed in this order because the evolution of a stream is governed only by the forces which act upon it and since these are well-determined then, in principle, following the evolution is straightforward.

In two recent reviews (Babadzhanov and Obrubov,1983, Williams and Fox,1983) references are given to many works on the subject and to its chronological development. For this reason we shall attempt to follow the development of ideas which lead to our current understanding of meteor stream and we shall make no attempt to give an exhaustive coverage of the literature.

2. THE PHYSICS APPLICABLE TO METEOR STREAM EVOLUTION

To a good approximation, each meteoroid in a meteor stream is moving on an ellipse with the Sun at one focus at any given instant. One process of spreading the meteoroids uniformly around the path of the stream follows from the formation process, whatever that formation process is. In no process will every meteoroid be given exactly the same initial velocity and neither will they all be ejected at exactly the same point. Consequently, all the meteoroids will be moving on slightly different ellipses, each with a slightly different period. The meteoroids will not therefore return to the ejection point at the same instant and a stream is formed with particles distributed throughout the volume enclosed by the stream envelope.

Each meteoroid is also subject to gravitational forces arising from the presence of the planets in the Solar System and these forces cause perturbations to the main orbit. As Jupiter is the most massive planet, perturbations due to it are always important. Also important are those due to the Earth, as the stream, by the very virtue of being detected, must pass very close to the Earth. The importance of the perturbations due to the other planets must be assessed individually before reaching a decision on inclusion or exclusion in a particular model. The inclusion of each additional planet increases very considerably the amount of calculation which needs to be carried out. Analytical expressions for the mean perturbation to an orbit are of course available and have been developed by Brouwer(1947) or Hagihara(1972) for example. These can give good results for the mean behaviour of a stream (that is the mean behaviour of a set of particles placed on an orbit which is the mean of all the known stream orbits). However much of the interest in meteor streams comes from the variation from the mean behaviour which individual meteoroids exhibit. For example, if a meteor stream has its period close to a resonance with Jupiter, then it is the same set of meteoroids that always experience a close encounter with Jupiter and hence experience large perturbations, while another set never get close to Jupiter and so suffers no appreciable perturbations. The mean behaviour predicted by the analytical theory may well be the average of the above two types of behaviour, but in reality, it misses all the interesting points in the evolution of the stream.

To deal with gravitational perturbations numerically requires a reasonable amount of time on a large computer. There is a requirement to know the position of each important planet at any given time so that the gravitational force due to it can be evaluated at any field point. This can be done, either by integrating the relevant n-body problem (Sun plus all the required planets), or by making use of a pre-existing Ephemeris tape such as the JPL tape. The former approach may be more efficient if only perturbations from a few planets are important, since searching through a large tape for the relevant data can be very time consuming. Having obtained the gravitational field, the equations of motion for each test particle representing the stream is then numerically integrated, preferably using a high order method. The methods in common use are Taylor, Runge-Kutta, Gauss-Jackson and Gauss-Radau. The efficiency of the various methods have been discussed by Fox(1984), who concluded that the choice between the above methods depended on the precise nature of the problem.

The individual meteoroids are small and so forces due to the existence of radiation from the Sun are also important. There are two effects which need to be considered. First, the radial component of the radiation field has the effect of weakening gravity. In consequence, any small meteoroid, released from a larger parent body will immediately move on a larger orbit. This effect was first discussed by Kresak(1974). Fox(1982) gives the following relations for the change in the semi major axis a and period P of the large parent to the corresponding elements a^* and P^* for a small meteoroid.

$$a^* = (ar(1-\beta))/(r-2a\beta)$$

$$P^* = P(1-\beta)(r/(r-2a\beta))^{3/2}.$$

Here r is the heliocentric distance of the meteoroid at the instant of ejection and β is the ratio of the magnitude of the forces due to radiation and gravity on the meteoroid. Numerically, for motion in the Solar System, β has a value $5.74 \times 10^{-5}/s\rho$, when all units are cgs, s being the radius and ρ the density of the meteoroid . It must be remembered that this effect is a one-off effect, that is the meteoroid has instantaneously a different orbit from the parent, even if the ejection velocity is zero, but this effect has no influence on subsequent orbits. They will all have a period P^* in the absence of other perturbations.

The second effect associated with the radiation field is the Poynting-Robertson drag (Robertson,1937). Here, angular momentum is lost from the meteoroid because it absorbs photons travelling in the rest frame of the Sun while it re-emitts photons isotropically in its own rest frame. These re-emitted photons therefore have some forward momentum associated with them in the rest frame of the Sun, and this momentum can only have come from the meteoroid. The resulting rate of loss of specific angular momentum is given by

$$GM_O\beta\theta/c,$$

where c is the velocity of light and G the gravitational constant.

The corresponding rate of decay of aphelion is approximately given by

$$1.5 \times 10^{-7}/s\rho,$$

assuming the value previously given for β.

The effect of radiation pressure on small particles, which leads to the two expressions above have been discussed in detail by Wyatt and Whipple(1950), Burns et al (1979), Williams(1983).

Taking s=0.5cm, which corresponds to the largest of the detected meteoroid sizes, and ρ=1gcm^{-3}, being about the highest density considered, a change in the aphelion distance of 0.1A.U., which is roughly the minimum dimension of a stream cross-section, will take place in about 3×10^5 years. For the smallest detected meteoroid, s=0.02cm and a typical density is 0.8gcm^{-1}, so that the same change in the aphelion occurs in 10^4 years. Thus for a time interval considerably less than 10^4 years, the Poynting-Robertson effect may be ignored, while it becomes important for all meteoroid sizes after 3×10^5 years. Of equal interest is the intermediate size range, where the Poynting-Robertson effect may cause a differentiation between the small and the large meteoroids to become apparent.

The remaining physical effect which needs to be considered is collisions. There are different types of collisions, with different consequences for the meteoroids concerned.

(1) Collisions between a meteoroid and the Earth, or any other planet, clearly occur since it is the consequence of such collisions that is the only observational evidence for the existence of meteor streams. As far as the individual meteoroid is concerned, such a collision is clearly catastrophic and the meteoroid ceases to exist. As far as the stream as a whole is concerned, one meteoroid is lost from its population, and whether or not this mechanism of meteoroid loss is important for the evolution of a stream depends on the ratio of the number of meteoroids lost to the number remaining in the stream. It is difficult to accurately determine either number, but reasonable bounds can be established for both. The total number of meteoroids is given by the

mass of a stream divided by the mean mass of a meteoroid and so the number must lie in the range 10^{16} to 10^{20}. On the other hand, in an average stream one observer sees the loss of up to 100 per hour for a period of up to 10 days. Scaling up to take account of the number vissible over the whole surface of the Earth, rather than the number vissible within the horizon of a single observer, gives a value for the total number lost per encounter with the Earth which is less than 10^{10}. This mechanism for the loss of meteoroids is not therfore important over time periods of only a few thousand years.

(2) Meteoroid-meteoroid collisions will occur, resulting in the possible loss of both meteoroids from the stream as their orbits may be drastically altered in the collision. Let the mean value of the meteoroid radius be s and let us also assume that the mean relative velocity in a collision, v, is equal to the mean orbital velocity. (This latter assumption clearly overestimates the relative velocity, since, meteoroids in a stream will be moving on nearly parallel orbits, and in consequence will also overestimate the number of collisions). Denote also the number of meteoroids per unit volume by n, the total mass of the stream by M, its average cross-section by A and its average period by P, then $M=mnAPv$, where m is the average mass of a meteoroid and is given by $3m=4\pi\rho s^3$. The mean free path, L, of a meteoroid is given by $4\pi ns^2 L=1$, and on average each meteoroid will experience a collision after travelling this distance, which takes a time L/v. Hence, the number of collisions per unit volume per unit time is nv/L. Each collision, by hypothesis, removes two meteoroids and so the fraction of meteoroids lost per orbit is $2vP/L$. Substitution from the above expressions gives the fraction lost as $6M/(As\rho)$. For the detectable meteoroids, $s=0.02$cm is the minimum radius which corresponds to the maximum loss and with $M=10^{15}$g and $A=10^{-2}$(A.U.)^2, the fraction lost per orbit is under 10^{-9}. Even if considerably smaller meteoroids are considered, this fraction cannot become significant over a few thousand orbits.

(3) Meteoroid-meteoroid collisions can occur which do not cause a loss of meteoroids but rather cause fragmentation which leads to a change in the size distribution of the meteoroids present in a stream. The expression for the collision rate is the same as found in (2) above except that we need to consider a smaller value of s. The minimum possible value of s is about 5×10^{-5}cm, since for smaller values, β exceeds unity and radiation pressure is stronger than gravity. Such meteoroids are driven directly out of the Solar System. Thus, the fraction involved in collision per orbit is increased by about 400. As the fraction of large meteoroids within a stream is small, then it is possible that this mechanism could lead to an errosion of the larger meteoroids on a realistic time scale but until more information is available on the mass distribution within streams, it is very difficult to quantify this effect. It may be that with the EURECA space experiment more data will become available in the near future.

At the current time, no quantitative account of any of the three types of collisions has been taken in any of the computer models that have been investigated.

3 METEOR STREAM EVOLUTION

From the earliest investigations, the aim of generating models for meteor stream evolution has been to match the predictions of the model to the observed data on meteor showers. The observational data consists of the following pieces of information: (i) the time of the year at which the stream is observed, (ii) the number of meteors per hour observed and the variation in this rate throughout the duration of the shower, (iii) the position of the radiant of the shower, and (iv) the date of the first recorded detection of the stream. The data given from (i) is an indication of the position of one of the nodes of the orbit, (ii) gives information about the stream density and the variation in this density along the path of the Earth through the stream, (iii) allows a determination of the orbital elements to be made and (iv) gives information regarding the motion of the mean stream relative to the Earth as a consequence of the perturbations acting on it.

One of the first streams on which any calculations were carried out was the Leonid. This stream is generally assumed to be associated with comet Tempel-Tuttle because of the similarity in their orbits. Magnificent displays of meteors had been seen in 1799, 1833 and 1866. It was postulated that the group of meteoroids causing these spectacular displays were close to the comet itself and so had the same orbital period as the comet. Hence, displays were seen when the comet passed close to the Earth. Calculation by Adams, Storey and Downey (see Lovell,1954) showed that no close encounter would ocurr in 1899, and so no display of meteors was to be expected. As predicted, no display was seen. No disply was seen in 1933 either, but in 1966 a very spectacular disply was observed. A second stream which gives occasional spectacular displays is the Draconid, which is associated with comet Giacobini-Zinner, and again these displays are seen only when the Earth and comet are close. Information on the relative positions of Earth and comet at each display can be found in Yeomans,(1981). In these two streams, which are presumably very young, and the meteoroids are still very close to the parent, simple models, where perturbations on only the single parent body are considered, were able to produce good agreement between theory and observations. However, as the stream evolves and the meteors spread around the orbit, more complex models, considering each meteoroid sepparatly are called for.

The first obvious development of a model is to include the perturbations from the planets on a set of slightly different orbits which represent the stream. By using such methods, Whipple and Hamid(1952) showed that the Taurid meteor stream and Encke's comet had very similar orbits 4700 years ago, while Zausaev (1972) showed that at no time in the past did the Quadrantid and the Delta Aquarid streams have similar orbits. Babadzhanov and Obrubov(1980) developed this model further by including the effects of radiation pressure. Their model was able reproduce the mean evolution of both the Quadrantid and the Geminid streams and predicted a behaviour in agreement with the observations of the corresponding showers. A further development was to produce models consisting of test particles rather than considering perturbations of orbits. Direct integration of test particles in model streams consisting of tens of particles were carried out by Levin *et al* (1972), Kazimirchak-Polonskaya *et al* (1972), Hughes *et al* (1979), Murray *et al* (1980). Hughes *et al*(1981) increased the number of test particles to a few hundred and by now, Fox *et al* (1983) can include many hundreds of thousands of test particles for some models of stream evolution. As a consequence of these models, a picture has emerged whereby it is clear that the general evolution of a meteor stream is governed by the physical processes already described, principally planetary perturbations and Poynting-Robertson drag.

A number of minor variations from the predicted mean behaviour has however maintained interestin the topic. For example, one anomaly in the Quadrantid stream is that the small radio meteors appear to show a different evolutionary behaviour from the visible meteors as far as the variation in the date of appearance is concerned (Hughes and Taylor, 1971). This was explained by Hughes *et al* (1981) in the following way. Radiation effects can cause a difference in the orbital parameters of meteoroids of different sizes, and in the case of the Quadrantids, this results in the small meteoroids having aphelia very close to Jupiter. By coincidence, this is also very close to a 2:1 resonance with Jupiter. Thus some of the small meteoroids suffer large perturbations, while others do not. The orbital parameter in which small changes are easiest to detect is the position of the node. Such changes results in a change in the annual time of appearance of the shower and explains the observed unpredictability of the small meteors.

There is therefore every reason to believe that the main mechanics of meteor stream evolution is understood and the interesting remaining question is whether we can use this understanding to gain an insight into the formation of meteor streams.

4   THE FORMATION OF METEOR STREAMS

The suggestion that meteor streams are associated with comets has been in circulation form some time `and it is this association that is developed further here. The suggestion for such an association has arrisen because of the similarity between meteor stream orbits and cometary orbits, and some well known pairings, such as the Leonids with comet Tempel-Tuttle or the Orionids with comet Halley, are well documented. We will not discuss the evidence for specific pairings further here, it can easily be found in the litteratrure. One of the major problems with this hypothesis of an association betweeen meteor streams and comets was the absence of comets associated with two of the richest regular meteor streams, namely the Quadrantids and the Geminids. Following the investigations of Hughes *et al*(1981), the reason for the failiure to find a comet associated with the Quadrantid stream became evident. The current Quadrantid stream passes very close to the orbit of Jupiter and so the meteoroids are subject to very large perturbations. Since the parent comet would not be on an identical orbit, it presumably did not pass quite so close to Jupiter and so experienced different perturbations. Consequently at the present time it could be on a very different orbit to the meteoroids we currently observe, and identification of the parent comet is therfore close to impossible.

The Geminid stream presents a very interesting problem. The actual orbit of the stream has a much smaller semi-major axis than any known cometary orbit and so any comet associated with the Geminid stream would have to be highly unusual. A number of standard calculation for the rate of change of node (Plavec,1950, Babadzhanov and Obrubov,1980, Fox *et al*,1982) which included all known physical effects, predicted a value of about -1.6⁰ per century and all the predictions were in excellent agreement with one another. However, the stream obstinately refuses to show any change in its appearance date over the last 150 years since regular observations have been available, and indeed, over 1000 years if the eleventh century identification of fireballs prove to be correct. Fox *et al*(1982) offered a possible solution to this dilemma resulting from the apparent contradiction between observations and theory. A stream with a very elongated projection of its cross-section onto the ecliptic could, if the mean motion of the stream due to perturbations happened to be in the general direction of the elongation, produce no apparent change in the position of the node. The Earth would in reality be passing through different parts of the stream, but would do so at the same time each year. The situation is illustrated with a simple sketch as Figure 1. It should be pointed out that Fox *et al*(1982) investigated all the likely ways, including general relativistic corrections, in which the calculated perturbations could be modified but could find none that were significant. Hence, the elongation of the projection of the cross-section seems to be the only acceptable way to reconcile observation and theory. The next question is obviously to discuss ways in which such an elongation of the stream projected cross-section could come about. Fox *et al*(1982), found an intellectually satisfying answer, which was that this was a consequence of the formation mode, where dust was released in random directions from a cometary nucleus. The ejection velocity is given by an expression derived by Whipple (1951), based on the icy agglomorate model for a cometary nucleus. In an extension of this model, Fox *et al*(1983) used 500 000 test particles to reprⱴsent the stream. With such a large number of particles, the authors were not only able to confirm the elongated shape, they could also estimate the number density of meteoroids encountered by the Earth during any passage through the stream. From this, it is possible to generate the expected Z.H.R. at all times during a shower. Spalding(1984), has gathered together most of the vissual observational evidence on the Geminid stream for the last decade, and has shown that the Z.H.R. profjlg is very skew, building up very slowly but decaying rapidly. It was very satisfying to find that that the theoretical model reproduced this charecteristic as well as the general shape. The important point is that this distribution of particles was generated through having a continuous ejection of material into the stream, with differing velocities at different points on the orbit. It is the particles released close to perihelion which generates the high density core which gives rise to the mapptwↄ in the predicted Z.H.R., while those ejected elswhere give the halo. Ejection of a large amount of dust at a single instant, as, for example, would occur as a consequence of a collision with the surface of an asteroidal parent, would not give rise to the same distribution. Hence, the evidence from the stream itself points to a cometary origin for the Geminid stream. The discovery by Green(I.A.U. circular 3878) using IRAS of object 1983TB seemed to be the final answer, in that what may be the elusive Geminid comet had been found. The orbit of 1983TB and that of the Geminids are indeed almost identical and the orbital evolution of both are discussed elswhere in this volume (see Fox, Williams and Hunt) and so we will not dwell further on the association here. However, some mysteries remain. For example, 1983TB is much more similar, judging by the observational evidence available to date, to an Apollo asteroid than to a comet. It is also rather a small object to be the parent of a very prolific stream like the Geminids, and it may be that rethinking the inter-relations between Apollo asteroids and comets is called for. Indeed, we may have to reconsider our ideas concerning cometary nucleii and entertain the possibility that somewhat larger lumps of solid, or semi solid, material than have hitherto been considered may be embedded within the conventional icy agglomerate model. We look forward with interest to observations of 1983TB in December 1984.

5 CONCLUSIONS

The study of the formation and evolution of meteor streams is at a very exiting stage at present. The basic phenomenon and the governing physics are well understood so that their general behaviour and evolution can be succesfully modeled with reasonable accuracy. However, when most streams are looked at in detail, some unexplained phenomenon or unusual behaviour seems to emerge. Most of these are associated with the formation stages rather than with the subsequent evolution and an interesting posibility is that we may obtain a deeper understanding of the structure and evolution of cometary nucleii through the study of meteor streams.

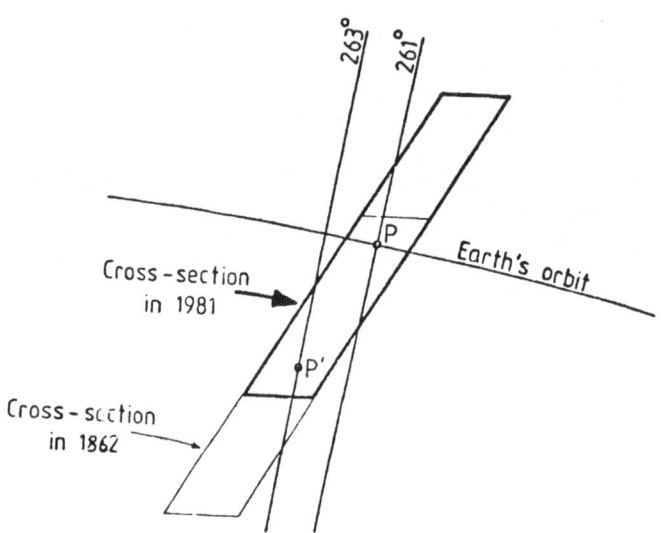

Figure 1. A schematic cross-section for the Geminid stream.

STREAM NAME	GEOCENTRIC RADIANT			DATE	MAXIMUM ZHR	
	λ_o	R.A.	Dec.		Visual	Radar
Quadrantids	282.7	230.1	+48.5	Jan 1-4	140	
Lyrids	31.7	271.4	+33.6	Apr 20-23	20	
η Aquarids	42.4	335.6	-1.9	Apr 21-May 12	30	
Daytime Arietids	77	44	+23	May 29-Jun 19		60
Daytime ζ Perseids	78	62	+23	Jun 1-17		40
Daytime β Taurids	96	86	+19	Jun 24-Jul 6		30
Phoenicids	109.6	31.1	-47.9	Jul 3-18		30
δ Aquarids	125	333.1	-16.5	Jul 21-Aug 29	30	
α Capricornids	127	307	-10	Jul 15-Aug 10	30	
Perseids	139	46.2	+57.4	Jul 23-Aug 23	70	
Daytime Sextantids	183.6	152	0	Sep 24-Oct 5		30
Orionids	208	94.5	+15.8	Oct 2-Nov 7	30	
Leonids	234.5	152.3	+22	Nov 14-20	10	
Geminids	261.0	112.3	+32.5	Dec 4-16	70	
Ursids	270.7	217.1	+75.9	Dec 17-24	20	

TABLE 1 Data on some important Meteor Streams.

I. P. WILLIAMS

REFERENCES

Astapovic,I.S. and Terentava,A.K.: 1968, in *Physics and Dynamics of Meteors*, Ed. Kresak,L. and Millman, P.M., Dortrecht.

Babadzhanov, P.B. and Obrubov, Y.V.: 1980, in *Solid Particles in the Solar System*, Ed. Halliday, I. and McIntosh, B.A., D. Reidel.

Babadzhanov, P.B. and Obrubov, Y.V.: 1983, in *Highlights in Astronomy*, Ed. West, R.M., D. Reidel.

Biot, E.: 1848, *Mem. pres. par divers savant d l'Ac. des Sci. de L'Inst. Nat. de France.* 10, 129.

Brouwer, D.: 1948, Astr. J., 52, 190.

Burns, J.A., Lamy, P.L. and Soter, S.: 1979, *Icarus*, 40, 1.

Cook,A.F.: 1973, in *Evolutionary and Physical properties of Meteors*, NASA SP-319.

Fox, K.: 1982, *Ph. D. Thesis*, London.

Fox, K.: 1984, *Celestial Mechanics*,33,127.

Fox, K., Williams, I.P. and Hughes, D.W.: 1982, *Mon. Not. R. astr. Soc.*, 200, 313.

Fox, K., Williams, I.P. and Hughes, D.W.: 1983, *Mon. Not. R. astr. Soc.*, 205, 1155.

Hagihara, Y.: 1972, *Celestial Mechanics*, M.I.T. London.

Hindley, K.B.: 1969, *J. Brit. astron. Soc.*, 79, 138.

Hughes, D.W.: 1978, in *Cosmic Dust*, Ed. McDonnell, J.A.M., J.Wiley, New York.

Hughes, D.W. and Taylor, I.W.: 1977, *Mon. Not. R. astr. Soc.*, 181, 517.

Hughes, D.W., Williams, I.P. and Murray, C.D.: 1979, *Mon. Not. R. astr. Soc.*, 189, 493.

Hughes, D.W., Williams, I.P. and Fox, K.: 1981, *Mon. Not. R. astr. Soc.*, 195, 625.

Kazimirchak-Polonskaya, E.I., Belyaev, N.A. and Terenteva, A.K.: 1972, in *Motion, Evolution of orbits and Origin of Comets*, Ed. Chebotarev, G.A., D. Reidel.

Kresak, L.: 1976, *Bull. Astr. Inst. Csl.*, 27, 35.

Kresakova, M.: 1974, *Bull. Astron. Inst. Czech.*, **25**, 20.

Levin, B.Y., Simonenko, A.N. and Sherbaum, L.M.: 1972, in *Motion, Evolution of orbits and Origin of Comets*, Ed. Chebotarev, G.A., D. Reidel.

Lovell, A. C. B.: 1954, *Meteor Astronomy*, O.U.P., Oxford.

Murray, C.D., Hughes, D.W. and Williams, I.P.: 1980, *Mon. Not. R. astr. Soc.*, **190**, 733.

Plavec, M.: 1950, *Nature*, **165**, 362.

Robertson, H.R.: 1937, *Mon. Not. R. astr. Soc.*, **97, 423.**

Spalding, G.H.: 1984, *J. Brit. Astron. Soc.*, **94**, 109.

Webster, A.R., Kaiser, T.R. and Pool, L.M.C.: 1960, *Mon. Not. R. astr. Soc.*, **133**, 309.

Whipple, F.L.: 1951, *Astroph. J.*, **113**, 464.

Whipple, F.L. and Hamid, S.E.D.: 1952, *Bull. Roy. Obs. Helwan*, 41.

Williams, I.P.: 1983, in *Dynamical Trapping and Evolution in the Solar System*, Ed. Markellos, V.V. and Kozai, D., D. Reidel.

Williams, I.P. and Fox, K.: 1983, in *Asteroids, Comets, Meteors*, Ed. Lagerqvist, C-I. and Rickman, H., Upsala University Press.

Yeomans, D.K.: 1981, *Icarus*, 47, 492.

Zausaev, A.F.: 1972, in *The Motion, Evolution of orbits and Origin Of Comets*, Ed. Chebotarev, G.A., D. Reidel.

THE TRANSITION BETWEEN LONG PERIOD COMETS, SHORT PERIOD COMETS AND
METEOROID STREAMS

David W. Hughes
Department of Physics
The University
Sheffield
S3 7RH

1. INTRODUCTION

It has long been realised that Jovian perturbation is the dominant cause
of the transition of long period comets (Period > 200 yr) into short
period ones (P < 200 yr). When the differences in the detectability of
comets in the two groups are taken into account it is clear that the
present day flux of long period comets is sufficient to provide the
present collection of short period comets in the inner solar system.
        The fact that meteoroid streams are produced by decaying short
period comets was first recognised around 1866 (see Hughes 1982a). The
magnificent display of Leonids in that year enabled the radiant position
and time of maximum rate to be easily calculated. Assuming the orbital
period to be 33.25 yr Le Verrier (1867) and Schiaparelli (1867)
published orbits for the meteoroid stream. The orbit of comet 1866 I,
which had been discovered by Guillaume Tempel, from Marseilles on
December 19, 1865 and independently by Horace P. Tuttle from Harvard,
Massachusetts on January 5, 1866, has been calculated and published by
Oppolzer (1867a). Almost to a man Peters (1867), Schiaparelli (1867)
and Oppolzer (1867b) realised that the comet and the stream had similar
orbits. Since that time many more examples have been put forward, two
famous ones being the Perseids and comet Swift-Tuttle (1862 III) and the
Eta Aquarids and Orionids both of which have comet Halley (1910 II) as
their parent. For more details see Cook (1973).
        Olivier (1925) was also convinced that the connection between
comet orbits and meteoroid stream orbits was too close to be fortuitous,
but he was still slightly worried, "The contrary fact has not yet been
stated, however, that some comets whose orbits come quite near the
earth's seem to furnish us with no attendant meteor streams". (In
passing, Olivier also warned against putting the comet first and
questioned "is it possible for a meteor stream to be condensed by the
perturbations of planets until, in extreme cases, the densest part
appears to be the head of a comet?" Sixty years later the overwhelming
opinion is that it is not).

A. Carusi and G. B. Valsecchi (eds.), Dynamics of Comets: Their Origin and Evolution, 129–142.
© 1985 by D. Reidel Publishing Company.

2. THE TRANSITION BETWEEN LONG AND SHORT PERIOD COMETS

Capture by Jupiter is the dominant mechanism. This is obvious from the peak in the aphelion distribution of short period comets around 5.0 AU. This capture process is rather selective and Everhart (1972) concluded that only a small set of long period comets, those with perihelia between 4 and 6 AU and inclinations between 0° and 9° were responsible for the large majority (~ 90%) of the short period comets. It was shown, however, that after this perturbation all values of perihelion distance (in the inner solar system) were equally as likely. The obvious lack of small perihelia short period comets is due to the rapid decay of this class of object. The quicker a comet decays, the quicker is formed the associated meteor stream, so this could have a converse effect on stream statistics.

The capture calculation runs as follows. Let $\dot{n}_{LP}$ be the flux of long period comets entering the $4 < q < 6$ AU capture area each year. The comets have random inclination. If only those with inclinations between 0 and $\theta°$ can be captured this represents a fraction $f = 0.5(1 - \cos \theta)$ of the flux. Everhart stipulated that θ was 9°. Fernandez and Ip (1983) relaxed this to $\theta = 30°$, which leads to $f = 0.067$. These comets lose, on average, an energy, ΔE, of 3×10^{-4} AU^{-1} per perihelion passage (see Fernandez 1981). So to go from a near parabolic orbit, $E \sim 0$, to a short period orbit (say $P \sim 13$ yr, $E_{SP} \sim - 0.18$ AU^{-1}) requires $(E_{SP}/\Delta E)^2 \simeq 4 \times 10^3$ passages. Unfortunately many of the comets in this group would be ejected by planetary perturbation long before this number of passages is reached. Everhart (1976) calculated that only about one in a hundred would be captured. Let this capture probability be f_c and introduce a coefficient α that takes into account the possibility of the comet being broken up and destroyed during its transition from a long to a short period orbit. The capture rate, $\dot{n}_{SP}$, of short period comets is then given by

$$\dot{n}_{SP} = \dot{n}_{LP} \times \alpha \times f_c \times f \qquad (1)$$

Fernandez and Ip took the number of new long period comets straddling Jupiter's orbit per year, $\dot{n}_{LP}$, to be 1.5. With $\alpha = 1$ (i.e. assuming no losses) $f_c = 10^{-2}$ and $f(\theta = 30°) = 0.067$ one gets $\dot{n}_{SP} = 10^{-3}$. [Using Hughes (1982b) we note that about 26 ± 3 long period comets are being discovered per decade, these having $0 < q < 1.5$ AU. Using this crude number would lead to an $\dot{n}_{LP}$ of about 3.5.]

To estimate the steady state numbers n_{SP} of short period comets in the inner solar system we need to have some idea as to their average lifetime, L_{SP}. (This is not the time since the origin of the comet, but the time since it was captured into the inner solar system). Kresak (1981), considering the decay of known short period comets, estimated that short period comets have a mean lifetime L_{SP} of 400 orbits. Taking the mean period to be 7 years this gave an L_{SP} of 2800 yr. So the steady state number of short period comets was given by

$$n_{SP} = \dot{n}_{SP} \times L_{SP} \simeq 2.8 \tag{2}$$

Fernandez and Ip (1983) stressed that, as this was two orders of magnitude below the observed number of short period comets, something was drastically amiss. They overcame this problem by proposing a second source of comets in addition to the Oort cloud, this second source being a belt of low inclination comets with semi-major axes of about 50 AU.

Maybe we don't need to be so inovatory. Firstly Kresak could have underestimated the cometary lifetime. Hughes and Daniels (1982) calculated that the members of the short period comet family have on average been past perihelion 800 times since capture. As the mean period is 7 years this corresponds to an average age, A_{SP}, of 6000 yr. The mean lifetime between capture and complete decay is, however, another matter. Unfortunately comets are not like people. If we measure the mean age of a cross section of people we would arrive at a value near 38 yrs, and we would be justified in doubling this to get a typical life duration of 76 yr. Comets, however, don't have a typical life duration. Short period comets can suffer from close approaches to Jupiter and the concomitant possibility of ejection from the inner solar system. Also as 'well behaved' comets seem to lose a fixed percentage of their mass each time they pass the Sun the more massive comets last longer than their slimmer brethren. Assume that the lifetime is β times the average age. Equation 2 thus becomes

$$n_{SP} = \dot{n}_{SP} \times \beta \times 6000 \simeq 6\beta \tag{3}$$

Astronomers since the dawn of scientific sophistication have seen about one hundred short period comets so if equation 3 is producing the correct answer β still seems to be uncomfortably large.

The problem really lies with the estimation of $\dot{n}_{LP}$ and this stems from the comparisons between the number of short period and long period comets that we have seen. Figure 1 shows the cumulative number of comets that are brighter than a specific absolute magnitude. Simply relying on the observations of astronomers during the period of scientific sophistication has produced detailed magnitude records of 104 short period and 522 long period comets. These two numbers must, however, be corrected before intercomparisons can be made. It can be seen from Figure 1 that the short period comet data breaks away from linearity when the absolute magnitude is greater than (i.e. the comets are fainter than) $H_{10} = 10.6$. This is equivalent to saying that during the historical period of scientific investigation we have recorded all the short period comets (SPC's) brighter than 10.6 absolute magnitude. As one progressively considers fainter and fainter comets more and more have slipped by unnoticed. Turning to the long period comets (LPC's) the knee of the curve occurs at $H_{10} = 5.8$.

One way of comparing the two comet groups is to extrapolate the LPC curve and slide the data set up until it breaks away from linearity at the same H_{10} value as the SPC curve does (this extrapolation is shown as a dashed line in figure 1). This is equivalent to saying that the Earth based astronomers have an equal probability of detecting and observing long period and short period comets. [There are obviously some minor problems with this bland

statement. Short period comets return every seven years (on average) so
we have more chances to discover an individual comet. Also the orbital
characteristics of the two groups, especially the inclination, differ
considerably and this could effect the detectability. The above
statement must be taken as a first approximation.]
   The initial data set contained 522 LPC's and 104 SPC's, a

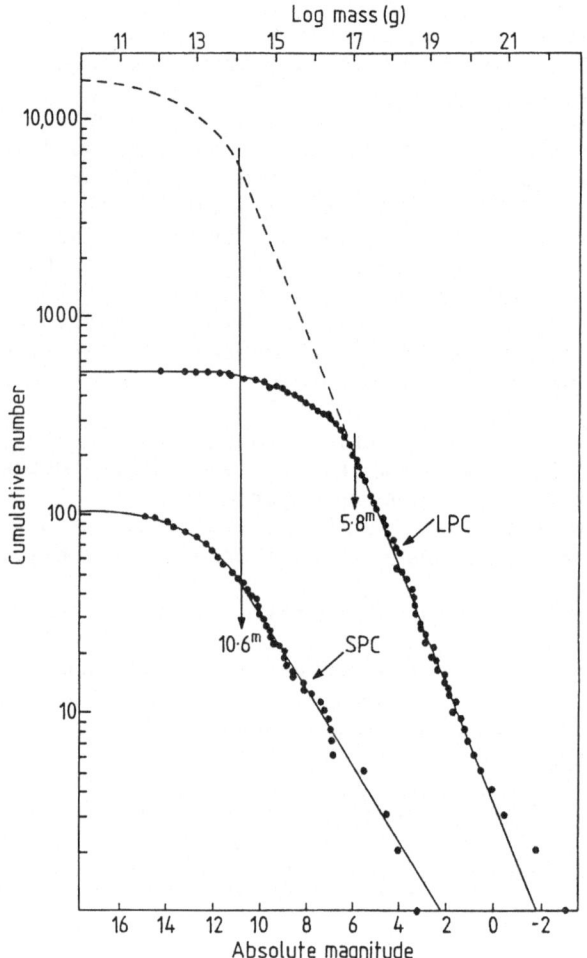

Figure 1 The cumulative number of short period (SPC) and long period
(LPC) comets plotted as a function of their absolute magnitude, H_{10}.
This is defined as being the apparent magnitude a comet would have if it
was 1 AU from the Sun and 1 AU from Earth as well. It is obtained by
substituting apparent magnitude values into the formula
$m = H_{10} + 10 \log r + 5 \log \Delta$. The upper abscissa is obtained by using
the relationship $\log M = 20.58 - 0.6 H_{10}$ where M is the mass of the
comet in grams.

ratio of 5.0. Using the intersection points on the ordinate of the extrapolated LPC data and the original SPC curve gives a ratio of 150. Due to the differing gradients (see Hughes et al 1982) of the two linear portions of the data sets in figure 1 this ratio will depend to an extent on the knee value used. In this rough analysis it is sufficient to note that the ratio between the numbers of long and short period comets is probably much closer to 150 ± 50 than to the value of 5 previously used. This change will be echoed in the $\dot{n}_{LP}$ value to be used in equation 1. The value of 1.5 yr^{-1} used by Fernandez and Ip must be replaced by a value of about 45. Using Kresak's value for the mean lifetime, and equation 2 gives n_{SP} = 84. Obviously the discrepancy between the predicted number of short period comets and the observed number has vanished.

The veracity of the above approach can be illustrated in another way. If we return to the original data set of 522 LPC's and 104 SPC's it is clear from figure 1 that the median LPC has an absolute magnitude of 6.4 and a mass of 5.5 x 10^{16} g. The median SPC has an H_{10} of 11.0 and a mass of 9.6 x 10^{13} g. No matter what feed system between the two groups is envisaged a mass difference between median members of a factor of 580 is surely untenable. Crudely assuming that a comet loses 1% of its mass at each close solar pass, a decay from mass m_i to mass m_f would take n passes where n is given by

$$\log \frac{m_f}{m_i} = n \log 0.99 \qquad (4)$$

The figures quoted above result in n being 630!

Using the extrapolated LPC data as shown in figure 1 results in the median mass of an LPC comet being only slightly larger than that of a SPC. The value of n obtained by using equation 4 turns out to be a single figure number - a result which is surely much more reasonable.

Returning to meteor streams it thus seems that the 'typical' short period comet in the inner solar system was captured by Jupiter in the not too distant past, has decayed somewhat but has by no means transferred all its dust mass to an infant meteor stream.

Unfortunately the mass of a comet is not an easy quantity to estimate. Hughes and Daniels (1980) briefly reviewed the problem and quoted four relationships that had been proposed between mass (M, in grams) and absolute magnitude H_{10}

$\log M = 21 - 0.4\ H_{10}$ (Allen, 1973)

$\log M = 19 - 0.4\ H_{10}$ (R. Newburn Jr, private communication)

$\log M = 19.39 - 0.6\ H_{10}$ (Whipple, 1975)

and $\log M = 20.30 - 0.6\ H_{10}$ (Öpik, 1973)

Hughes (1984) has reinvestigated the problem and concludes tentitatively that

$$\log M = 20.58 - 0.6 H_{10} \qquad (5)$$

This relationship has been used to mark out the upper ordinate of figure 1. Use of the other relationships can change the median mass of the short period comets by up to two orders of magnitude!

The mass of the meteoroids in a meteor stream can be estimated from the influx per year to the Earth and the volume of the stream. Hughes (1974, 1983) found that the mass of dust in the Quadrantids,

Perseids and Geminids was 5×10^{13} g, 2×10^{15} g and 9×10^{14} g
respectively. McIntosh and Hajduk (1983) obtained a value of 5×10^{17} g
for the Orionid and Eta Aquarid streams of Comet Halley. The meteor
stream dust probably constituted about 20 per cent of the mass of the
material that had decayed from the comet so these values are reasonably
consistent with the cometary masses given above.

3. A COMPARISON BETWEEN METEOROID STREAM AND SHORT PERIOD COMET ORBITS

Cook (1973) produced a working list of meteoroid streams and from this
one can conclude that there are around 48 easily observable independent
meteoroid streams intersecting the Earth's orbit at the present time
(see also Drummond 1981). The distribution of the orbital parameters of
these streams are shown in Figure 2. The dashed lines indicate the
median values (the median has half the values smaller and half the
values larger and is much less sensitive to odd extreme values than, for
example, the mean). As is to be expected the distribution of Ω and ω
seems to be random.

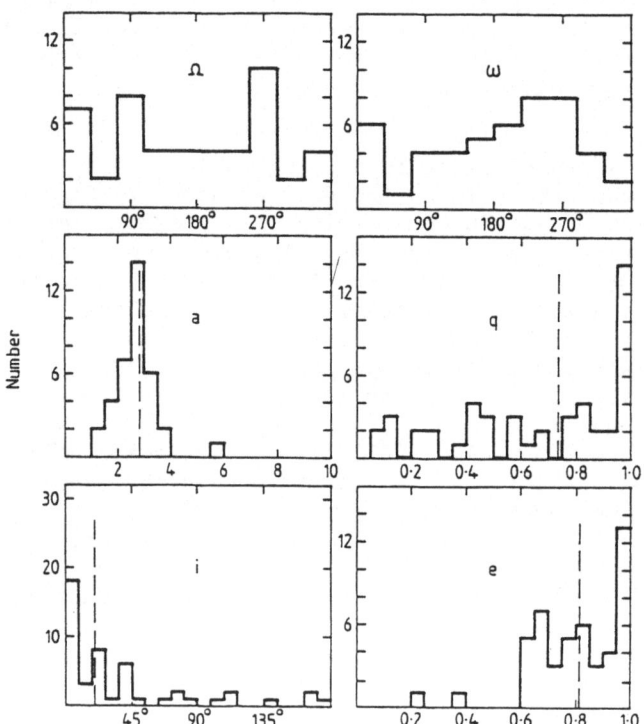

Figure 2   The distribution of the orbital parameters (longitude of
ascending node, Ω, argument of perihelion, ω, semi-major axis a, AU,
perihelion distance q, AU, inclination, i, and eccentricity, e) of the
48 independent meteoroid streams listed by Cook (1973). The dashed
lines indicate the median values of the parameters and these are
$a = 2.84$ AU, $q = 0.73$ AU, $i = 19.2°$ and $e = 0.813$.

It is instructive to compare stream and comet orbits. The
orbital parameters of the 123 known short period comets have been taken
from Marsden (1979). The inclination of the orbits of the short period
comets and meteoroid streams are shown in Figure 3A. The median
inclinations are 12.4° and 19.2° respectively and it is interesting to
note that there is a higher percentage of meteoroid streams with
i < 2.5° than comets.

The distribution of semi-major axes are shown in Figure 3B and
the differences between the two groups are now much more striking. the
median value for comets is 3.73 AU and for streams is lower at 2.75 AU.
It must be stressed, however, that meteoroid stream statistics vary
considerably as a function of the strength of the meteor shower. This
was clearly pointed out by Kresak (1968) who differentiated between the
major meteor showers with high flux rates (the Quadrantids, Lyrids, Eta
and Delta Aquarids, Perseids, Orionids, Taurids, Leonids, Geminids and
Ursids) and the host of minor showers which are hardly distinguishable
from the sporadic background and are usually recognised in photographic
and radar surveys simply by comparing individual meteor orbits. The ten
major showers mentioned above have median values of inclination, semi-
major axis and eccentricity of 76°, 8.0 AU and 0.93 respectively. The
transition from major to minor shower seems to be characterised by a
considerble decrease in all three parameters to values which are roughly
9°, 2.1 AU and 0.7. This is illustrated graphically in Figure 4.

Perihelion distribution is shown in Figure 5. Notice that

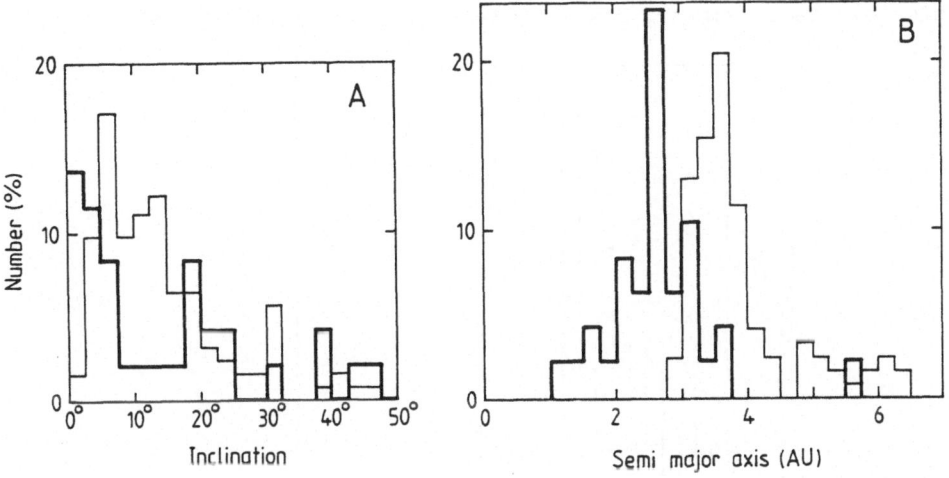

Figure 3(A) The distribution of the orbital inclinations of meteoroid
streams (thick lines) and short period comets (thin lines). Seven
percent of the comets lie off the diagram and 29 percent of the streams
do likewise. (B) The distribution of the semi-major axes of meteoroid
streams (thick lines) and short period comets (thin lines). Seventeen
percent of the comets and 27 percent of the meteoroid streams lie off
the diagram.

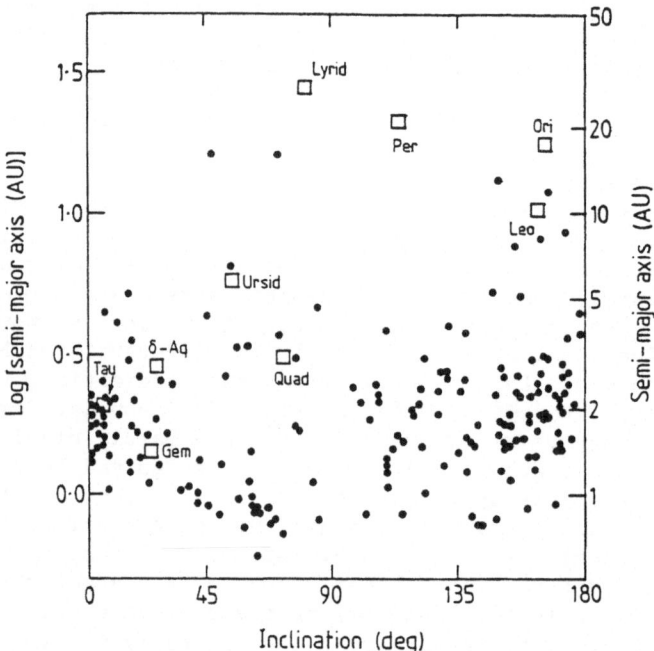

Figure 4   The distribution of the semi-major axes (shown on a
logarithmic scale) and the inclinations of meteoroid streams, □ the
nine major showers, ● the minor showers (taken from the Northern
Hemisphere radio survey, Kasceev, Lebedinec and Lagutin, 1967).

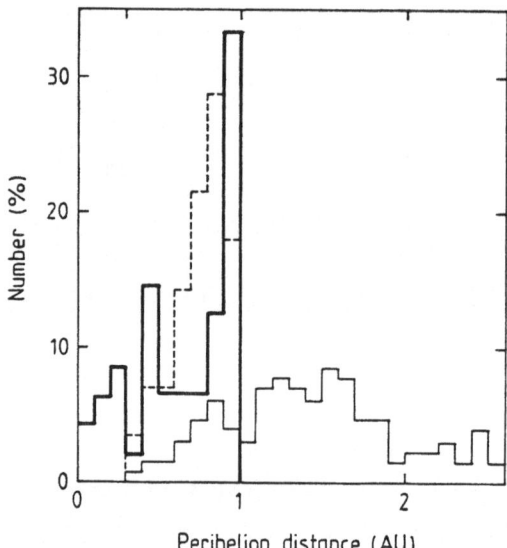

Figure 5   The distribution of the perihelion distances of meteoroid
streams (thick lines) and short period comets (thin lines). The dashed
line represent the distribution of the short period comets that have q <
1.0 AU.

there are no observed short period comets with q < 0.3 AU whereas about
19 percent of the streams fall into this catagory. These q < 0.3 AU
comets decay so quickly that our chances of detecting them are much
reduced.

Obviously only q < 1.0 AU must be considered because meteor
streams outside this ramge cannot be detected from Earth orbit.
Concentrating on q < 1.0 AU the median q for comets is 0.73 AU which
fortuitously turns out to be the same as the value obtained for
streams.

4. DISCUSSION

Kresak (1973) divided the evolution of a meteoroid stream into three
principle phases. (I) - that influenced by the orbital evolution of the
parent comet, (II) - differential accelerations, experienced at the time
of separation from the cometary nucleus and III subsequent stream
evolution. Phase I is governed by major planetary perturbations and
also the non-gravitational effects. Now planetary perturbations tend to
produce a random walk in orbital parameters whereas non- gravitational
forces are much more effective in producing a long term change in one
specific sense. Assuming that the spin axis direction of the cometary
nucleus does not vary, the value of the reciprocal of the semi-major
axis undergoes regular, unidirectional steps and the total change
produced in this quantity is simply a function of the number of times
the comet passes close to the Sun.

Non-gravitational forces are responsible for changing the
orbits of many of the short-period comets. Weissman (1979) suggests for
example that they have decreased the semi-major axis of the periodic
comet Encke and have pulled it away from the influence of Jupiter. This
effect can thus greatly increase the inner Solar system lifetime of
these short-period comets and there is the possibility that they could
lose all their ices without violent disruption and so evolve into Apollo
or Amor-type asteroids. In stage III the meteoroids are small and are
influenced by planetary perturbation and also by the dynamical effects
of solar radiation and interplanetary magnetic fields.

Why do some streams have extant parent comets whereas others
do not? If streams last longer, as observable entities, than comets, it
is to be expected that the majority of streams would be comet-less.
also the closer a comet gets to the Sun the faster it decays. So, as
Kresak (1968) pointed out, streams with small perihelion distances are
less likely to have extant parents than the others. A division of the
streams into two groups, those without extant comets (NC) and those with
extant comets (C) (see figure 6) shows that about 56% of the C group
have perihelia in the 0.9 < q < 1.0 AU range whereas only 22% of the NC
group lie in this range. And this isn't the only difference between
these two groups. The NC group tend to have lower inclinations with 62%
in the range 0 < i < 20° as opposed to 31% in the C group. This trend
is obviously following the one between major and minor streams. The
semi major axes also differ slightly. The NC group has about 87% of the
semi-major axes in the 1 < a < 4 AU range whereas the C group only has

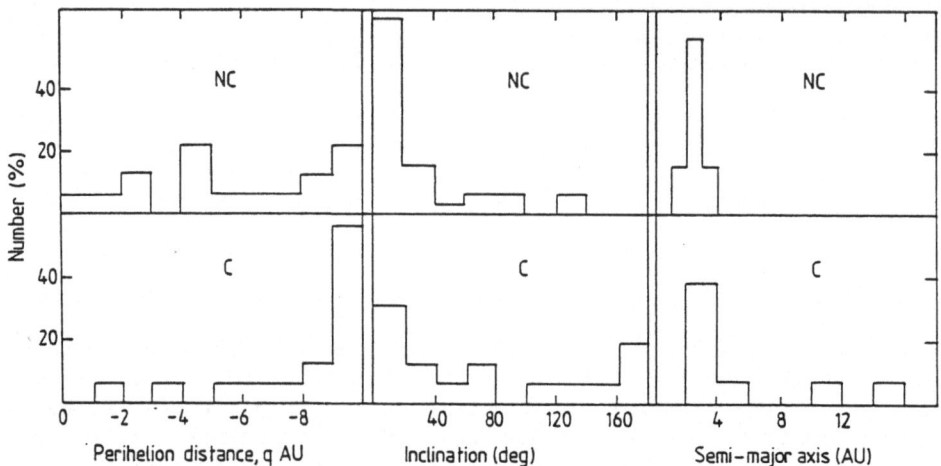

Figure 6   The orbital parameters of meteor streams, the streams having
been divided into two groups, those with no extant parent comets (NC)
and those with extant parent comets (C).

37.5% of its members in that range. So the C group has in the main,
larger semi-major axes.

5. SELECTION

The effect of observational selection is severe. Let us take meteor
streams first. To be seen at all they must intersect the Earth's orbit,
so q must be less than 1 AU. We also have a better chance of seeing
wide streams, such as the Perseids, Taurid, Geminid and Halley streams,
than narrow ones, like the Quadrantids and Draconids. The probability
of encounter also increases for streams close to the ecliptic, that is
with inclination near 0° and 180°. Radiant position is important too,
the summer day-time streams being difficult to observe visually because
the radiant is approximately in the solar direction.
        The size of the orbit is significant. For streams of similar
dust masses the flux per unit time at any intersection will be inversely
proportional to the orbital period. This favours the streams that lie
well within Jupiter's orbit, added to which, they have a special
advantage in as much as Jovian perturbation does not change the orbital
parameters significantly, a change which can quickly sweep a stream node
away from the Earth's orbit. Streams which have aphelia way beyond
Jupiter suffer due to the decrease in the spatial density of the
meteoroids.
        The activity of a shower, as observed from a given location,
decreases with the increasing zenith distance of the radiant and becomes
very low when the radiant passes below the horizon.
        Even today the majority of our information about meteoroid
streams comes from observations in the latitude zone between 35°N and
55°N. This effect is clearly shown in figure 7A. Notice that the

'cometary' streams seem to be predominantly northern hemisphere.The observations that have been made from the southern hemisphere indicate that this is not entirely due to observational selection, no major southern streams having been found.

       The geometry of the intersection between the meteoroid stream and the Earth is of paramount importance. It is well known that the geocentric velocities of meteoroids lie in the range 11 to 74 km sec^{-1}, the lower limit being set by particles which just manage to catch up with the Earth and the upper ocurring when the particle and our planet have a head-on collision. Thus shower meteoroids have encountered velocities dispersed over a range of over 1 to 6. More significant is the fact that the kinetic energy per unit mass of the incident meteoroid varies over a range of 1 to 44. Both the meteor luminosity and the electron line density are direct functions of the kinetic energy of the meteoroid. So streams with high geocentric velocities, such as the Leonids, Orionids and Perseids are over impressive because, with a specific piece of equipment such as the eye, we can detect much lower mass meteoroids than we could in the low geocentric velocity minor streams (see Figure 7B). In sharp contrast to the above, our ability to detect comets does not depend drastically on their geocentric velocity.

       Kresák (1968) points out a further problem. The internal dispersion of stream meteoroid heliocentric velocities leads to the radiants being distributed over a finite area of the sky. This area increases considerably for radiants elongated at between 100 and 180 degrees of the apex, thus making visual meteors from these streams much

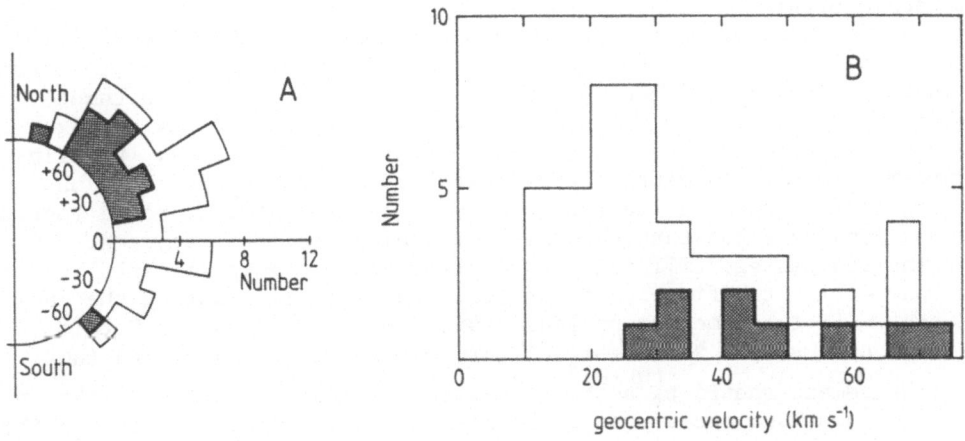

Figure 7(A) The distribution of declinations of observed meteor streams. Hatched region – the C set, bold histogram, the total group.
(B) The distributions of meteoroid geocentric velocities for the 48 meteor streams. The bold histogram represents the nine most active streams: Quadrantid, April Lyrid, Delta Aquarid, Perseid, Orionid (plus Eta Aquarid), Taurid, Leonid, Geminid and Ursid.

less easy to differentiate from the visual sporadic background.
Retrograde streams are thus greatly favoured over the stream meteoroids
which are trying to overtake the Earth from the direction of the
antapex. To quote Kresak "This is one of the reasons why an
identification of minor streams based on the orbital elements and not on
the radiant co- ordinates enhances the number of diffuse direct short
period streams".

6. CONCLUSIONS

One important point must be stressed. Both meteoroid streams and comets
are notorious for moving around the solar system. The
Quadrantids (see Williams et al 1979) and Geminids (see Fox et al 1982)
are both sweeping past the Earth's orbit and have only been seen, as
meteor showers from Earth, since the early nineteenth century. Their
orbital parameters, with the exception of the semi-major axes, are
changing quickly. There is no reason why the parent comet should
undergo perturbations of a similar magnitude so even though they started
in the same place the stream and comet can quickly separate as time
passes. This is probably the only satisfactory explanation as to why
two out of three of the streams in Cook's list do not have recognisable
parents. Figure 7 illustrates that the smaller the perihelion distance
and inclination of the stream the greater is the chance of the comet
being lost. [It is worth noting that the sporadic background is much
denser at low inclinations, so that more spurious streams may be
mistakenly recognised here due to chance coincidences between sporadic
meteor orbits].

The ages of the stream and the comet are also important.
Meteor streams are in the main not produced instantaneously (the term
'in the main' is used because there is the possibility of a comet
suffering a catastrophic break up, disappearing, and transferring a
large percentage of its mass to its offspring the stream). Usually
stream formation is gentle, typically one percent of the comet mass
being lost at each perihelion passage. Drummond (1981b) noted that a
comet loosing a constant thickness of nucleus material at each
perihelion passage will inject 270 times as much material into its
stream during the first tenth of its life as it will during the last
tenth. [If the cometary nucleus started out with a radius R this
fraction is simply $\frac{R^3 - (0.9R)^3}{(0.1R)^3}$]. This means that masive and thus
bright comets should be associated with dense meteor streams, the decay
of the comet and the stream going hand in hand. The corollary to this
is that streams such as the Quadrantids and Geminids must have parent
comets somewhere. It is simply a matter of recognising them. A rough
illustration of this decay process is given in Table I. It is obvious
from this table that the stream has reached about 75 percent of its
final mass by the time the decaying comet has increased in magnitude by
only unity.

We obviously have streams without comets, do we have comets
without streams? Drummond (1981b) disagreed with Olivier (1925) and
answers this question with a qualified negative. To quote his paper

"every short period comet that has approached the Earth's orbit to within 0.08 AU has produced recognisable meteor showers except P/Lexell, P/Finlay, P/Grigg-Skjellerup and P/Denning - Fujikawa". The first two are in unstable orbits, the third is marginal and Drummond (1981b) drew attention to the favourable observing conditions for detecting meteors from it in April 1982. The fourth has a radiant less than 30° from the Sun and of negative declination. P/Lexell was so violently perturbed by Jupiter in 1767 and 1779 that its radiant is now probably rather inactive.

Many long period comets have come close to the Earth's orbit but very few (five according to Drummond 1981b) have produced meteor streams.

Table I

This table illustrates the decay of a comet which loses a constant thickness of nucleus material at each perihelion passage. The comet starts out with a mass of M, a radius of R and an absolute magnitude of H. A fraction 'a' of its mass goes into the meteoroid stream (a ~ 0.3). It has been assumed that the brightness of the comet is proportional to its surface area [following equation 5].

Percentage of cometary life that has passed	Radius of nucleus	Mass of comet	Mass of stream	Magnitude of comet
0	R	M	0	H
10	0.9R	0.729M	0.271Ma	H+0.23
20	0.8R	0.512M	0.488Ma	H+0.48
30	0.7R	0.343M	0.657Ma	H+0.77
40	0.6R	0.216M	0.784Ma	H+1.1
50	0.5R	0.125M	0.876Ma	H+1.5
60	0.4R	0.064M	0.936Ma	H+2.0
70	0.3R	0.027M	0.973Ma	H+2.6
80	0.2R	0.008M	0.992Ma	H+3.5
90	0.1R	0.001M	0.999Ma	H+5.0
100	0	Q	Ma	∞

REFERENCES

Allen, C. W., 1973, Astrophysical Quantities (3rd Edition), Athlone Press, University of London.
Cook, A. F., 1973, in Evolutionary and Physical Properties of Meteoroids, IAU Col. 13, 183-191, NASA SP319.
Drummond, Jack D., 1981a, Icarus, 45, 545, 1981.
Drummond, J. D., 1981b, Icarus, 47, 500.
Everhart, E., 1972, Astrophysical Lett., 10, 131.

Everhart, E., 1976, in The Study of Comets, IAU Col 25, 445,
        NASA SP 393.
Fernandez, J. A., 1981, Astron. Astrophys., 96, 26.
Fernandez, J. A. and Ip, W. H., 1983, in Asteroids, Comets and Meteors,
        ed. C-I, Lagerkvist and H. Rickman, Univ. of Uppsala Press,
        p.387.
Fox, K., Williams, I. P. and Hughes, D. W., 1982, Mon. Not., R. astr.
        Soc., 199, 313.
Hughes, D. W., 1974, Space Research XIV, 709.
Hughes, D. W. and Daniels, P. A., 1980, Monthly Notices R. Astr. Soc.,
        191, 511.
Hughes, D. W. and Daniels, P. A., 1982, Monthly Notices R. Astr. Soc.,
        198, 573.
Hughes, D. W., 1982a, Vistas in Astronomy, 26, 325-345.
Hughes, D. W., 1982b, J. Brit. astr. Assoc., 92, 61.
Hughes, D. W., 1983, in Asteroids, Comets, Meteors, Ed. C.-I. Lagerkvist
        and H. Rickman, Univ. of Uppsala Press, p.239.
Hughes, D. W., 1984, Monthly Notices R. Astr. Soc., in press.
Kasceev, B. L., Lebedinec, V. N. and Lagutin, M. F., 1967, Rezultaty
        Issted. MGP-Issled. Meeorov No. 2, 1.
Kresak, L., 1968, in Physics and dynamics of meteors, IAU Symposium 33,
        D. Reidel, 391.
Kresak, L., 1973, in Physical Properties of Meteoroids, IAU Col 13,
        NASA SP 319, p.331.
Kresak, L., 1981, Bull. astr. inst. Czech., 32, 321.
Le Verrier, U. J. J., 1867, Sur les étoiles filantes due 13 Novembre et
        du 10 Août, Comptes Rendus, 64, 94-99.
Marsden, B. G., 1979, Catalogue of Cometary orbits, 3rd Edition,
        Smithsonian Astrophysical Observatory, 1979.
McIntosh, B. A. and Hajduk, A., 1983, Mon. Not. R. astr. Soc., 205,
        931.
Olivier, C. P., 1925, Meteors, Williams and Wilkins, Baltimore, p.199.
Opik, E. J., 1973, Astrophys. Space Sci., 21, 307.
Oppolzer, T von, 1867a, Bahnkestimmung des Cometen I. 1866, Astron.
        Nach., 68, 241 .
Oppolzer, T von, 1867b, Schreiben des Herrn Dr. Th. Oppolzer an den
        Herausgeber, Astron. Nach., 68, 333.
Peters, C. F. W., 1867, Bernerkung über den Sternschnuppenfall vom 13
        November und 10 August 1866, Astron. Nach., 68, 287.
Schiapparelli, G. V., 1867, Sur la relation qui existe entre les comètes
        et les étoiles filantes, Astron. Nach., 68, 331.
Weissman, P. R., 1979, Astrophysical J., 84, 580.
Whipple, F. L., 1975, Astr. J., 80, 525.
Williams, I. P., Murray, C. D. and Hughes, D. W., 1979, Mon. Not. R.
        astr. Soc., 189, 483.

THE PAST AND FUTURE OF 1983 TB AND ITS RELATIONSHIP TO THE GEMINID METEOR STREAM.

Ken Fox, Iwan P. Williams and J. Hunt,
Theoretical Astronomy Unit,
School of Mathematical Sciences,
Queen Mary College, Mile End Road,
London. E1 4NS, U.K.

ABSTRACT. It is now well known that object 1983 TB, discovered by IRAS, has an orbit very similar to that of the Geminid meteor stream. Calculations show that this orbit crossed over the orbit of Venus about 500 years ago. We will describe calculations tracing the history of both the object and the stream through this interaction with Venus and the present interaction with the Earth.

1. INTRODUCTION

Mathematical models of the formation of meteor streams, simply by the ejection of material away from a cometary nucleus, are never able to produce a stream wide enough to match the length of the shower which is actually observed. This implies either that during formation the parent comet spiralled away from its Keplerian orbit under the influence of non-gravitational forces, as proposed by Weissman (1979) or that after formation the stream was dispersed by gravitational perturbations, radiation effects or some other unknown mechanism. This paper investigates the effects of gravitational perturbations on the evolution of the Geminid meteor stream cross-section.

Table I

Orbital element	Geminid Stream	1983 TB
Argument of perihelion	324.8°	321.6839°
Ascending node	260.3°	265.0332° 1950
Inclination	23.6°	22.0339°
Eccentricity	0.896	0.8902365
Perihelion distance	0.140 AU	0.1395603 AU
Period	1.57 years	1.434 years
Aphelion distance	2.56 AU	2.4033679 AU

143

A. Carusi and G. B. Valsecchi (eds.), Dynamics of Comets: Their Origin and Evolution, 143–148.
© 1985 by D. Reidel Publishing Company.

The mean orbital elements of the Geminid meteor stream and asteroid 1983 TB (MPC 8678) are listed in Table I. As can be seen, they follow virtually identical orbits. Fox et. al. (1984) have concluded that 1983 TB is not just an Apollo asteroid but is the degassed remains of the parent comet of the Geminid meteor stream. Fox et. al. (1982,1983) described a mechanism for the formation of the Geminid meteor stream by the ejection of particles away from the nucleus of a hypothetical comet, travelling on the present day Geminid orbit. This model was able to explain many of the observed features of the Geminid shower, in particular its skew rate profile. However, the model was not entirely realistic as it would have produced a shower of less than two days duration compared to that observed (Spalding (1984)) of about eight days.

Babadzhanov and Obrubov (1980) have studied the long term behaviour of the Geminid meteor stream using a secular perturbation technique. They have shown that the stream is undergoing rapid orbital evolution, in the past 4000 years alone it has crossed over the orbits of Earth, Venus and Mercury twice at either its descending or ascending node. Such crossings must disrupt the stream causing it to spread, but whether this disruption is significant is not easy to estimate. Secular perturbation methods are only able to describe the average behaviour of an orbit's evolution. In particular they cannot describe the disruption of a stream as its orbit crosses over the orbit of any perturbing planet, as a singularity occurs in the equations when the orbits intersect. Therefore the only way to accurately study this disruption is to use the direct approach and integrate the full equations of motion.

2.   THE FORMATION MODEL

The same procedure is used here as that used by Fox et. al. (1982), except that this time the orbit used for the parent comet is that of asteroid 1983 TB. The orbit of 1983 TB was integrated backwards 500 years so that it was now interior to the orbit of Venus. Twenty particles were then ejected from equally spaced points around this orbit with speeds v, given by Whipple's formula , below, taking typical values of s=0.1 cm and ρ=0.8 g/cm^3 for the particles' radius and density respectively and a cometary radius R_C of 5 km, a value which is slightly larger than the observed radius of 1983 TB.

$$v = \left[\frac{1}{s \, \rho \, r^{9/4}} - 0.013 \, R_C \right]^{1/2} R_C^{1/2} \times 656 \text{ cms}^{-1}.$$

The equations of motion of these twenty particles were then integrated forwards in time one thousand years, at which stage virtually all their orbits had crossed over the orbits of Venus and the Earth. Perturbations by Venus, Earth and Jupiter were included in the calculations. The equations were integrated by a Runge-Kutta-Nystrom method of Dormand and Prince (1978) as Fox (1984) has shown this to be one of the best methods available for systems like this.

3. THE EVOLUTION OF THE THEORETICAL STREAM

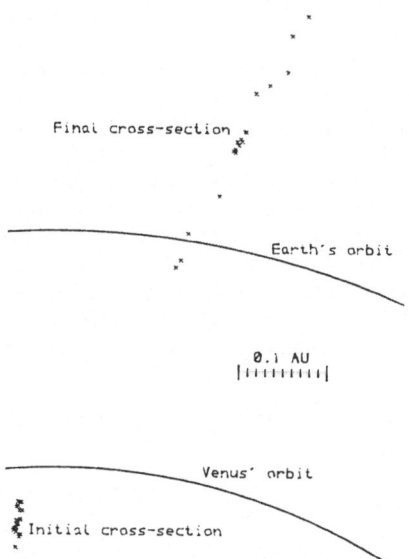

Figure 1. In the plane of the ecliptic, the initial and final positions of the
descending nodes of the particles in the theoretical stream.

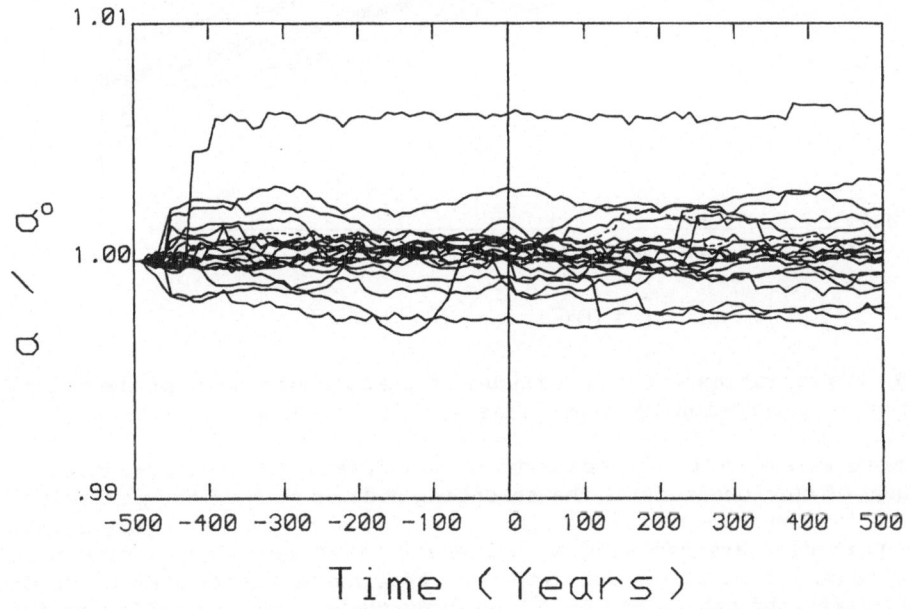

Figure 2. The variation of the normalised semi-major axes of the twenty
particles as a function of time. The variation of the semi-major axis of 1983
TB is also shown as a dashed line. Time = 0 is September 23rd 1983.

Figure 1 shows the initial state of the theoretical stream cross-section 500 years ago, interior to the orbit of Venus, and how over the one thousand year period the cross-section has evolved across both the orbits of Earth and Venus. This movement is virtually entirely due to perturbations by Jupiter. One of the best ways of seeing how the stream has been disrupted is to look at the changes in the semi-major axes of the particles in the stream as a function of time. Normalising the semi-major axes by dividing by their initial value enables relative changes to be seen more clearly. Figure 2 is a plot of these normalised semi-major axes over the one thousand year period. It is evident that many of the particles are affected by crossing over Venus' orbit during the first one hundred years of the integration by the large relative changes in their semi-major axis. The disruption caused by the Earth is not so obvious, because the stream soon spreads out and the crossing point occurs at different times for each particle. However, further spreading can be seen during the last five hundred years due to this cause.

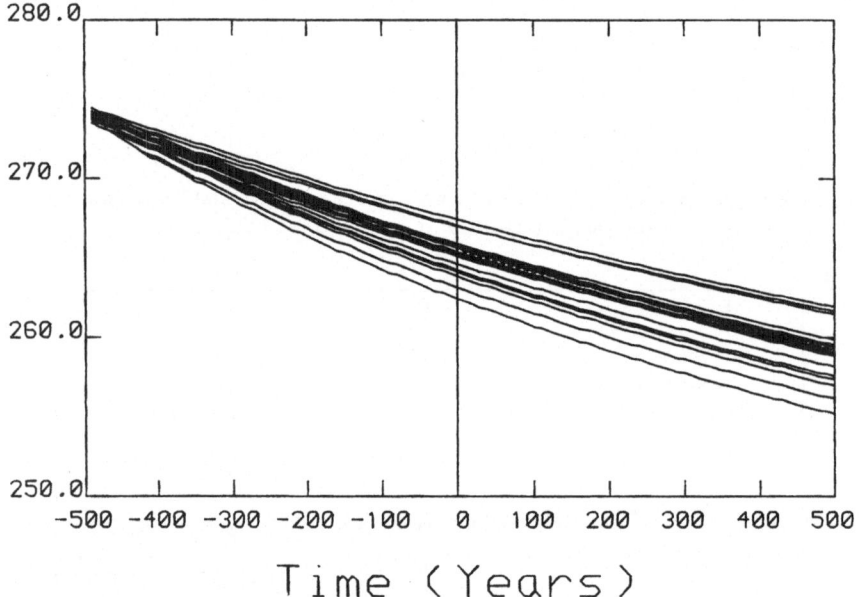

Figure 3. The variation of the longitudes of the ascending node of the twenty particles as a function of time. Time = 0 is September 23rd 1983.

Of more relevence to the observed shower duration is the variation in the change of the longitude of the ascending node of each of the particles. Figure 3 shows the change in this angle as a function of time. It is clear that all the particles are evolving at different rates causing the stream to slowly spread. This is mainly due to the fact that the particles started out on slightly different orbits and thus Jovian perturbations are different for each orbit. There is virtually no sign of any major disruption caused by the Earth and Venus in the evolution of this angle.

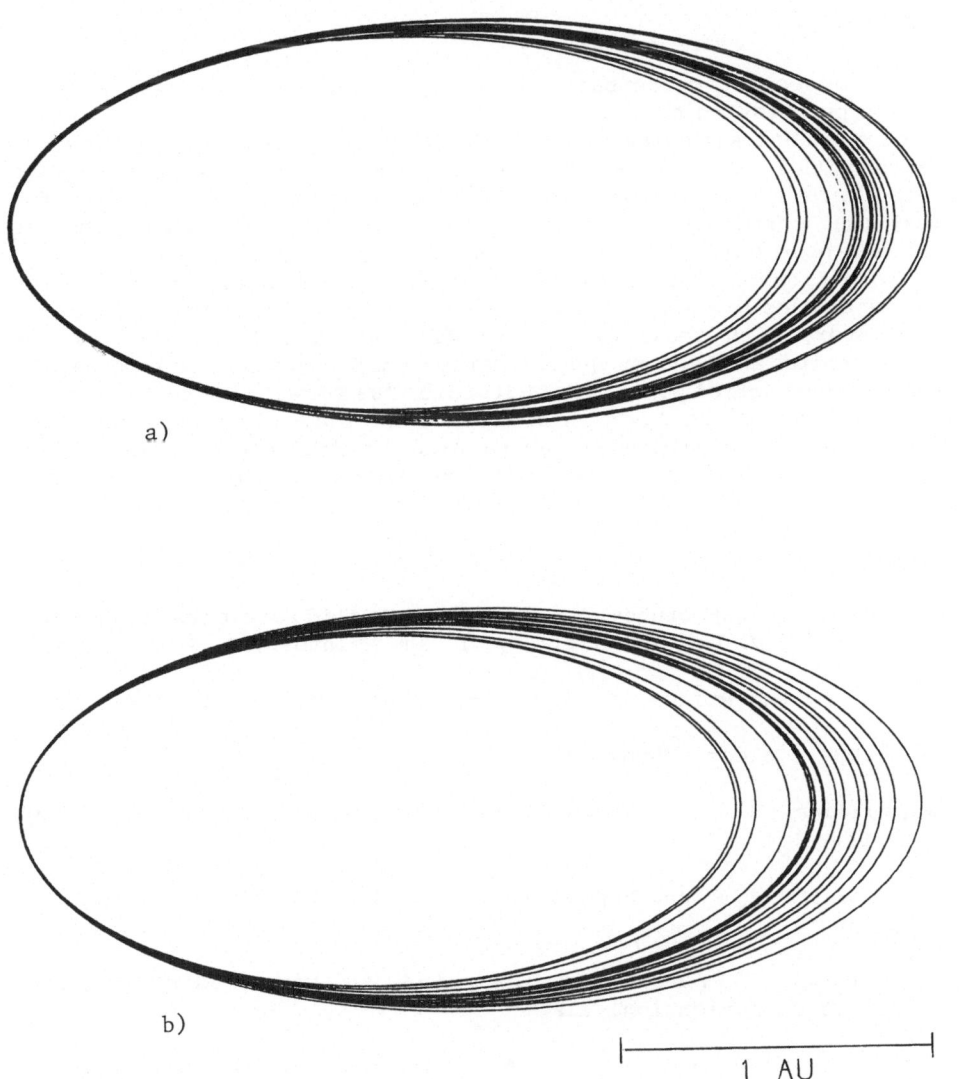

a)

b)

├────────────────┤
       1 AU

Figure 4. A comparison of a) the orbits of actual observed Geminid meteors with b) those produced by the theoretical model.

Finally a comparison should be made of the actual shapes of the theoretical orbits compared to those observed. Jacchia and Whipple (1961) obtained twenty high precision orbits of photographic Geminid meteors. These ellipses, with pericentres alligned are shown in Figure 4, alongside the elliptical orbits produced by the theoretical model at time = 0. As can be seen the orbits are very similar.

4. CONCLUSIONS

Due to secular perturbations by Jupiter and to a lesser extent, disruption by crossing over the orbits of the inner planets, the Geminid meteor stream is slowly spreading out. The lack of any close approaches to or resonances with Jupiter means that the spreading is slow. From calculations presented here it would seem that thousands of years are necessary for gravitational perturbations to spread out the stream enough to match the present day observations of the shower.

The formation mechanism used here produces an asymmetric distribution of semi-major axes in the ejected particles' orbits about that of the parent body. However, with realistic values for the comet's radius and the particles' radius and density the asymmetry is not so great as to immediately explain the non-central position of asteroid 1983 TB relative to the Geminid meteor stream. The crossing of the orbit's of the inner planets causes small jumps in the orbital elements of any individual particle within the stream. This mechanism is probably sufficient to explain this non-central position.

REFERENCES

Babadzhanov, P. B. and Obrubov, Y. V.: 1980, in 'Solid Particles in the Solar System', p. 157, ed. Halliday, I. and McIntosh, B. A., D. Reidel.

Dormand, J. R. and Prince, P. J.: 1978, Celest. Mech., 18, 223.

Fox, K.: 1984, Celest. Mech., 33, 127.

Fox, K., Williams, I. P. and Hughes, D. W.: 1982, Mon. Not. R. astr. Soc., 200, 313.

Fox, K., Williams, I. P. and Hughes, D. W.: 1983, Mon. Not. R. astr. Soc., 205, 1155.

Fox, K., Williams, I. P. and Hughes, D. W.: 1984, Mon. Not. R. astr. Soc., 208, Short Communication, 11P.

Jacchia, L. G. and Whipple, F. L.: 1961, Smithson. Contrib. Astrophys., 4, 97.

Spalding, G. H.: 1984, J. Brit. astron. Assoc., 94, 3.

Weissman, P. R.: 1979, Astr. J., 84, 580.

INTERRELATIONS BETWEEN COMETS AND ASTEROIDS

Hans Rickman
Astronomical Observatory
Box 515
S-75120 Uppsala
Sweden

ABSTRACT. Various aspects of comet/asteroid distinctions and interrelations are reviewed with emphasis on recent work and paying special attention to the following problems: characteristics of cometary activity at large heliocentric distance and uniqueness of comet P/Schwassmann-Wachmann 1 with respect to physical properties, the rôle of Trojans and other small bodies in the outer planetary system concerning comet/asteroid classification, possibilities for physical evolution of comets into asteroids, orbital and dynamical overlap of the comet and asteroid populations, and the cometary versus asteroidal origin of Earth-approaching asteroids. With regard to these latter questions it is argued that recent discoveries indicate a more substantial probability for Jupiter family comets to develop into asteroidal objects than earlier believed, and several examples of cometary association for newly discovered Apollo-Amor asteroids are also referred to. However, the fractional cometary contribution to the traditional Apollo-Amor asteroid population (aphelia far inside Jupiter's orbit) apparently can not yet be reliably estimated.

1. INTRODUCTION

Comets and asteroids, the two major types of interplanetary objects, are both considered to carry important information on the origin and evolution of the Solar System. It is thus essential to understand their interrelations as well as distinctions and the possible evolutionary links connecting them. Many reviews have been devoted to these problems (e.g. Kresák 1977, 1979, 1980, 1983; Degewij and Tedesco 1982) and the present paper will mainly deal with new results, attempting to evaluate the impact of many recent findings, both observational and theoretical. Our approach will be to consider, in succession, a number of different aspects of the comet/asteroid distinction, i.e., the observational, physico-chemical, genetical and dynamical aspects.

A. Carusi and G. B. Valsecchi (eds.), Dynamics of Comets: Their Origin and Evolution, 149–172.
© *1985 by D. Reidel Publishing Company.*

2. OBSERVATIONAL AND PHYSICO-CHEMICAL DISTINCTIONS

Observationally, the distinction between asteroids and comets is quite clear. An asteroid always appears stellar (angular diameter < 1"), but a comet develops a diffuse coma due to gas production from a solid nucleus (Whipple 1950), giving rise to an expanding atmosphere entraining grains of dust and ice. Nonetheless, comets sometimes appear stellar, especially at relatively large distances from the Sun, and there are a number of examples of asteroidal designations given to objects identical to known comets or later recognized as comets. Some of these were recently found by Nakano (1984) including a pre-discovery observation of comet P/Smirnova-Chernykh as asteroid 1967 EU. There are also examples of preliminary cometary designations which have been changed into asteroidal (such as 1977t Lovas = 1977 YA), since the activity reported at the discovery observation failed to be confirmed. Such cases are rare, but they should not always be regarded only as curiosities, since they may provide clues as to specific requirements for the onset of cometary activity or the existence of variable or sporadic activity among comets.

Of more profound significance than the observational classification would be a comet/asteroid distinction referring to the physico-chemical properties of the objects. This may be formulated simply as asteroids being dominated by refractory materials (metals, silicates) and comets containing large quantities of volatile compounds (ices). For reviews of the general constitution of asteroids, see e.g. Chapman et al. (1978) or Gaffey and McCord (1979), and for cometary nuclei, see e.g. Whipple and Huebner (1976), Delsemme (1977a) or Greenberg (1982).

Certainly, asteroids are not completely deprived of volatiles. The $3\mu m$ absorption due to H_2O has been identified in the reflectance spectrum of Ceres (Lebofsky 1978; Lebofsky et al. 1981), and Larson et al. (1979) found that such data for both Ceres and Pallas were consistent with surface minerals containing water of hydration. The shapes of the reflectance curves characteristic of RD-type asteroids have a likely analogue among the C1 and C2 carbonaceous chondrites (Degewij and van Houten 1979) containing a hydrated clay mineral matrix (see Dodd 1981) and their colours resemble kerogen-containing, low-temperature carbonaceous condensates (Gradie and Veverka 1980). Indications of volatile material contained in asteroids are as yet restricted to the C and RD types, and thus they become more important as more remote groups of asteroids are considered.

On the other hand, it is clear that for an object to be observed as a comet, it must also have physico-chemical characteristics of the cometary type, i.e., large quantities of ice must be present near its surface. However, even with an icy surface there is no guarantee that an object will display cometary activity. This depends on the orbital parameters, mainly the perihelion distance, and perhaps also on the evolutionary status of the object. Typical variations with heliocentric distance of sublimation rates of various ices as induced by absorption of solar radiation were found by Delsemme and Miller (1971) assuming an isothermal cometary nucleus with the mean insolation for a spherical body. Recent work has considered different improvements of this approximation, such as e.g. the use of latitude-dependent mean diurnal insolations (Cowan and A'Hearn 1979) or treatment of heat flow inside the nucleus (Weissman and Kieffer 1981; Rickman and Froeschlé 1983). Lacking a systematic exploration of the influence of such effects on the expected activity limit, we can only estimate this limit within a factor two, where the lower bound is set by the isothermal approximation and the upper bound corresponds to a subsolar point unaffected by heat conduction. For a H_2O-dominated nucleus, then, using visual and infrared albedos = 0.1, the normalizing distance r_o entering into the g(r) function in the standard expression for the nongravitational force (Marsden et al. 1973; Marsden 1974) would fall between 2.8 AU and 5.6 AU. For different choices of the albedos this range would be modified (Marsden et al. 1973), and for other cosmochemically likely substances such as CO_2 or CO it occurs much further from the Sun (Marsden 1974; see Delsemme 1985). According to Delsemme, the principal reason to believe comets in general to be dominated by H_2O ice is the tendency for cometary activity to follow the behaviour predicted for this case. Nevertheless, cometary activity at large distances from the Sun is still a relatively unexplored phenomenon, and the limited observations at hand appear to allow a wide range of theoretical interpretations.

3. P/SCHWASSMANN-WACHMANN 1 AND THE JUPITER-SATURN RESERVOIR

It is of interest in this connection to consider the trans-Jovian intermediate reservoir of comets suggested by Kresák (1972a) and evidenced by numerical simulations of cometary capture such as those performed by Everhart (1972, 1977) and Fernández and Ip (1983a). The number of potentially active comets in this reservoir, as restricted to the Jupiter-Saturn region, was estimated using Monte Carlo simulations by Rickman and Vaghi (1976) and Froeschlé and Rickman (1980) to be $\sim 10^4$. "Potentially active" means that at least upon reduction of the perihelion distance during capture into the Jupiter family the objects will display cometary activity.

Comet P/Schwassmann-Wachmann 1 has an orbit typical of the Jupiter-Saturn reservoir and must be regarded as one of its members. Thus it is unique at least in the respect that it is, as yet, the only member discovered in an orbit of this class. The question as to whether it is also unique with respect to its violent and frequent outbursts (5-8 magnitudes; Whipple 1980) is closely connected to the size of its nucleus. Estimates of the diameter have been quoted as 20-25 km (Kresák 1979) with a large uncertainty depending on the range of possible albedos (see Roemer 1966). Recently Cruikshank and Brown (1983) found a value of appr. 40 km corresponding to a geometric albedo $p_v = 0.13$.

If P/Schwassmann-Wachmann 1 is indeed the only member of the Jupiter-Saturn reservoir exhibiting the phenomenon of frequent, violent outbursts, then there is nothing strange about the fact that no other member has yet been discovered. It is reasonable to expect that even among 10^4 objects, the nucleus of P/Schwassmann-Wachmann 1 is one of the biggest. This can be seen by a comparison with the sample of long-period comets which passed perihelia inside Jupiter's orbit during the last 400 years. Using an influx rate of Oort cloud comets of 5 yr^{-1} per AU of perihelion distance (Weissman 1980a, 1982), we get a contribution of 10000 objects. The number of passages of old long-period comets is more difficult to estimate, since observational selection dependent on the perihelion distance is more serious. Taking Everhart's (1967) estimate of 8000 passages during the 127 years from 1840 to 1967 with perihelia inside 4 AU, we get appr. 25000 such passages, and a conservative extrapolation to 5 AU increases this by 20% to 30000. Furthermore, if intrinsically much fainter objects (in general not possible to observe) are also included, the old comets certainly outnumber the new ones by a large factor (see e.g. Kresák 1982; Weissman 1982). Hence as a very conservative estimate we may consider our long-period comet sample to be at least as large as the Jupiter-Saturn reservoir. According to the analysis by Kresák (1979), among the long-period comets there are only two - comets 1729 and 1882 II - that have indisputably an intrinsic brightness high enough for their nuclei to be as large as, or possibly somewhat larger than, the above-quoted values for P/Schwassmann-Wachmann 1. Some uncertainties are nevertheless connected with these two objects: comet 1882 II being a Sun-grazer, probably a fragment of a very big parent comet (Marsden 1967), and comet 1729 having a large perihelion distance (4.05 AU) and most likely being a new comet from the Oort cloud. In both these cases there are reasons to believe that standard methods for the translation from observed to absolute brightness and from absolute brightness to nuclear diameter may give misleading results.

Thus we definitely have no reason to suspect that there should be any cometary nuclei much bigger than that of P/Schwassmann-Wachmann 1 in the Jupiter-Saturn reservoir. In the absence of outbursts, with the low albedo values recently considered for cometary nuclei (e.g. Veeder and Hanner 1981; Hartmann et al. 1982; Whipple 1983a; Weissman 1984), they may all easily have escaped detection. The problem then is to explain why P/Schwassmann-Wachmann 1 should be unique among all the objects in similar orbits to display the outbursts. The mechanism causing these outbursts has not yet been identified, and none of the different suggestions that appear plausible at present (Froeschlé et al. 1983) is able to explain such a uniqueness. This holds for the volatile pocket hypothesis involving sudden exposures of CH_4 (Whitney 1955) or CO_2 or CO (Cowan and A'Hearn 1982) on the surface of the nucleus, the phase transition hypothesis invoking an amorphous-cubic phase transition proceeding in isolated bursts (Patashnick et al. 1974; Froeschlé et al. 1983; Klinger 1983a,b), and the free-radical hypothesis where the energy feeding the outbursts is supposed to originate in the recombination of free radicals trapped in the ice (Haser 1955; Donn and Urey 1956; for a suggestion concerning ion molecular clusters, see Shul'man 1983). Likewise, the model recently proposed by Hartmann et al. (1982) where outbursts occur due to gas pressure accumulating in a dusty regolith would not predict P/Schwassmann-Wachmann 1 to be unique.

Obviously, we must consider the possibility that P/Schwassmann-Wachmann 1 is not at all unique in the Jupiter-Saturn reservoir producing frequent outbursts by a factor 100 or more in brightness. Although the evolutionary aspects of the above hypotheses have not yet been worked out, at least the phase transition hypothesis would predict outbursts to be a commonplace phenomenon for a certain range of cometary

orbits characterized by orbital mean temperatures (Klinger 1983a) near
the value of P/Schwassmann-Wachmann 1 (120 K) and moderate eccentricities. There is one further important restriction, namely, that the
comet should never have made any visit into the Jupiter family, thus
allowing amorphous ice to exist near the surface of the nucleus. Similarly, Hartmann et al. (1982) indicate a rather broad range of orbital
motions (relatively circular orbits at 4-7 AU from the Sun) for their
mechanism to be effective.

Taking such restrictions into account, only a subset of the 10^4
members of the Jupiter-Saturn reservoir already considered would be candidates for outbursts, but apparently this subset may still contain a
large number of comets. If this is so, we must again consider the question why P/Schwassmann-Wachmann 1 is the only of these as yet discovered. It is impossible at present to give a detailed answer to this
question, since there may well be a considerable spread in the outburst
amplitudes for different comets depending e.g. on dust/gas ratios.
However, as a reasonable estimate we would have to assume that all
the other comets displaying outbursts must have nuclear diameters at
least approximately ten times lower than P/Schwassmann-Wachmann 1. If
the value by Cruikshank and Brown (1983) is indeed trusted as referring
to light scattered by the solid nucleus of P/Schwassmann-Wachmann 1,
this assumption appears reasonable. It is interesting to note in this
connection that Festou and Atreya (1983) found a possible production of
H atoms from P/Schwassmann-Wachmann 1 using Ly-α observations. The
derived production rate was $\gg 10^{30}$ s^{-1} and made these authors conclude
that the nucleus of P/Schwassmann-Wachmann 1 is at least ten times
bigger than those of other comets, whose gas production rates have
been measured near the Earth. Parent molecules have not been identified:
the detection of CO$^+$ in P/Schwassmann-Wachmann 1 by Cochran et al.
(1980) and Larson (1980) strongly indicates the presence of CO (Festou
and Atreya 1983), but the presence of an atomic hydrogen coma would show
that a hydrogen-bearing molecule must also be produced in large quantities from the nucleus.

4. ACTIVITY LIMITS AND INTERNAL HEAT SOURCES

Thus, even though the activity exhibited by P/Schwassmann-Wachmann 1 may
be of a somewhat different kind as compared to other observed comets, it
appears possible that objects dominated by H_2O ice may have an effective
activity limit well beyond Jupiter's orbit under certain circumstances.
Further work is needed to settle this question, both theoretically and
observationally.

Obviously, the discussion of activity limits is closely related to
the long-standing question of internal heat sources in cometary nuclei,
as shown by the above examples of suggested mechanisms for the outbursts
of P/Schwassmann-Wachmann 1. There is little doubt that such internal
heat sources may exist in most if not all comets, but much work remains
to be done in order to estimate quantitatively their importance. For the
time being, variations in cometary activity at large distances from the
Sun may be explained both by compositional differences and differences

in the action of internal heat sources. One example is given by the well-known fading of long-period comets after their first apparition as newcomers from the Oort cloud (e.g. Oort and Schmidt 1951; Kresák 1982; Bailey 1985) which appears particularly serious for comets with large perihelion distances (Marsden and Sekanina 1973; Marsden et al. 1978). One possible explanation is that the nuclei of Oort cloud comets have a particularly volatile surface layer, either due to a chemical inhomogeneity (Marsden and Sekanina 1973) or due to effects of long-term cosmic-ray bombardment (Whipple 1977; see also Brown et al. 1982; Johnson et al. 1983). Another possibility is that the amorphous–cubic phase transition sweeps through the surface layer of a nucleus coming directly from the Oort cloud, thereby increasing the gas production significantly at large distances from the Sun (Smoluchowski 1981).

In conclusion, there is apparently a broad interval of heliocentric distance where objects with icy surfaces may or may not develop cometary activity. The inner boundary may be put at $r \simeq 5$ AU, as evidenced by the relatively inert icy surfaces of most Galilean satellites. The outer boundary may be at least as far as $r \simeq 15$ AU, judging from observations of the tails of some new comets from the Oort cloud (Sekanina 1973a) or from recent observations of comet P/Halley (West 1983). According to Hartmann et al. (1981, 1982), the surface of asteroid 2060 Chiron may be at least partly icy, as expected from the cometary nature of its dynamical behaviour (Scholl 1979; Oikawa and Everhart 1979).

5. ICY INTERIORS, ROCKY SURFACES

Next, let us consider the possible existence of objects whose surfaces are asteroidal but whose bulk compositions are of the cometary type involving large quantities of ice. Two different types of observed asteroids may belong to this category. One of them is best represented by the Trojans, but may count members among all classes of outlying asteroids in stable orbits. The other type includes the Aten, Apollo and Amor groups of Earth-crossing or Earth-approaching asteroids (Shoemaker et al. 1979).

The general background to these ideas is the conception that the ice/dust mixture characteristic of a cometary nucleus may turn into an inert dust layer as the ice sublimates from the surface (see Mendis 1984). The reason in the case of a typical comet passing perihelion within 3 AU of the Sun would be that there is a critical size of a dust grain where the pressure of the sublimating gas is just sufficient to carry the grain off the gravitational field of the nucleus (Whipple 1951; Whipple and Huebner 1976), and all grains larger than this critical size are thus bound to fall back to the nucleus. Even if this concerns a very minor fraction of the grains, they may accumulate on the surface as time goes on into an inert mantle seriously affecting the sublimation of the underlying ice (Dobrovolsky and Markovich 1972; Mendis and Brin 1977, 1978; Brin and Mendis 1979; Brin 1980). This idea was recently developed by Horányi et al. (1983) into the "friable-sponge model" of the dust mantle. Depending on e.g. the dust/ice ratio of the cometary material, this mantle may develop into an insulating crust terminating the cometary activity altogether (Brin 1980; Horányi et al. 1983; Mendis 1985), until possibly the crust is broken by a meteoroid impact.

Nevertheless, there may be significant local variations in the build-up of this crust making the nucleus spotty, or strong latitudinal variations caused by the rotation of the nucleus or seasonal insolation effects such as seem to be present on P/Encke (Whipple and Sekanina 1979). Evidence for spottedness or localized activity among the nuclei of short-period comets in general is abundant (e.g. Whipple 1977, 1980, 1982, 1983a,b; Sekanina 1981a,b).

This discussion refers to comets with reasonably small perihelion distances, as already indicated, and thus the resulting asteroidal objects with cometary interiors might be found among the Aten-Apollo-Amor asteroids. Let us also consider the possible evolution of objects consisting of a dust/ice mixture moving in orbits in the Jupiter-Saturn region! For the expected composition with H_2O dominating the ice, the sublimation flux in most cases will be extremely small, so that a very large fraction of the dust grains remain on the surface. If the object can be trapped for a long time in the inner parts of this region, e.g. as a temporary Jovian satellite, it might be possible to obtain a crust thick enough to protect the object even after a reduction of the perihelion distance. This speculation may be worthy of further consideration since it conforms to a dynamically attractive evolutionary path for supplying Earth-approaching asteroids from the cometary source (Kresák 1979; see below).

The Trojan asteroids constitute a group of objects which may be quasi-permanently trapped near Jupiter's orbit (see Greenberg and Scholl 1979) ever since the early stages of evolution of the Solar System. If this is so, then comparison with the estimated bulk composition of Ganymede and Callisto indicates that the Trojans should have been formed out of a material containing a substantial fraction of ice. Possibly this holds true even for their present structure (Hartmann et al. 1982): their interiors may contain large quantities of ice, while since a long time their surface layers have been outgassed by sublimation. Collisional events (see Hartmann 1979) might temporarily cause some rejuvenation, exposing ice at the surfaces, but no long-lasting gas production can be expected. Some evidence against a bulk composition dominated by ice for the Trojan asteroids is, however, found in the estimated density of 2.5 g/cm^3 for 624 Hektor, the largest Trojan (Weidenschilling 1980). This value is derived on the assumption that Hektor is an equilibrium binary system, which seems to be the most likely explanation of its large-amplitude light variations (see Farinella et al. 1982). An alternative model of a very elongated Jacobi ellipsoid would imply a density of only 1 g/cm^3 but would be dynamically unstable.

A very important recent finding is the association between the surface material on RD asteroids and cometary dust, i.e., the refractory constituent of cometary nuclei. This was predicted from cosmogonic considerations by Gradie and Veverka (1980), and subsequently strong observational support for this idea has been found by Hartmann et al. (1982) and by Hartmann and Cruikshank (1984). Degewij and Tedesco (1982) find a preponderance of RD-type reflectivity spectra among Trojans in the

two clouds together (72%), and Degewij and van Houten (1979) formulated
the question: "Are the RD-type objects extinct cometary nuclei?" The
idea of a genetic association between Trojans and comets, however, dates
much further back. Rabe (1971, 1974) considered dynamical evidence in
support of such an association, but Kresák (1979) showed that this evidence can not be upheld. Future observations are needed in order to
settle the question if the Trojans have icy interiors.

6. GENETIC CLASSIFICATION

In the preceding sections we have paid some attention to different possibilities for comets to appear as asteroids, i.e., for objects observationally classified as asteroids to have the physico-chemical properties
of cometary nuclei. One example is given by the "future comets" that are
bound to display activity upon a major reduction of the perihelion distance, and a possible representative of this group is 2060 Chiron. Another example may be extinct cometary nuclei deactivated by gradual
build-up of an insulating crust of dust. Possible representatives may be
found among asteroids in "cometary orbits" (see below) or more generally
among Earth-approaching asteroids.
      The evolution of a comet into an asteroid may, however, proceed in
a different way, if the nucleus contains an inner core of refractory
material (see Whipple 1977; Wilkening 1979; Degewij and Tedesco 1982).
After exhaustion of the volatile material by sublimation, then, there
would remain an object having asteroidal characteristics both observationally and physico-chemically. Obviously, though, it would be genetically distinct from usual asteroids. Hence there is also some interest in
a genetic comet/asteroid distinction referring to the physico-chemical
properties with which the object was formed.
      It is not yet known with certainty what fraction, if any, of the
cometary nuclei may contain refractory cores. To some extent, different
theories for the origin of comets yield different predictions about
this, depending e.g. on the assumed place of formation. As remarked by
Kresák (1982): "Different authors put the origin of comets practically
anywhere between the asteroid belt and the interstellar medium interacting with the outskirts of the Solar System". Nevertheless, it is fair
to say that, at present, very few scientists favour the asteroid belt as
the place of origin of comets. This idea goes back to Oort's (1950)
classical paper but was also considered by van Flandern (1977) in an
hypothesis invoking a recent cometary origin. Convincing arguments that
comets must have formed (and stayed) at larger distances from the Sun
were given by Delsemme (1977b).
      Generally speaking, the rocky-core/icy-mantle structure of cometary
nuclei appears more plausible, the closer to the Sun one imagines the
comets to have formed, since the higher maximum temperatures reached in
the Solar nebula (Cameron and Pine 1973; Cameron 1978a) make it more
likely to have a significant time lag between the condensation of
refractories and volatiles during the subsequent cooling (see Greenberg
1983). Another possible reason for large-scale differentiation of cometary nuclei might be radiogenic heating by short-lived isotopes such as

[26]Al, leading to melting of the ice, as indicated e.g. by Degewij and Tedesco (1982). The obvious requirement of rapid condensation and grain accretion under this hypothesis again makes birthplaces closer to the Sun more favourable to the development of rocky cores.

The ultimate aim of the stepwise succession of observational, physico-chemical and genetic comet/asteroid distinctions so far considered is, of course, to be able to place constraints on the cosmogonic processes representing the origin of each separate class of objects. Unfortunately, the serious uncertainty concerning the origin of comets makes it impossible to say to what extent the asteroidal and cometary birthplaces may join to each other in a continuous manner, thus e.g. allowing to consider distant asteroids and outer Jovian satellites as being in some sense intermediate objects between the two classes. A complete and systematic review of the theories of cometary origin was given by Delsemme (1977a).

Of the more recent issues, let us briefly consider the Uranus-Neptune accretion zone mechanism of comet formation, the possibility of a dense inner "core" of the Oort cloud and the suggestion of Oort cloud dissipation and replacement by molecular cloud encounters. Fernández (1980a, 1982) and Fernández and Ip (1981) have presented dynamical evidence in favour of the Uranus-Neptune region as the place of formation of comets, earlier suggested e.g. by Kuiper (1951), Safronov (1972) and Whipple (1972). In particular, Fernández and Ip (1983b) found that capture of short-period comets (orbital periods < 13 yr) from a reservoir in the Uranus-Neptune region by gravitational interactions with the giant planets is much better able to maintain the presently observed population than corresponding captures from near-parabolic, Oort cloud orbits. Due to the very long time-scale for scattering cometesimals from the Uranus-Neptune region (Öpik 1973; Ip 1977; Fernández 1980b), the formation of the Oort cloud by this mechanism would still be taking place at present (Fernández and Ip 1981).

However, arguments have also been advanced for cometary formation at much larger distances ($\sim 10^4$ AU) from the Sun (Biermann and Michel 1978; Biermann 1981; see also Greenberg 1985), and intermediate possibilities involve the formation of a very large number of comets at heliocentric distances $\sim 10^2$ AU (e.g. Cameron 1962, 1978b). At least in this latter case the Oort cloud would be expected to have a dense inner "core", as argued also by Hills (1981). This core has attracted much attention recently (see Bailey 1983a; Weissman 1984) since it may offer an explanation to many different phenomena, such as otherwise unmodelled perturbations on the outer planets (Bailey 1983b; see also Whipple 1964, 1972), supply of short-period comets (Fernández 1980a; cf. Fernández and Ip 1983b), a low-temperature sky background detected by IRAS (Low et al. 1984; Bailey 1984 ; cf. Bailey 1983a,c), or replenishment of the outer parts of the Oort cloud after dissipation by encounters with Giant Molecular Cloud Complexes (van den Bergh 1982; Bailey 1983d; Weissman 1984). These encounters have been found to make the outer parts of the Oort cloud dynamically unstable over a time-scale $\sim 10^8$ years (Napier and Staniucha 1982; Clube and Napier 1982; Napier 1982), and gravitational capture of new comets from these star-forming regions was proposed as a source for replenishment (Clube 1983; Clube and Napier 1984; see also

Valtonen and Innanen 1982; Valtonen 1983). For further discussion of these issues, see Clube and Napier (1983) and Weissman (1983).

Let us briefly return to the possibility of identifying 2060 Chiron genetically as a comet. It must be noted that its diameter of several hundred km (see Kowal 1979) would necessarily make it a very unusual comet, as seen from the above discussion of the sizes of cometary nuclei. The cometary identification of Chiron appears more plausible on the hypothesis of the Uranus–Neptune accretion zone as the origin of comets than it does if even more remote birthplaces are imagined. A much larger number of comets is naturally expected to move in Chiron-like orbits, if the source is very close, and thus the existence of a giant object appears less unlikely. If the diameter-frequency relation for cometary nuclei in the vicinity of 10^2 km has the same slope as that for asteroids presented by Zellner (1979), then a value of 40 km for the diameter of P/Schwassmann–Wachmann 1 as the biggest among 10^4 objects would imply an estimate of 10^7 objects in the Saturn–Uranus region as the maximum number out of which Chiron could be expected to be the biggest. The Oort cloud should then contain some $10^4 - 10^5$ Chiron-sized comets and, of course, possibly some even bigger ones. Such an estimate is in no conflict with the estimated mass of the Oort cloud (see Weissman 1982), but apparently a large fraction of this mass may be contributed by exceptionally big objects.

In conclusion to the discussion of cometary birthplaces, we note that the vast majority of authors recently put the origin of comets far outside Jupiter's orbit. Hence to the extent that primordial objects remain in the Jupiter–Saturn region (these will have to be locked into stable resonances or satellite motions; see Lecar and Franklin 1973, 1974; Everhart 1973a,b; Froeschlé and Scholl 1979), they might in fact be genetically unrelated to both asteroids and comets.

7. DYNAMICAL CLASSIFICATION

In the previous sections we have come across several possibilities for comets to develop into asteroids or at least to have an asteroidal appearance. In order to identify such cases among the multitude of observed objects, it has proved essential to use the orbital properties (e.g. Kresák 1977, 1979). In the analysis by Kresák (1979) three different parameters were considered: the aphelion distance (Q), the minimum approach distance to Jupiter (ρ), and the Tisserand invariant (T).

In particular, T turned out to provide a clear separation of asteroids and comets in general. To quote from Kresák: "...the definition of a cometary orbit as one of T < 3 without resonance, and of an asteroidal orbit as one which either has T > 3 or librates around a simple resonance ratio, sets a very good dividing line between the two populations". The stability of T has been disputed (see Everhart 1976), and of course it is an invariant only in the circular, restricted 3-body problem (Sun–Jupiter–object). However, perturbations ΔT resulting from Jupiter's orbital eccentricity have been shown to be small (e.g. Froeschlé and Rickman 1981), and if one limits attention to typical objects under observation, having perihelia within or near Jupiter's orbit,

Jupiter is indeed the dominant perturber of their orbits, so that the
ΔT:s caused by other planets are almost always relatively small. These
facts have been stressed by Kresák (1972, 1977, 1980) and the empirical
separation of comets and asteroids with respect to the dividing line at
T = 3 remains a fact, proving the importance of T under most circumstances in question.

Very few objects definitely crossed this boundary, as of the beginning of 1979. Most conspicuously among the comets, a group of three
objects (P/Oterma, P/Smirnova-Chernykh and P/Gehrels 3) was found in
temporary motion near the 3/2 resonance with relatively small aphelion
distances and values of T between 3 and 3.05, following low-velocity
encounters with Jupiter ("quasi-Hilda type motions"; see Kresák 1979).
In the case of P/Oterma, this motion took place during 1937-63 and
was thus already terminated (Kazimirchak-Polonskaya 1967; Marsden 1970a;
Carusi et al. 1981). Additional examples are known from orbital integrations of other short-period comets outside the observed time interval,
and the phenomenon of temporary captures into low-eccentricity orbits
near the 3/2 resonance appears to be quite common (Kresák 1979).

Among the asteroids there were only two cases of T < 2.9 occurring
without any libration in mean longitude or critical argument to protect
the object from encountering Jupiter. One of these was 1373 Cincinnati
with present-day osculating elements a = 3.4 AU; Q = 4.5 AU; i = 39°,
which turns out to avoid encounters with Jupiter due to libration of the
argument of perihelion around 90° (Kozai 1962; Marsden 1970b; Froeschlé
and Scholl 1979). The other case was 944 Hidalgo, the only asteroid
known at that time to approach Jupiter rather closely (minimum distance
= 0.38 AU in 1673; Marsden 1970b) and generally considered the primary
candidate for being an extinct cometary nucleus (Degewij and Tedesco
1982).

Indeed, the minimum approach distance to Jupiter has been shown to
provide another interesting distinction between short-period cometary
and asteroidal orbits: comets tend to approach Jupiter closely while
asteroids tend not to approach Jupiter (Marsden 1970b), and as of 1979
only Hidalgo among the asteroids had ρ < 1 AU, while among the short-
period comets (P < 20 yr) only P/Encke, P/Arend-Rigaux and P/Neujmin 1
had ρ > 0.8 AU. As remarked by Marsden (1970b), the latter two comets
are known for their low level of activity thus making them the most
asteroidal comets both regarding physical appearance and dynamical
behaviour. As remarked by Kresák (1979), both Hidalgo, P/Arend-Rigaux
and P/Neujmin 1 have typically cometary values of T, and they can not be
expected to settle into stable orbits of the asteroidal type. Over a
time-scale $\sim 10^3$ yr their motions are indeed relatively stable, in spite
of the approaches to Jupiter by Hidalgo, and due to resonance librations
by the two comets (Marsden 1970b), but consideration of a longer time
interval might well change this situation.

Apparently the occurrence of relatively stable motion of a short-
period comet over $\sim 10^3$ yr gives a possibility for the object to develop
asteroidal characteristics, probably by growth of an inert crust on the
nucleus. Such a process should be strongly dependent on the perihelion
distance, but the details of this dependence have not yet been worked
out. Let us remark in this connection that the long-term perturbations

of Hidalgo's orbit as estimated analytically by Kozai (1979) allow its perihelion distance to drop as low as 1.1 AU.

A couple of comets have been found in fairly stable orbits near the 1/1 resonance with Jupiter, involving temporary mean longitude libra- tions. This holds for P/Slaughter-Burnham (Marsden 1970b; Rabe 1972) with a present perihelion distance of 2.5 AU and P/Boethin (Benest et al. 1980, 1982, 1983) with a present perihelion distance of 1.1 AU. How- ever, these objects may approach Jupiter rather closely (P/Slaughter- Burnham to 0.29 AU in 2075 and P/Boethin to 0.5 AU in 1909), and hence the long-term behaviour of their motions may be affected by serious uncertainties.

8. NEW CANDIDATES FOR EXTINCT COMETS

The most interesting feature to be noticed at present in connection with the dynamical comet/asteroid distinction is the recent addition of a number of asteroids in cometary orbits. In particular, for Mars-crossing or Mars-tangent objects (perihelion distance q < 1.67 AU) there was a very clear separation of comets from asteroids in 1979 such that all asteroids except 6344 P-L with Q = 4.21 AU (determined from very few observations and seriously uncertain) had Q < 4.1 AU, while all comets except P/Encke had Q > 4.6 AU. This situation has now changed drastical- ly. Table I lists some orbital data for newly discovered asteroids with q < 1.67 AU and Q > 4 AU, and it is readily seen that nine of these have Q $\gtrsim$ 4.3 AU. In fact, three are even Jupiter-crossers (1982 YA, 1983 SA and 1984 BC), and while the quality of the orbits of 1982 YA and 1984 BC is inferior, the orbit of 1983 SA is already quite well-determined. Furthermore, the Tisserand invariants of these three objects are deeply inside the cometary domain (T < 2.9), and five more are situated in the interval 2.95 < T < 3.00. Obviously, with regard to Kresák's classifica- tion as quoted above, it is of interest to examine whether the objects with T < 3 librate around simple resonance ratios.

Indeed a preliminary investigation in the elliptic restricted three-body problem Sun-Jupiter-object (Hahn and Rickman 1984) shows such librations to exist in four cases, as indicated in Table I. Two of these refer to the above-mentioned Jupiter-crossers, and 1982 YA is thus protected from approaching Jupiter to within 1 AU during a considerable time by libration at the 5/3 resonance, while the libration of 1983 SA is the second one known, after 279 Thule, at the 4/3 resonance. This latter libration is also confirmed by Benest et al. (1985) using more complete dynamical models, and in both investigations it is found to be broken after < 1000 yr in the future, whereafter encounters to within less than 0.4 AU of Jupiter occur. 1984 BC, on the other hand, as yet does not appear to librate and shows moderately close encounters with Jupiter. The closest encounters are, however, found for 1983 XF (also with a well-determined orbit) in connection with a large-amplitude lib- ration at the 2/1 resonance. After the termination of this libration, in both the quoted investigations, very close encounters with Jupiter are found.

Table I. Orbital properties of ten recently discovered asteroids with Q > 4 AU according to the investigation by Hahn and Rickman (1984). The number of observations and the observational arc refer to the orbit treated in this investigation, and the minimum distance to Jupiter (ρ) and the librational property correspond to the motion over appr. ± 1000 yr.

Asteroid	T	Q(AU)	q(AU)	i($^\circ$)	No. of obs.	Obs. arc (days)	ρ(AU)	Libr. around
1979 VA	3.08	4.29	0.98	2.8	49	88	1.16	
1981 FD	2.99	4.79	1.69	2.6	18	40	3.14	2/1
1981 VA	2.96	4.29	0.63	22.0	23	49	1.54	
1982 TA	3.09	4.07	0.53	12.1	46	213	1.19	
1982 YA	2.38	6.29	1.12	34.6	11	27	1.02	5/3
1983 LC	2.98	4.50	0.77	1.5	12	19	0.91	
1983 SA	2.31	7.25	1.21	30.8	54	174	0.51	4/3
1983 VA	2.98	4.36	0.80	16.2	6	68	0.85	
1983 XF	2.98	4.78	1.45	4.2	35	100	0.01	2/1
1984 BC	2.78	5.30	1.55	22.5	7	32	0.30	

Approach distances to Jupiter significantly smaller than 1 AU have been found for all the five objects discovered in 1983 and 1984 (see Table I). Thus Hidalgo is no longer unique in this respect. We have three new first-rank candidates for being asteroids of cometary origin (1983 SA, 1983 XF and 1984 BC) and three more, only somewhat less certain cases (1982 YA, 1983 LC and 1983 VA). Of the other asteroids, we remark that 1981 FD appears to add to the Griqua group (see Franklin et al. 1975; Kresák 1979; Schubart 1979), as indicated already by Bowell and Marsden (1981).

Recently another Apollo asteroid was also discovered using the IRAS satellite, providing even more clearcut evidence for a cometary association from the dynamical point of view. This is 1983 TB, the asteroid of the Geminid meteors (Whipple 1983c; see also Hughes 1983). The idea of possible associations of asteroids with meteor streams is an old one (see Sekanina 1973b, 1976; Kresák 1977), and recently Drummond (1982) suggested several such associations, the most likely cases involving asteroids 2101 Adonis and 2201 Oljato.

However, for 1983 TB there can be no reasonable doubt about its association with the Geminid stream. Although no parent comet was known for this stream, it was considered highly probable that such a comet had earlier existed, being now extinct (e.g. Kresák 1973). Thus 1983 TB could be this extinct comet, but one important problem still remains to be solved: how can an active comet be transferred into the orbit of the Geminid stream having the very high value of T = 4.27 and the low aphelion distance Q = 2.6 AU? There is not yet any satisfactory answer to this often posed question except for, possibly, the simple observation that P/Encke seems somehow to have managed at least part of the required evolution (T = 3.00; Q = 4.10 AU). Certainly, nongravitational forces may be involved (Sekanina 1971), and the stable, favourable orientation of the spin axis found for P/Encke in the past (Whipple and Sekanina 1979) may indeed have been essential for reducing Q to its present value. However, it must also be noted that this orientation is changing dramatically in such a way that the evolution of Q will soon be reversed (Whipple and Sekanina 1979). We hence can not be sure that a stable settling into an asteroidal orbit will in fact occur even for comet P/Encke. Furthermore, the value of Q = 2.6 AU for the Geminids appears too small to be produced by nongravitational effects (Sekanina 1971). We must pay attention to the possibility that genuinely asteroidal meteor streams may exist as a result of collisional fragmentation or release of ejecta clouds by minor impacts (Degewij and Tedesco 1982; Drummond 1982).

Let us briefly mention another property expected to reveal possible ex-comets among asteroids (Kresák 1977), i.e., the existence of nongravitational forces affecting the orbital motion. It was recently claimed that such effects may exist for some asteroids (Ziolkowski 1983). However, the true nature of the effects in question has not yet been fully worked out (see e.g. Marsden 1970b, 1984).

Further observational studies of the physical nature of Aten-Apollo-Amor asteroids are obviously needed. In view of the results by Gradie and Veverka (1980), Hartmann et al. (1982) and by Hartmann and Cruikshank (1984), an important indicator of cometary origin would be an RD-type reflectivity spectrum. This has not yet been found (McFadden 1983; McFadden et al. 1984), but many of the above-mentioned candidates for cometary origin remain to be examined. Statistics of rotation rates for Earth-approaching asteroids appears to indicate a bimodal distribution, suggesting the existence of both cometary and asteroidal contributions (Debehogne et al. 1983; Harris 1983). However, to associate rapid spin of a group of Earth-approaching asteroids with an origin in the main belt may not be justified, since a recent analysis by Farinella et al. (1984) indicates no clear difference between the spin rates of comets and small main-belt asteroids. A significant fraction of the Apollo-Amor objects (5 out of 21 observed photometrically; Farinella, priv. comm.) have a highly elongated shape. However, comparison of these statistics with the shape distribution of small main-belt asteroids (Binzel and Mulholland 1983; Binzel 1984; Lagerkvist 1983a,b) or with that of fragments produced in laboratory impact experiments (Capaccioni et al. 1984) is complicated by the likely existence of various selection effects. At present no conclusion regarding the importance of the cometary contribution appears possible from such data.

9. ORIGIN OF THE EARTH-APPROACHING ASTEROIDS

One of the outstanding issues regarding comet-asteroid evolution is the problem of the origin of Earth-approaching asteroids. Reviews of work performed in this field have been given e.g. by Shoemaker et al. (1979) and Wetherill (1979). In brief, the Earth-approaching asteroids often have typically asteroidal orbits as far as the Tisserand invariant is concerned. In principle there are evolutionary tracks of the coplanar Tisserand criterion connecting some orbits of Apollo asteroids with the main belt, but the problem is that in the absence of close encounters it appears impossible under most circumstances to produce the necessary increases of eccentricity. Even if gravitational captures by the action of Mars are invoked, any transfer mechanism from the main asteroid belt appears unable to explain the high ratio of Apollo to Amor objects as inferred from observational statistics (Shoemaker et al. 1979). This difficulty adds to the problem of accounting for the required infeed rate of appr. 15 objects per 10^6 yr (Wetherill 1979), derived under the assumption of a steady state for the population of Earth-approaching objects.

Such arguments led Öpik (1963) to suggesting comets as a source for Apollo asteroids, and this idea has remained a popular one, especially since it conforms well to ideas about the evolution of cometary nuclei, as discussed above. Specifically, Wetherill (1976, 1979) found dynamical evidence in favour of a cometary origin for most Apollo asteroids. This cometary source would evidently be identifiable with the Mars-crossing Jupiter family (q $\lesssim$ 1.5 AU; Q $\lesssim$ 8 AU). However, the dynamical lifetimes of objects in such orbits are limited to $\sim 10^5$ yr due to Jovian perturbations, mostly at close encounters (e.g. Froeschlé and Rickman 1981; Carusi et al. 1979). This interval may be shorter than the typical one during which the object is observable as a comet. This observable lifetime is estimated to be several hundred revolutions for the cometary orbits in question (Kresák 1981a,b; Fernández 1981) as derived from observational and orbital statistics, and even longer (Weissman 1980b) if standard models for the sublimation from cometary nuclei are to be trusted. Processes decreasing Q and increasing T are needed in order to capture comets from the source in question into typical Apollo asteroid orbits as described above, and evidently they must work rapidly in order not to be disturbed by Jovian perturbations of the cometary orbit.

One possibility is that active comets are transferred into Encke-type orbits by nongravitational forces whereafter their activity may terminate, and the extinct nuclei appear as Apollo asteroids (Kresák 1979). The likelihood of occurrence of this process needs to be further investigated. The other alternative is that extinct comets moving in unstable orbits (Q λ 4.5 AU) are gravitationally captured by the terrestrial planets at near-collisions, so that Q is suddenly decreased by a large amount. At least in the second case it would be justified to compare the observed vs. expected numbers of both Apollo-Amor asteroids with: (q < 1.3 AU; Q $\lesssim$ 4 AU) and corresponding extinct comets with: (q < 1.3 AU; Q λ 4.2 AU). Such a comparison was carried out by Rickman and Froeschlé (1980) on the basis of a Monte Carlo simulation of the distribution of extinct comet orbits. The absence, at the time of writing of that paper, of any observed Apollo-Amor asteroid with a safely determined Q well in excess of 4 AU, combined with a large number of expected extinct comets, led these authors to the conclusion that most extinct comets are non-existent, i.e., that no more than several percent of the Jupiter family comets may develop into sizeable asteroidal bodies at the end of their activity (cf. Kresák 1980; Whipple 1981). By using a similar argument for high-inclination comets, Nakamura (1983) arrived at the same estimate.

We may now add two comments to this discussion. Firstly, the conclusion by Rickman and Froeschlé (1980) holds under the assumption that the majority of observed Apollo-Amor asteroids are indeed extinct comet nuclei. On the other hand, if these objects should be mainly collisional fragments of main-belt asteroids, they might have a much higher albedo than the extinct nuclei of the Jupiter family, and the lack of observations of such nuclei might to some extent result from this albedo difference. Secondly, Table I shows that the number of zero observed Apollo-Amor asteroids with $Q \gtrsim 4.2$ AU used by Rickman and Froeschlé has now increased to six! In fact, four of these belong to the above-mentioned group of candidates for cometary origin judging from their orbital evolutions. The conclusion by Rickman and Froeschlé would now be changed into an estimate that almost 10% of the short-period comet nuclei develop into sizeable asteroidal objects, and by the albedo effect just mentioned this could in fact be taken as a lower limit. However, it must be emphasized that the statistical material underlying these estimates is still extremely poor, and that further serious sources of uncertainty exist in the necessary estimates of the steady-state number of extinct comets and the discovery probability of such an object.

Evidently, the question of the cometary vs. asteroidal origin of the usual Earth-approaching asteroids with $Q \lesssim 4$ AU is still far from being satisfactorily answered. The above arguments give some evidence against a major cometary contribution. Another piece of evidence pointing in the same direction is the difficulty in identifying Apollo objects both as extinct comets and as the source of stony meteorites (Levin and Simonenko 1981).

Unfortunately, the orbital inclinations (i) of Earth-approaching asteroids do not yet appear to provide any clearcut evidence regarding their origin. Four known objects have $i > 50^\circ$, and three of these belong to the Apollo group. This might possibly be indicative of a cometary origin for these asteroids, but it must be noted that as yet there is no statistically significant difference between the i-distributions of Apollos and Amors. Furthermore, it is not yet clear to what extent a dynamical transfer from the main asteroid belt would lead to smaller inclinations than a transfer from the Jupiter family of comets.

In this connection one should also note the recent work by Wisdom (1982, 1983). By application of an algebraic mapping of phase space onto itself, motions near the 3/1 resonance with Jupiter could be tracked over very long time intervals, and large sudden increases in eccentricity were often found. These eccentricity jumps are similar to those earlier found by Scholl and Froeschlé (1977) at the 3/1, 5/2 and 2/1 resonances by numerical integration using Schubart's (1964) averaging method. However, by extension to a longer time span this phenomenon now appears more wide-spread for near-resonant orbits. Thus Mars-crossers and perhaps even Earth-approachers may result from the 3/1 resonance, and this possibility appears to increase considerably the efficiency of gravitational transfer from the main belt into Apollo-Amor orbits, as compared with existing estimates. The orbit of the recently discovered asteroid 1984 AB (a = 1.58 AU, e = 0.076, i = $14^\circ.8$ computed by Marsden using an identification with 1975 XL4 by Bardwell; see MPC 8679) is of great interest in this connection, being quite similar to the orbit of Mars. This kind of orbit is indeed to be expected as an intermediate stage in a capture by Mars of an object coming from the main belt, since the Tisserand parameter with respect to Mars has a value near 3 and the encounter speed is thus relatively low. Future studies of the orbital evolution of this object may indicate whether the idea here outlined can be upheld.

Regarding dynamical possibilities for a cometary origin of Apollo-Amor objects, an interesting suggestion by Kresák (1979; see also Carusi et al. 1981) is to follow the T = 3 evolutionary track, perhaps even with T somewhat above 3, via a quasi-Hilda type motion whereby Q reaches relatively low values at an early stage of capture, and possibly further so that low enough perihelion distances may be reached for an efficient action of nongravitational forces. This kind of evolution should be studied further. Some attention has already been paid to it since it is closely connected with temporary satellite captures by Jupiter (Carusi and Valsecchi 1981, 1983). Indeed, two of the three above-mentioned quasi-Hilda type comets (P/Oterma and P/Gehrels 3) have experienced such satellite captures lasting for short but significant time intervals (Chebotarev 1967; Carusi and Valsecchi 1979, 1981, 1982; Rickman 1979; Rickman and Malmort 1981).

In conclusion, the problem of the origin of Earth-approaching asteroids is not yet solved. It appears at present that there are some indications of a mixture of two disparate populations among the 'usual' Apollo-Amor asteroids ($Q \lesssim 4$ AU), corresponding perhaps to the two sources classically considered. The recent discovery of a number of Apollo-Amor and Mars-crossing asteroids in 'unusual', cometary orbits ($Q \gtrsim 4.2$ AU) strengthens the evidence for evolution of Jupiter-family comets into asteroidal objects. However, much work remains to be done in order to clarify the dynamical transfer mechanisms from both the asteroidal and cometary sources.

J.A. FERNANDEZ: For the estimate of the conversion rate of short-period comets into Apollo-Amor asteroids it is necessary to know the dynamical lifetime, t_{dyn}, of AA objects. Have you considered any particular value of t_{dyn} in your study?

H. RICKMAN: For estimating that a certain fraction (at least 10% according to my discussion) of short-period comets with $q < 1.3$ AU develop into Apollo-Amor asteroids, no knowledge of t_{dyn} is required. When it comes to estimating what fraction of such objects may be stabilized from Jovian perturbations by reduction of the aphelion distance, too little is known at present to give any quantitative figure. Assuming that the majority of AA asteroids do come from the cometary source, we would require a conversion rate supplying $\gtrsim 10$ new objects per 10^6 yr, and this corresponds to an estimate of $\sim 10^8$ yr for t_{dyn}, where collisions as well as dynamical ejections are taken into account.

P.R. WEISSMAN: I would be very cautious about accepting Cruikshank and Brown's radius for P/Schwassmann-Wachmann 1. They base their estimate on the magnitude of the comet when it is quiescent at around m = 18, assuming there is no coma at that time. But IRAS has looked at Schwassmann-Wachmann 1 and found that it is very bright in the infrared even during quiescent periods, indicating that there is always a very substantial coma. Thus the estimate of the radius by Cruikshank and Brown is likely too large.

H. RICKMAN: I fully agree that one should not take it for granted that the 20-μm observations by Cruikshank and Brown pertain to the solid nucleus without any coma. Hence it might indeed be preferable to consider their estimate of the nuclear diameter as an upper limit. However, the existence of a visual brightness threshold during quiescent periods, below which the comet appears never to fall, speaks against the presence of an optically thick dust coma on such occasions.

P.R. WEISSMAN: Another example of an Apollo asteroid that is likely an extinct comet is 2201 Oljato. Chris Russel at UCLA has detected disturbances in the solar wind associated with close approaches of this asteroid to Venus, using the Pioneer-Venus spacecraft. He interprets this as some sort of outgassing debris stream in the asteroid's orbit. Also, Lucy McFadden has found a brightening of this object in the ultraviolet which she interprets as Rayleigh scattering from a cloud of fine particles around the asteroid. Thus, Oljato may be another extinct cometary nucleus like 1983 TB.

H. RICKMAN: Indeed Oljato is one of the most promising candidates for being an extinct comet. Another piece of evidence in support of this is its possible meteor stream association suggested by Drummond.

REFERENCES

Bailey, M.E.: 1983a, in Asteroids, Comets, Meteors, eds. C.-I. Lager-
    kvist and H. Rickman, Univ. of Uppsala, p. 383
Bailey, M.E.: 1983b, Nature 302, 399
Bailey, M.E.: 1983c, Mon. Not. Roy. Astron. Soc. 205, 47p
Bailey, M.E.: 1983d, Mon. Not. Roy. Astron. Soc. 204, 603
Bailey, M.E.: 1984, Mon. Not. Roy. Astron. Soc., in press
Bailey, M.E.: 1985, this volume.
Benest, D., Bien, R., Rickman, H.: 1980, Astron. Astrophys. 84, L11
Benest, D., Bien, R., Rickman, H.: 1982, in Sun and Planetary System,
    proc. IAU VI European Meeting in Astronomy, eds. W. Fricke and G.
    Teleki, D. Reidel Publ. Co., p. 397
Benest, D., Bien, R., Rickman, H.: 1983, in Dynamical Trapping and
    Evolution in the Solar System, proc. IAU Coll. 74, eds. V.V. Mar-
    kellos and Y. Kozai, D. Reidel Publ. Co., p. 107
Benest, D., Bien, R., Rickman, H.: 1985, this volume.
van den Bergh, S.: 1982, J. Roy. Astron. Soc. Can. 76, 303
Biermann, L.: 1981, Phil. Trans. Roy. Soc. London A303, 351
Biermann, L., Michel, K.W.: 1978, the Moon and the Planets 18, 447
Binzel, R.P.: 1984, Icarus 57, 294

Binzel, R.P., Mulholland, J.D.: 1983, Icarus 56, 519
Bowell, E., Marsden, B.G.: 1981, IAU Circular No. 3592
Brin, G.D.: 1980, Astrophys. J. 237, 265
Brin, G.D., Mendis, D.A.: 1979, Astrophys. J. 229, 402
Brown, W.L., Lanzerotti, L.J., Johnson, R.E.: 1982, Science 218, 525
Cameron, A.G.W.: 1962, Icarus 1, 13
Cameron, A.G.W.: 1978a, in Protostars and Planets, ed. T. Gehrels, Univ.
    of Arizona Press, p. 453
Cameron, A.G.W.: 1978b, in The Origin of the Solar System, ed. S.F.
    Dermott, John Wiley & Sons, p. 49
Cameron, A.G.W., Pine, M.R.: 1973, Icarus 18, 377
Capaccioni, F., Cerroni, P., Coradini, M., Farinella, P., Flamini, E.,
    Martelli, G., Paolicchi, P., Smith, P.N., Zappalà, V.: 1984, Nature
    308, 832
Carusi, A., Kresák, L., Valsecchi, G.B.: 1981, Astron. Astrophys. 99,
    262
Carusi, A., Pozzi, F., Valsecchi, G.B.: 1979, in Dynamics of the Solar
    System, proc. IAU Symp. 81, ed. R.L. Duncombe, D. Reidel Publ. Co.,
    p. 185
Carusi, A., Valsecchi, G.B.: 1979, in Asteroids, ed. T. Gehrels, Univ.
    of Arizona Press, p. 391
Carusi, A., Valsecchi, G.B.: 1981, Astron. Astrophys. 94, 226
Carusi, A., Valsecchi, G.B.: 1982, in The Comparative Study of the
    Planets, eds. A. Coradini and M. Fulchignoni, D. Reidel Publ.
    Co., p. 131
Carusi, A., Valsecchi, G.B.: 1983, in Asteroids, Comets, Meteors, eds.
    C.-I. Lagerkvist and H. Rickman, Univ. of Uppsala, p. 331
Chapman, C.R., Williams, J.G., Hartmann, W.K.: 1978, Annu. Rev. Astron.
    Astrophys. 16, 33
Chebotarev, G.A.: 1967, Analytical and Numerical Methods of Celestial
    Mechanics, Elsevier Publ. Co., p. 239-240
Clube, S.V.M.: 1983, in Asteroids, Comets, Meteors, eds. C.-I. Lager-
    kvist and H. Rickman, Univ. of Uppsala, p. 369
Clube, S.V.M., Napier, W.M.: 1982, Quart. J. Roy. Astron. Soc. 23, 45
Clube, S.V.M., Napier, W.M.: 1983, Highl. of Astron., vol. 6, ed. R.M.
    West, D. Reidel Publ. Co., p. 355
Clube, S.V.M., Napier, W.M.: 1984, Mon. Not. Roy. Astron. Soc., in press
Cochran, A.L., Barker, E.S., Cochran, W.D.: 1980, Astron. J. 85, 474
Cowan, J.J., A'Hearn, M.F.: 1979, the Moon and the Planets 21, 155
Cowan, J.J., A'Hearn, M.F.: 1982, Icarus 50, 53
Cruikshank, D.P., Brown, R.H.: 1983, Icarus 56, 377
Debehogne, H., de Sanctis, G., Zappalà, V.: 1983, Icarus 55, 236
Degewij, J., Tedesco, E.F.: 1982, in Comets, ed. L.L. Wilkening, Univ.
    of Arizona Press, p. 665
Degewij, J., van Houten, C.J.: 1979, in Asteroids, ed. T. Gehrels, Univ.
    of Arizona Press, p. 417
Delsemme, A.H.: 1977a, in Comets, Asteroids, Meteorites, proc. IAU Coll.
    39, ed. A.H. Delsemme, Univ. of Toledo, p. 453
Delsemme, A.H.: 1977b, in Comets, Asteroids, Meteorites, proc. IAU Coll.
    39, ed. A.H. Delsemme, Univ. of Toledo, p. 3
Delsemme, A.H.: 1985, in proc. NATO Adv. Res. Workshop Ices in the Solar

System, eds. D. Benest, A. Dollfus, J. Klinger, R. Smoluchowski, D. Reidel Publ. Co., in press

Delsemme, A.H., Miller, D.C.: 1971, Planet. Space Sci. 19, 1229

Dobrovolsky, O.V., Markovich, M.Z.: 1972, in The Motion, Evolution of Orbits, and Origin of Comets, proc. IAU Symp. 45, eds. G.A. Chebotarev, E.I. Kazimirchak-Polonskaya and B.G. Marsden, D. Reidel Publ. Co., p. 287

Dodd, R.T.: 1981, Meteorites: a Petrologic-Chemical Synthesis, Cambridge Univ. Press

Donn, B., Urey, H.C.: 1956, Astrophys. J. 123, 339

Drummond, J.D.: 1982, Icarus 49, 143

Everhart, E.: 1967, Astron. J. 72, 1002

Everhart, E.: 1972, Astrophys. Lett. 10, 131

Everhart, E.: 1973a, Astron. J. 78, 316

Everhart, E.: 1973b, Astron. J. 78, 329

Everhart, E.: 1976, in The Study of Comets, proc. IAU Coll. 25, eds. B. Donn, M. Mumma, W. Jackson, M.F. A'Hearn and R. Harrington, NASA SP-393, p. 445

Everhart, E.: 1977, in Comets, Asteroids, Meteorites, proc. IAU Coll. 39, ed. A.H. Delsemme, Univ. of Toledo, p. 99

Farinella, P., Paolicchi, P., Zappalà, V.: 1982, Icarus 52, 409

Farinella, P., Paolicchi, P., Zappalà, V.: 1985, this volume

Fernández, J.A.: 1980a, Mon. Not. Roy. Astron. Soc. 192, 481

Fernández, J.A.: 1980b, Icarus 42, 406

Fernández, J.A.: 1981, Astron. Astrophys. 96, 26

Fernández, J.A.: 1982, in Sun and Planetary System, proc. IAU VI European Regional Meeting in Astronomy, eds. W. Fricke and G. Teleki, D. Reidel Publ. Co., p. 371

Fernández, J.A., Ip, W.-H.: 1981, Icarus 47, 470

Fernández, J.A., Ip, W.-H.: 1983a, Icarus 54, 377

Fernández, J.A., Ip, W.-H.: 1983b, in Asteroids, Comets, Meteors, eds. C.-I. Lagerkvist and H. Rickman, Univ. of Uppsala, p. 387

Festou, M.C., Atreya, S.K.: 1983, in Cometary Exploration, vol. I, ed. T.I. Gombosi, Hungarian Acad. Sci., p. 203

van Flandern, T.C.: 1977, in Comets, Asteroids, Meteorites, proc. IAU Coll. 39, ed. A.H. Delsemme, Univ. of Toledo, p. 475

Franklin, F.A., Marsden, B.G., Williams, J.G., Bardwell, C.M.: 1975, Astron. J. 80, 729

Froeschlé, Cl., Klinger, J., Rickman, H.: 1983, in Asteroids, Comets, Meteors, eds. C.-I. Lagerkvist and H. Rickman, Univ. of Uppsala, p. 215

Froeschlé, Cl., Rickman, H.: 1980, Astron. Astrophys. 82, 183

Froeschlé, Cl., Rickman, H.: 1981, Icarus 46, 400

Froeschlé, Cl., Scholl, H.: 1979, Astron. Astrophys. 72, 246

Gaffey, M.J., McCord, T.B.: 1979, in Asteroids, ed. T. Gehrels, Univ. of Arizona Press, p. 688

Gradie, J., Veverka, J.: 1980, Nature 283, 840

Greenberg, J.M.: 1982, in Comets, ed. L.L. Wilkening, Univ. of Arizona Press, p. 131

Greenberg, J.M.: 1983, in Asteroids, Comets, Meteors, eds. C.-I. Lagerkvist and H. Rickman, Univ. of Uppsala, p. 259

Greenberg, J.M.: 1985, this volume

Greenberg, R., Scholl, H.: 1979, in Asteroids, ed. T. Gehrels, Univ. of
     Arizona Press, p. 310

Hahn, G., Rickman, H.: 1984, Icarus, in press

Harris, A.W.: 1983, Bull. Amer. Astron. Soc. 15, 828

Hartmann, W.K.: 1979, in Asteroids, ed. T. Gehrels, Univ. of Arizona
     Press, p. 466

Hartmann, W.K., Cruikshank, D.P.: 1984, Icarus 57, 55

Hartmann, W.K., Cruikshank, D.P., Degewij, J.: 1982, Icarus 52, 377

Hartmann, W.K., Cruikshank, D.P., Degewij, J., Capps, R.W.: 1981, Icarus
     47, 333

Haser, L.: 1955, C. R. Acad. Sci. Paris 241, 742

Hills, J.G.: 1981, Astron. J. 86, 1730

Horányi, M., Gombosi, T.I., Cravens, T.E., Kecskeméty, K., Nagy, A.F.,
     Szegö, K.: 1983, in Cometary Exploration, vol. I, ed. T.I. Gombosi,
     Hungarian Acad. Sci., p. 59

Hughes, D.W.: 1983, Nature 306, 116

Ip, W.-H.: 1977, in Comets, Asteroids, Meteorites, proc. IAU Coll. 39,
     ed. A.H. Delsemme, Univ. of Toledo, p. 485

Johnson, R.E., Lanzerotti, L.J., Brown, W.L., Augustyniak, W.M., Mussil,
     C.: 1983, Astron. Astrophys. 123, 343

Kazimirchak-Polonskaya, E.I.: 1972, Astron. Zh. 44, 439

Klinger, J.: 1983a, Icarus 55, 169

Klinger, J.: 1983b, in Asteroids, Comets, Meteors, eds. C.-I. Lagerkvist
     and H. Rickman, Univ. of Uppsala, p. 205

Kowal, C.T.: 1979, in Asteroids, ed. T. Gehrels, Univ. of Arizona Press,
     p. 436

Kozai, Y.: 1962, Astron. J. 67, 591

Kozai, Y.: 1979, in Dynamics of the Solar System, proc. IAU Symp. 81,
     ed. R.L. Duncombe, D. Reidel Publ. Co., p. 231

Kresák, L.: 1972a, in The Motion, Evolution of Orbits, and Origin of
     Comets, proc. IAU Symp. 45, eds. G.A. Chebotarev, E.I. Kazimirchak-
     Polonskaya and B.G. Marsden, D. Reidel Publ. Co., p. 503

Kresák, L.: 1972b, Bull. Astron. Inst. Czech. 23, 1

Kresák, L.: 1973, in Evolutionary and Physical Properties of Meteoroids,
     proc. IAU Coll. 13, eds. C.L. Hemenway, P.M. Millman and A.F. Cook,
     NASA SP-319, p. 331

Kresák, L.: 1977, in Comets, Asteroids, Meteorites, proc. IAU Coll. 39,
     ed. A.H. Delsemme, Univ. of Toledo, p. 313

Kresák, L.: 1979, in Asteroids, ed. T. Gehrels, Univ. of Arizona Press,
     p. 289

Kresák, L.: 1980, the Moon and the Planets 22, 83

Kresák, L.: 1981a, Bull. Astron. Inst. Czech. 32, 19

Kresák, L.: 1981b, Bull. Astron. Inst. Czech. 32, 321

Kresák, L.: 1982, in Sun and Planetary System, proc. IAU VI European
     Regional Meeting in Astronomy, eds. W. Fricke and G. Teleki, D.
     Reidel Publ. Co., p. 361

Kresák, L.: 1983, Highl. of Astron., vol. 6, ed. R.M. West, D. Reidel
     Publ. Co., p. 377

Kuiper, G.P.: 1951, in Astrophysics, ed. J.A. Hynek, McGraw-Hill, p. 357

Lagerkvist, C.-I.: 1983a, Highl. of Astron., vol. 6, ed. R.M. West,

D. Reidel Publ. Co., p. 371

Lagerkvist, C.-I.: 1983b, in Asteroids, Comets, Meteors, eds. C.-I.
    Lagerkvist and H. Rickman, Univ. of Uppsala, p. 11

Larson, H.P., Feierberg, M.A., Fink, U., Smith, H.A.: 1979, Icarus
    39, 257

Larson, S.M.: 1980, Astrophys. J. 238, L47

Lebofsky, L.A.: 1978, Mon. Not. Roy. Astron. Soc. 182, 17p

Lebofsky, L.A., Feierberg, M.A., Tokunaga, A.T., Larson, H.P., Johnson,
    J.R.: 1981, Icarus 48, 453

Lecar, M., Franklin, F.A.: 1973, Icarus 20, 422

Lecar, M., Franklin, F.A.: 1974, in The Stability of the Solar System
    and of Small Stellar Systems, proc. IAU Symp. 62, ed. Y. Kozai,
    D. Reidel Publ. Co., p. 37

Levin, B.J., Simonenko, A.N.: 1981, Icarus 47, 487

Low, F.J., Beintema, D.A., Gautier, T.N., Gillett, F.C., Beichman, C.A.,
    Neugebauer, G., Young, E., Aumann, H.H., Boggess, N., Emerson,
    J.P., Habing, H.J., Hauser, M.G., Houck, J.R., Rowan-Robinson, M.,
    Soifer, B.T., Walker, R.G., Wesselius, P.R.: 1984, Astrophys. J.
    278, L19

Marsden, B.G.: 1967, Astron. J. 72, 1170

Marsden, B.G.: 1970a, Astron. J. 75, 75

Marsden, B.G.: 1970b, Astron. J. 75, 206

Marsden, B.G.: 1974, Annu. Rev. Astron. Astrophys. 12, 1

Marsden, B.G.: 1985, this volume

Marsden, B.G., Sekanina, Z.: 1973, Astron. J. 78, 1118

Marsden, B.G., Sekanina, Z., Everhart, E.: 1978, Astron. J. 83, 64

Marsden, B.G., Sekanina, Z., Yeomans, D.K.: 1973, Astron. J. 78, 211

McFadden, L.A.: 1983, Ph. D. dissertation, Univ. of Hawaii

McFadden, L.A., Gaffey, M.J., McCord, T.B.: 1984, Icarus 59, 25

Mendis, D.A.: 1985, in proc. NATO Adv. Res. Workshop Ices in the Solar
    System, eds. D. Benest, A. Dollfus, J. Klinger and R. Smoluchowski,
    D. Reidel Publ. Co., in press

Mendis, D.A., Brin, G.D.: 1977, the Moon 17, 353

Mendis, D.A., Brin, G.D.: 1978, the Moon and the Planets 18, 77

Nakamura, T.: 1983, in Dynamical Trapping and Evolution in the Solar
    System, proc. IAU Coll. 74, eds. V.V. Markellos and Y. Kozai, D.
    Reidel Publ. Co., p. 97

Nakano, S.: 1984, Nakano Note Nos. 445-448, 453 (Minor Planet Circulars
    8694-8695)

Napier, W.M.: 1982, in Sun and Planetary System, proc. IAU VI European
    Regional Meeting in Astronomy, eds. W. Fricke and G. Teleki, D.
    Reidel Publ. Co., p. 375

Napier, W.M., Staniucha, M.: 1982, Mon. Not. Roy. Astron. Soc. 198, 723

Oikawa, S., Everhart, E.: 1979, Astron. J. 84, 134

Oort, J.H.: 1950, Bull. Astron. Inst. Netherlands 11, 91

Oort, J.H., Schmidt, M.: 1951, Bull. Astron. Inst. Netherlands 11, 259

Öpik, E.J.: 1963, Adv. Astron. Astrophys. 2, 219

Öpik, E.J.: 1973, Astrophys. Space Sci. 21, 307

Patashnick, H., Rupprecht, G., Schuerman, D.W.: 1974, Nature 250, 313

Rabe, E.: 1971, in Physical Studies of Minor Planets, proc. IAU Coll.
    12, ed. T. Gehrels, NASA SP-267, p. 407

Rabe, E.: 1972, in The Motion, Evolution of Orbits, and Origin of Co-
    mets, proc. IAU Symp. 45, eds. G.A. Chebotarev, E.I. Kazimirchak-
    Polonskaya and B.G. Marsden, D. Reidel Publ. Co., p. 55
Rabe, E.: 1974, in Asteroids, Comets, Meteoric Matter, proc. IAU Coll.
    22, eds. C. Cristescu, W.J. Klepczynski and B. Milet, Ed. Acad.
    Rep. Soc. Romania, p. 165
Rickman, H.: 1979, in Dynamics of the Solar System, proc. IAU Symp. 81,
    ed. R.L. Duncombe, D. Reidel Publ. Co., p. 293
Rickman, H., Froeschlé, Cl.: 1980, the Moon and the Planets 22, 125
Rickman, H., Froeschlé, Cl.: 1983, in Cometary Exploration, vol. I, ed.
    T.I. Gombosi, Hungarian Acad. Sci., p. 75
Rickman, H., Malmort, A.M.: 1981, Astron. Astrophys. 102, 165
Rickman, H., Vaghi, S.: 1976, Astron. Astrophys. 51, 327
Roemer, E.: 1966, Mém. Soc. Roy. Sci. Liège 12, Sér. 5, 23
Safronov, V.S.: 1972, in The Motion, Evolution of Orbits, and Origin of
    Comets, proc. IAU Symp. 45, eds. G.A. Chebotarev, E.I. Kazimirchak-
    Polonskaya and B.G. Marsden, D. Reidel Publ. Co., p. 329
Scholl, H.: 1979, Icarus 40, 345
Scholl, H., Froeschlé, Cl.: 1977, in Comets, Asteroids, Meteorites,
    proc. IAU Coll. 39, ed. A.H. Delsemme, Univ. of Toledo, p. 293
Schubart, J.: 1964, SAO Special Report No. 149
Schubart, J.: 1979, in Dynamics of the Solar System, proc. IAU Symp. 81,
    ed. R.L. Duncombe, D. Reidel Publ. Co., p. 207
Sekanina, Z.: 1971, in Physical Studies of Minor Planets, proc. IAU
    Coll. 12, ed. T. Gehrels, NASA SP-267, p. 423
Sekanina, Z.: 1973a, Astrophys. Lett. 14, 175
Sekanina, Z.: 1973b, Icarus 18, 253
Sekanina, Z.: 1976, Icarus 27, 265
Sekanina, Z.: 1981a, Annu. Rev. Earth Planet. Sci. 9, 113
Sekanina, Z.: 1981b, Astron. J. 86, 1741
Shoemaker, E.M., Williams, J.G., Helin, E.F., Wolfe, R.F.: 1979, in
    Asteroids, ed. T. Gehrels, Univ. of Arizona Press, p. 253
Shul'man, L.M.: 1983, in Cometary Exploration, vol. I, ed. T.I. Gombosi,
    Hungarian Acad. Sci., p. 55
Smoluchowski, R.: 1981, Astrophys. J. 244, L31
Valtonen, M.J.: 1983, preprint
Valtonen, M.J., Innanen, K.: 1982, Astrophys. J. 255, 307
Veeder, G.J., Hanner, M.S.: 1981, Icarus 47, 381
Weidenschilling, S.J.: 1980, Icarus 44, 807
Weissman, P.R.: 1980a, Nature 288, 242
Weissman, P.R.: 1980b, Astron. Astrophys. 85, 191
Weissman, P.R.: 1982, in Comets, ed. L.L. Wilkening, Univ. of Arizona
    Press, p. 637
Weissman, P.R.: 1983, Highl. of Astron., vol. 6, ed. R.M. West, D. Rei-
    del Publ. Co., p. 363
Weissman, P.R.: 1984, Science, in press
Weissman, P.R., Kieffer, H.H.: 1981, Icarus 47, 302
West, R.M.: 1983, the Messenger, No. 32, p. 1
Wetherill, G.W.: 1976, Geochim. Cosmochim. Acta 40, 1297
Wetherill, G.W.: 1979, Icarus 37, 96
Whipple, F.L.: 1950, Astrophys. J. 111, 375

Whipple, F.L.: 1951, <u>Astrophys. J. 113</u>, 464
Whipple, F.L.: 1964, <u>Proc. Nat. Acad. Sci. U.S.A. 51</u>, 711
Whipple, F.L.: 1972, in <u>The Motion, Evolution of Orbits, and Origin of Comets</u>, proc. IAU Symp. 45, eds. G.A. Chebotarev, E.I. Kazimirchak-Polonskaya and B.G. Marsden, D. Reidel Publ. Co., p. 401
Whipple, F.L.: 1977, in <u>Comets, Asteroids, Meteorites</u>, proc. IAU Coll. 39, ed. A.H. Delsemme, Univ. of Toledo, p. 25
Whipple, F.L.: 1980, <u>Astron. J. 85</u>, 305
Whipple, F.L.: 1981, in <u>Comets and the Origin of Life</u>, ed. C. Ponnamperuma, Univ. of Maryland, p. 1
Whipple, F.L.: 1982, in <u>Comets</u>, ed. L.L. Wilkening, Univ. of Arizona Press, p. 227
Whipple, F.L.: 1983a, in <u>Cometary Exploration</u>, vol. I, ed. T.I. Gombosi, Hungarian Acad. Sci., p. 95
Whipple, F.L.: 1983b, <u>Highl. of Astron.</u>, vol. 6, ed. R.M. West, D. Reidel Publ. Co., p. 323
Whipple, F.L.: 1983c, IAU Circular No. 3881
Whipple, F.L., Huebner, W.F.: 1976, <u>Annu. Rev. Astron. Astrophys. 14</u>, 143
Whipple, F.L., Sekanina, Z.: 1979, <u>Astron. J. 84</u>, 1894
Whitney, C.: 1955, <u>Astrophys. J. 122</u>, 190
Wilkening, L.L.: 1979, in <u>Asteroids</u>, ed. T. Gehrels, Univ. of Arizona Press, p. 61
Wisdom, J.: 1982, <u>Astron. J. 87</u>, 577
Wisdom, J.: 1983, <u>Icarus 56</u>, 51
Zellner, B.: 1979, in <u>Asteroids</u>, ed. T. Gehrels, Univ. of Arizona Press, p. 783
Ziolkowski, K.: 1983, in <u>Asteroids, Comets, Meteors</u>, eds. C.-I. Lagerkvist and H. Rickman, Univ. of Uppsala, p. 171

ON THE ROTATION OF COMETARY NUCLEI AND SMALL ASTEROIDS

P. Farinella[1], P. Paolicchi[2] and V. Zappala'[3]
1 - Department of Mathematics, University of Pisa, Italy
2 - Institute of Astronomy, University of Pisa, Italy
3 - Astronomical Observatory of Torino, Pino Torinese, Italy

ABSTRACT. The spin rate distributions of comet nuclei and small asteroids are compared, and it is shown that Whipple's (1982) finding of a
faster average rotation for the asteroid sample was due to observational
biases. In fact, the presently available rotational data do not exhibit
any clear differentiation among comet nuclei and asteroids, except
possibly for a higher abundance of short rotational periods among the
Apollo-Amor objects.

Do cometary nuclei rotate (in a statistical sense) like the small asteroids and, in particular, like the Apollo-Amor objects among which a
significant fraction of extinct cometary nuclei is widely believed to be
present ? Clearly, an answer would represent an important observational
test on the theories about the origin and the subsequent physical and
dynamical evolution of comets (e.g., see Rickman, 1985). The first
attempt to collect and compare systematically the available data on this
issue was carried out by Whipple (1982), in the frame of a thorough
review of the present knowledge on the rotation of comet nuclei. He
showed that for a sample of 47 comets, whose rotational periods were
mostly determined by the halo method (Whipple, 1981), the rotation was
on the average considerably slower than for a comparable sample of 41
small asteroids (of diameter less than 40 km) extracted from the data
set of Harris and Burns (1979). Whipple's tentative conclusion was in
the sense of inferring a different origin of comets and small asteroids,
i.e., accretion at low relative velocity for comets and collisional
fragmentation for asteroids.

   We now believe that this conclusion is not justified, because it is
vitiated by a strong observational bias affecting the asteroid sample
used for the comparison. A substantial part of this sample (more than
2/3 of the objects) was formed by asteroids whose spin period had
been determined from photographic observations, which are usually

173

A. Carusi and G. B. Valsecchi (eds.), Dynamics of Comets: Their Origin and Evolution, 173–178.
© *1985 by D. Reidel Publishing Company.*

carried out on a single night and therefore exclude a priori the determination of periods longer than about 6 hours. For this reason the
statistical analyses of asteroid rotation rates performed by Tedesco and
Zappala' (1980) and by Farinella et al. (1981), who tried to minimize
the most important selection effects, excluded from the samples under
scrutiny all the objects observed photographically. Today new lightcurve
data obtained by photoelectric photometry has become available for a
significant number of small asteroids, mainly as a result of observational programs carefully designed by Harris and Young (1983) and by Binzel
and Mulholland (1984) to prevent any bias in favour of short periods.
Comparing this new data with the former one, Binzel (1984) has concluded
that the photographic technique produces indeed a strong bias against
the long periods, thus decreasing artificially the sample variances in
the distributions of rotational frequency; as a consequence, photographic and photoelectric data should not be combined in statistical
studies, lest the resulting trends are physically meaningless.

       We have also to stress that Whipple's results on the spin periods
of comet nuclei are affected by several sources of uncertainty. As
pointed out by Whipple himself (1982, pp.233-235), a number of ambiguities and faults are unescapable when the halo method is applied to
observational data, and quite often the resulting periods can be seen
only as reasonable guesses. Moreover the weaknesses of the method are
such that in several cases the periods are either quite accurate, or
completely wrong (e.g., by a factor two), so that one cannot be confi-
dent that statistically the errors average out. In spite of these
problems, we think that the importance of the issue for the understanding of comet/asteroid interrelations justifies the attempt to analyse
the presently available data in the best possible way, i.e., trying to
eliminate at least the known biases.

       This consideration has led us to remake Whipple's analysis, by
employing an "unbiased" asteroid data set formed only by objects whose
lightcurves have been observed photoelectrically; this set has been
extracted from the data file on asteroid rotational properties compiled
by one of us (V.Z.) at the Turin Observatory (and available on request).
Weights have been assigned in the following way : among asteroids,
objects whose period is uncertain (or for which only lower limits for
the period have been derived from lightcurve data) have been weighed
0.5, while objects with well determined periods have been weighed 1.0;
for comet nuclei, we have simplified Whipple's procedure by changing
into 0.5 and 1.0 all Whipple's weights $\lesssim$ and $>$ 0.5 respectively. In
Figure 1, above the horizontal axis the resulting spin rate distribution
is shown for the comet sample, including also a (dotted) histogram
referring to comets classified by Whipple (1982) into types III, IV and
V, whose shorter orbital periods suggest a longer time of activity;

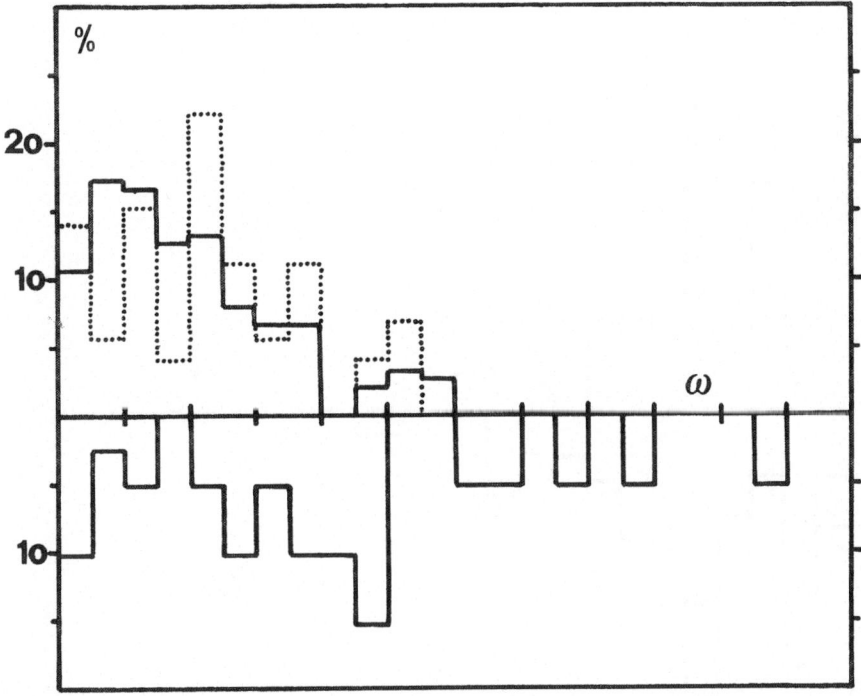

Figure 1. The spin rate distribution is shown for cometary nuclei (above) and for Apollo-Amor asteroids (below). The unit of the horizontal axis is 1 rotation/day, the width of each histogram bin is 0.5 rotations/day, while the vertical axis gives the percentage of objects contained in the different bins. The dotted histogram refres to comets of types III, IV and V according to Whipple's (1982) classification.

below the axis, the corresponding distribution for 21 AA asteroids is shown. In Figure 2, the same comet distribution (above) is compared with that of 68 main-belt asteroids of diameter smaller than 50 km (below); the lower dotted histogram refers to 38 small asteroids which do not belong to dynamical families according to Williams' (1979) classification (we recall that a possibly significant faster rotation of family asteroids has been confirmed by statistical analyses performed by Dermott et al. (1984) and Binzel (1984); this is related perhaps to a different colli-sional history between the two groups). Surprisingly enough, we can see from the Figures that cometary nuclei and main-belt small asteroids do not display distributions that are impressively different (as in the case of Whipple's comparison), implying that the inference about different origins cannot be drawn by this kind of evidence. On the contrary, the AA asteroids appear to behave in a peculiar way, and in particular they

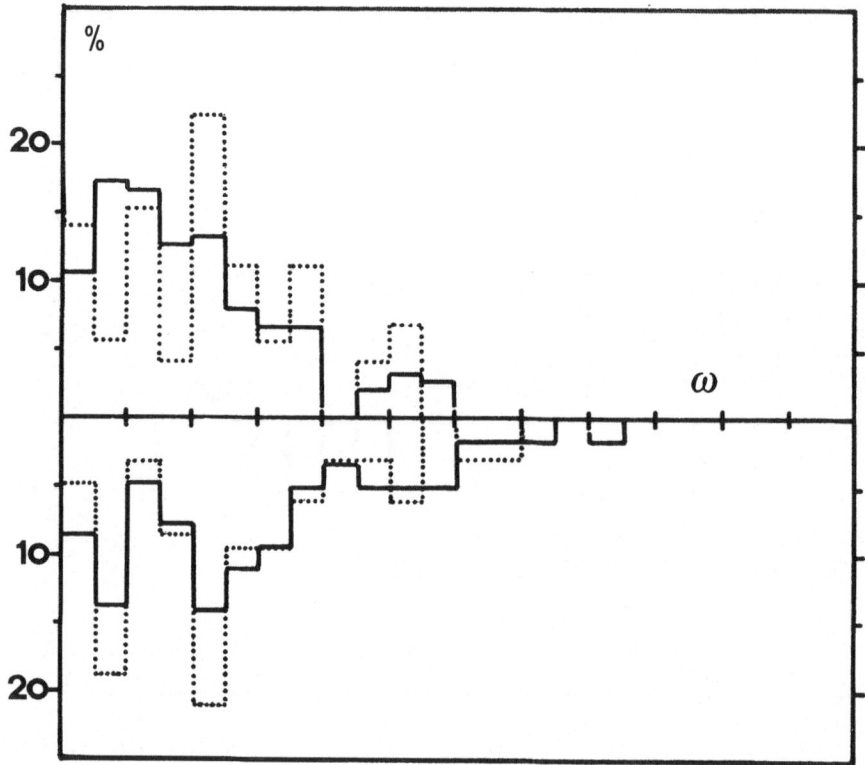

Figure 2. The same comet distribution of Figure 1 (above) is compared
with the spin rate distribution for a sample of main-belt, photoelectrically observed asteroids of diameter smaller than 50 km. The dotted histogram in the lower part of the Figure refers to nonfamily asteroids. The
units are the same as in Figure 1.

present a tail of fast rotators which has no apparent counterpart either
among main-belt small asteroids or among comet nuclei. We recall that
the AA objects have also a peculiar distribution of shapes, as deduced
from the lightcurve amplitudes, in the sense that they include a significant fraction of very elongated, cigar-like shapes which are very unusual
in the main belt and also among fragments from hypervelocity impact experiments (Capaccioni et al., 1984; Catullo et al., 1984). The idea that
these peculiar features may be due to some process related to aging of
cometary nuclei is attractive, but finds only a weak support from the
statistically unsignificant difference between the spin rate distributions
of "older" and "younger" comets. On the other hand, few rotational data
are available for main-belt asteroids of size comparable to that of the
AA objects (typically, of the order of 1 km), so that we cannot exclude

that a similar size dependence of the rotational properties is a feature of the whole asteroid population. Indeed, Binzel (1984) has presented some evidence indicating that, even excluding photographic observations, below a diameter of 120 km an inverse correlation between spin rate and size exists in the main belt, so that AA and main-belt objects of the same size might really behave in a similar way. Rotational data are also consistent with the widely accepted "intermediate" hypothesis that we are observing a mixture of two classes of AA objects, one of cometary and one of asteroidal origin.

As a conclusion, we have to stress that at the present stage every result must be taken with caution, because the statistics are poor and because subtle but significant biases could still be present. Moreover, our understanding of the origin and evolution of both asteroid and comet rotation is not such that we can draw any sound inference from the observational data. Our analysis only shows that the presently available data does not exhibit clear differentiations between cometary nuclei and asteroids, except possibly for a higher abundance of short periods among AA objects, a phenomenon of difficult interpretation (if real). Further studies are clearly needed both to make available larger and unbiased samples for the statistics, and to define better the physical mechanisms which, affecting the rotational properties, could be related to any evolutionary relationship or difference between comets and asteroids.

ACKNOWLEDGMENTS

We thank M. Fulchignoni and R. Greenberg for helpful remarks on the manuscript, and D.K. Yeomans for a comment at the meeting. This work has been supported in part by the National Research Council of Italy (C.N.R.).

REFERENCES

BINZEL, R.P. (1984). 'The rotation of small asteroids'. Icarus 57, 294 - 306.

BINZEL, R.P., AND J.D. MULHOLLAND (1984). 'A photoelectric lightcurve survey of small main belt asteroids'. Icarus 56, 519 - 533.

CAPACCIONI, F., P. CERRONI, M. CORADINI, P. FARINELLA, E. FLAMINI, G. MARTELLI, P. PAOLICCHI, P.N. SMITH, AND V. ZAPPALA' (1984). 'Shapes of asteroids compared with fragments from hypervelocity impact experiments'. Nature 308, 832 - 834.

CATULLO, V., V. ZAPPALA', P. FARINELLA, AND P. PAOLICCHI (1984). 'Analysis of the shape distribution of asteroids'. Astron. Astrophys., in press.

DERMOTT, S.F., A.W. HARRIS, AND C.D. MURRAY (1984). 'Asteroid rotation rates'. Icarus 57, 14 - 34.

FARINELLA, P., P. PAOLICCHI, AND V. ZAPPALA' (1981). 'Analysis of the spin
    rate distribution of asteroids'. Astron. Astrophys.104, 159 - 165.
HARRIS, A.W., AND J.A. BURNS (1979).'Asteroid rotation: I. Tabulation and
    analysis of rates, pole positions and shapes'. Icarus 40, 115 - 144.
HARRIS, A.W., AND J.W. YOUNG (1983). 'Asteroid rotation: IV. 1979 obser-
    vations'. Icarus 54, 59 - 109.
RICKMAN, H. (1985). 'Interrelations between comets and asteroids'. This
    volume.
TEDESCO, E.F., and V. ZAPPALA' (1980). 'Rotational properties of asteroids:
    Correlations and selection effects'. Icarus 43, 33 - 50.
WHIPPLE, F.L. (1981). 'On observing comets for nuclear rotation'. In
    Modern Observational Techniques for Comets (J.C. Brandt, J.M. Green-
    berg, B. Donn and J. Rahe, Eds.), pp. 191 - 200, Jet Propulsion
    Laboratory, Pasadena.
WHIPPLE, F.L. (1982). 'The rotation of comet nuclei'. In Comets (L.L.
    Wilkening, Ed.), pp. 227 - 250, University of arizona Press, Tucson.
WILLIAMS, J.G. (1979). 'Proper elements and family memberships of the
    asteroids'. In Asteroids (T. Gehrels, Ed.), pp. 1040 - 1063, Univer-
    sity of Arizona Press, Tucson.

THE ORBITAL EVOLUTION OF METEOR STREAMS AT THE 2/1 RESONANCE WITH JUPITER.

Cl. Froeschlé, Observatoire de Nice, and

H. Scholl, Astronomisches Rechen-Institut, Heidelberg

We investigated the orbital evolution of Quadrantid-like meteor streams situated in the vicinity of the 2/1 resonance with Jupiter. For the starting orbital elements we took the values of the orbital elements of the Quadrantid meteor stream except for the semi-major axis which was varied between $a = 3.22$ and $a = 3.34$ AU. We considered these meteor streams as a ring and we investigated the resonant effect on the dispersion of this ring over a period of 13 000 years. Only gravitational forces due to the Sun and due to Jupiter were taken into account.

The main results are the following : Different parts of this ring evolved in a different way according to their starting values for the resonant argument $\dot{\sigma}$. For the 2/1 resonance, σ is defined by $\sigma = -(\tilde{\omega} + \ell - 2\ell_J)$, where $\tilde{\omega}$ and ℓ are the longitude of perihelion and the mean longitude respectively of a ring particle and where ℓ_J denotes Jupiter's mean longitude.

For $\sigma = 0°$, a ring particle is situated at its perihelion, and for $\sigma = 180°$, a ring particle is situated at its aphelion during a conjunction with Jupiter.

The dispersion is not uniform with respect to σ. We can say that the dispersion rate is fastest for ring pieces starting with σ

A. Carusi and G. B. Valsecchi (eds.), Dynamics of Comets: Their Origin and Evolution, 179–180.
© *1985 by D. Reidel Publishing Company.*

close to 180° while the dispersion rate is slowest for σ close
to 0°. Therefore, the following effect occurs : After a few
thousand years, only a piece of the ring persists while the
rest of the ring has already dispersed.

After 13 000 years, the perihelion distances range from almost 0
to about 3 AU.

In addition, it also happens that a ring piece starts to precess
rapidly in such a way that finally the particles of this ring piece
encounter the particles of the remainder of the ring in almost
opposite directions with relative velocities exceeding 50 km/s.
These high velocity encounters between particles of the same
original ring occur in the range between 0.3 and 1 AU from the Sun.
Further investigations have to be made in order to find out wether
or not this peculiar behaviour might be an important mechanism
for the destruction of meteor stream particles.

DYNAMICAL EFFECTS OF COMETARY BOMBARDMENT OF SATURN'S RINGS AND MOONS

Jack J. Lissauer
NAS-NRC Resident Research Associate
Space Science Division
NASA-Ames Research Center
Moffett Field, CA 94035
U.S.A.

EXTENDED ABSTRACT. Saturn's ring particles and airless moons are exposed to a large flux of interplanetary debris, principally comets and comet dust. Collisions with this debris are responsible for both physical and dynamical changes in objects orbiting about Saturn. Physical changes include cratering of large bodies and catastrophic disruption of small bodies. Dynamical changes, which are analyzed in this paper, include orbital alteration (principally of ring particles) and changes in spin state (which are only important for moons, as ring particle spins are continually altered by mutual collisions).
      Saturn's rings are rapidly being eroded by impacts of hypervelocity meteoroids in cometary orbits. Ejecta from these impacts will, in most cases, remain in orbit about Saturn and eventually be reaccreted by the rings, possibly at a different radial location. The resulting mass transport has been suggested as the cause of some of the features observed in Saturn's rings (see Durisen 1984 for a review). Previous attempts to model this transport have used numerical simulations which have not included effects of angular momentum transport coincident with this mass transport. I have developed an analytic model for ballistic mass transport in Saturn's rings. The model includes the effects of angular momentum advection and shows that the net material movement due to the combined effects of mass and angular momentum transport is roughly half that calculated when angular momentum advection is ignored. See Lissauer (1984) for further details.
      All of Saturn's mid-sized moons are rotating synchronously with their orbital period; thus, the same hemisphere of these moons always faces the planet, and the same point is always at the center of the satellite's leading hemisphere (the apex). The satellites orbit Saturn with velocities ranging from 14 km/sec for Mimas to 8.5 km/sec for Rhea and 3.3 km/sec for Iapetus. These speeds are a significant fraction of the encounter velocities between comets and the Saturn system (~ 10-25 km/sec); thus, due to a type of "windshield effect" (more raindrops hit the windshield of a moving car than hit the rear window), more comets will collide with the moons' leading hemispheres than with their trailing hemispheres; also, higher relative velocities between comets

181

A. Carusi and G. B. Valsecchi (eds.), Dynamics of Comets: Their Origin and Evolution, 181–182.
© *1985 by D. Reidel Publishing Company.*

and the moons will lead to larger craters for impacts by comets of a
given mass on the leading hemispheres. The combination of these effects
suggests that regions of the satellites' surfaces near the apex should
be much more heavily cratered than regions near the antapex (Shoemaker
and Wolfe 1982). Such a major cratering asymmetry has not been observed (Plescia and Boyce 1983). A similar situation exists for
Jupiter's moons Ganymede and Callisto (Passey and Shoemaker 1982).

McKinnon (1981) suggested that stochastic reorientation of the
moons by impact-induced "spinup" during the establishment of the
cratering record tended to equilibrate the crater densities between
hemispheres. I have re-examined the dynamics of this problem, and I
conclude that most impacts large enough to have caused "spinup"
would have catastrophically disrupted the moons in question (Lissauer
1985).

REFERENCES.

Durisen, R. H., (1984). Transport Effects due to Particle Erosion
Mechanisms. In Planetary Rings, R. Greenberg and A. Brahic, eds.
Univ. of Arizona Press, Tucson, 416-446.

Lissauer, J. J. (1984). Ballistic Transport in Saturn's Rings: An
Analytic Theory. Icarus 57, 63-71.

Lissauer, J. J. (1985). Can Cometary Bombardment Disrupt Synchronous
Rotation of Planetary Satellites? In preparation for submission to
J. Geophys. Res..

McKinnon, W. B. (1981). Reorientation of Ganymede and Callisto by
Impact, and Interpretation of the Cratering Record. Eos Trans. AGU
62, 318.

Passey, Q. R. and Shoemaker, E. M. (1982). Craters and Basins on
Ganymede and Callisto: Morphological Indications of Crustal
Evolution. In Satellites of Jupiter, D. Morrison, ed., Univ. of
Arizona Press, Tucson, 379-434.

Plescia, J. B. and Boyce, J. M. (1983). Crater Numbers and Geological Histories of Iapetus, Enceladus, Tethys and Hyperion. Nature
301, 666-670.

Shoemaker, E. M. and Wolfe, R. (1982). Cratering Time Scales for
the Galilean Satellites. In Satellites of Jupiter, D. Morrison, ed.,
Univ. of Arizona Press, Tucson, 277-339.

SECTION IV

DYNAMICS OF COMETS: NUMERICAL MODELLING

AN EFFICIENT INTEGRATOR THAT USES GAUSS-RADAU SPACINGS

Edgar Everhart
Physics Department and Chamberlin Observatory
University of Denver
Denver, Colorado 80208   USA

ABSTRACT. This describes our integrator RADAU, which has been used by
several groups in the U.S.A., in Italy, and in the U.S.S.R. over the
past 10 years in the numerical integration of orbits and other problems
involving numerical solution of systems of ordinary differential equations. First- and second-order equations are solved directly, including
the general second-order case. A self-starting integrator, RADAU proceeds by sequences within which the substeps are taken at Gauss-Radau
spacings. This allows rather high orders of accuracy with relatively few
function evaluations. After the first sequence the information from
previous sequences is used to improve the accuracy. The integrator
itself chooses the next sequence size. When a 64-bit double word is
available in double precision, a 15th-order version is often appropriate, and the FORTRAN code for this case is included here. RADAU is
at least comparable with the best of other integrators in speed and
accuracy, and it is often superior, particularly at high accuracies.

1. INTRODUCTION

The first description of RADAU was by Everhart (1974a). A more full
treatment is in an unpublished technical report, Everhart (1974b),
which has had a wide distribution, and there have been improvements
suggested by experience in the past 10 years. Our method is related to
the implicit Runge-Kutta procedures of Butcher (1964), which also use
Gauss-Radau spacings, but the algorithms of the two methods are quite
different. RADAU can easily be written to a very high order of accuracy.
In special tests we have used it in 31st order.

It has been used sucessfully in determining the original orbits for
some 224 comets, Marsden, Sekanina, and Everhart (1978), Everhart and
Marsden (1983). Starting with the comet in its accurate osculating orbit
near perihelion, they integrated in 15th order backwards through the
solar system, taking into account the perturbations of every planet on
every other planet and on the comet. It was a 10-body integration near
the sun, with 30 simultaneous second-order equations. As the comet moved
outward in this barycentric integration, the masses of the inner planets
were one by one added to the sun's mass until at large distances only

A. Carusi and G. B. Valsecchi (eds.), Dynamics of Comets: Their Origin and Evolution, 185–202.

the 5 outer planets and the comet were included. This 6-body integration
was continued until the comet was 60 AU away from the sun, at which
place the original barycentric orbit was determined. The formal accuracy
was far more than was justified by the accuracy of the starting elements. Such effects as the close encounters of Comet Bowell with both
Saturn and Jupiter were accurately handled. The perturbations of Saturn
on Jupiter, which changed the position of Jupiter slightly, made a
significant difference on the computed original orbit of this comet.

The results of the first paper on this, Everhart (1974a), showed
the enormous advantages of using Gauss-Radau spacings. In 1973 we were
developing an integrator for the equation $y'' = F(y,t)$ for applications
in celestial mechanics. We expanded y'' in an empirical time series
through terms in t^3, and the corresponding expansion for y was to t^5.
The idea was to examine a sequence of overall length T. With uniform
spacing of substeps it was evident that the force function F would have
to be evaluated at 0, .3333T, .6667T, and at T. This did result in a
5th-order integrator, with errors of the order of T^6 in y. Thus if one
cut the sequence size in half one should expect to see 1/64th the error.
Then some rather complicated algebra suggested that if the substeps were
taken at 0., 0.21234T, 0.59053T, and 0.91441T, then the integrator
would be 7th order, instead. When this was tried the results were aston-
ishing. Errors dropped by a factor of 100 at the same number of function
calls. It was now a 7th-order integrator. Soon thereafter, this was
discussed with W. H. Goodyear, who suggested that Gaussian spacings had
just been re-invented. It turned out that the spacings found here were
those used in Gauss-Radau quadratures. They were camouflaged a bit because for quadratures one uses the range -1 to +1, instead of 0 to +1 as
here. The special spacings of Gaussian quadratures allow one to get
almost double the order of accuracy that one gets with constant
spacings, comparisons being made at the same number of function calls.
The present integrator extends the advantages of Gaussian spacings to
the integration of differential equations.

The detailed example of the next section describes an expansion
that would give 6th-order results at constant spacings, but which is a
9th-order integrator with Gauss-Radau spacings. In our most practical
version we use a time expansion through t^9 in y (and through t^7 in y''),
achieving 15th-order accuracy with these spacings.

The present method solves simultaneous differential equations of
first or second order directly. We define several classes:

Class I $y' = F(y,t)$

Class IIS (special) $y'' = F(y,t)$

Class II   (general) $y'' = F(y',y,t)$

The method can be extended also to solve general 3rd- or 4th-order
differential equations directly. The theory is given for solving one
equation. Extension to any number of such simultaneous equations
involves simple loops as seen in the examples of Sec. 4.

2. THE ALGORITHM

We illustrate the method in solving the general class II equation, $y'' = F(y',y,t)$, in 9th order of accuracy.

2.1 Expansions

Since y' and y are implicit functions of the independent variable t, one may expand y'' (which is F) as a truncated series,

$$y'' = F = F_1 + A_1 t + A_2 t^2 + A_3 t^3 + A_4 t^4, \tag{1}$$

about the initial value $t_1 = 0$, where F has the value F_1. At the end of a sequence of length T the problem is to find $y(T)$. Equation (1) is not a Maclaurin series expansion in t where the error is comparable with the first term neglected. With Gaussian spacings the error in the truncated series can be a millionth the size of the last term included.

Using $h = t/T$ and $B_1 = A_1 T$, $B_2 = A_2 T^2$, $B_3 = A_3 T^3 \dots$, one has

$$y'' = F = F_1 + B_1 h + B_2 h^2 + B_3 h^3 + B_4 h^4. \tag{2}$$

After F_1 is found at $h_1 = 0$, other values F_2, $\dots$, F_5 are developed at suitable spacings h_2, $\dots$, h_5 within the interval of h between 0 and 1. We expand F in a way that incorporates these spacings h_n, introducing a set of coefficients G and writing

$$F(h) = F_1 + G_1 h +$$
$$+ G_2 h(h-h_2) + G_3 h(h-h_2)(h-h_3) + G_4 h(h-h_2)(h-h_3)(h-h_4) . \tag{3}$$

This truncates at each location h_n. Thus $F_3 = F_1 + G_1 h_3 + G_2 h_3(h_3 - h_2)$. Abbreviating with $r_{nj} = 1/(h_n-h_j)$ and $r_{n1} = 1/h_n$ one finds

$$
\begin{aligned}
G_1 &= (F_2-F_1)r_{21} , \\
G_2 &= ((F_3-F_1)r_{31}- G_1)r_{32} , \\
G_3 &= (((F_4-F_1)r_{41}- G_1)r_{42}- G_2)r_{43} , \\
G_4 &= ((((F_5-F_1)r_{51}- G_1)r_{52}- G_2)r_{53}- G_3)r_{54} .
\end{aligned} \tag{4}
$$

The B's and G's are related through Eqs. (2) and (3). This gives

$$
\begin{aligned}
B_1 &= c_{11}G_1 + c_{21}G_2 + c_{31}G_3 + \dots , \\
B_2 &= \phantom{c_{11}G_1 +} c_{22}G_2 + c_{32}G_3 + \dots , \\
B_3 &= \phantom{c_{11}G_1 + c_{21}G_2 +} c_{33}G_3 + \dots ,
\end{aligned} \tag{5}
$$

where the recurrence relationships for the c's are

$$c_{jj} = 1 \quad ,$$

$$c_{j1} = -h_j c_{j-1,1} \quad , \qquad\qquad j > 1 \quad , \tag{6}$$

$$c_{jk} = c_{j-1,k-1} - h_j c_{j-1,k} \quad , \quad k < j \quad .$$

The reverse relationships are also needed. These are

$$G_1 = d_{11}B_1 + d_{21}B_2 + d_{31}B_3 + \cdots \quad ,$$

$$G_2 = \qquad\qquad d_{22}B_2 + d_{32}B_3 + \cdots \quad , \tag{7}$$

$$G_3 = \qquad\qquad\qquad\qquad d_{33}B_3 + \cdots \quad ,$$

with the recurrence relationships

$$d_{jj} = 1 \quad ,$$

$$d_{j1} = h_2 d_{j-1,1} = h_2^{j-1} \qquad\qquad , \quad j > 1 \quad , \tag{8}$$

$$d_{jk} = d_{j-1,k-1} + h_{k+1} d_{j-1,k} \quad , \quad k < j \quad .$$

In this dimensionless form the c-, d-, and r-values do not depend on the sequence length T and are calculated but once for a given integration order. Evaluating these constants uses simple loops and takes much less than one second.

2.2 Predictors and Correctors

The predictors $y_n(h_n)$ and $y_n'(h_n)$ at each substep n within a sequence are found by integrating Eqs. (1) or (2). These are

$$y_n = y_1 + y_1' h_n T +$$

$$+ h_n^2 T^2 (F_1/2 + h_n(B_1/6 + h_n(B_2/12 + h_n(B_3/20 + h_n B_4/30)))) \quad , \tag{9}$$

$$y_n' = y_1' +$$

$$+ h_n T (F_1 + h_n(B_1/2 + h_n(B_2/3 + h_n(B_3/4 + h_n B_4/5)))) \quad . \tag{10}$$

At the end of the sequence, where h = 1 and t = T, the correctors are

$$y(T) = y_1 + y_1' T +$$

$$+ T^2 (F_1/2 + B_1/6 + B_2/12 + B_3/20 + B_4/30) \quad , \tag{11}$$

$$y'(T) = y_1' + T(F_1 + B_1/2 + B_2/3 + B_3/4 + B_4/5) \quad . \tag{12}$$

The system is implicit; the B-values are not known when they are first needed.

2.3 Procedure

After the first sequence, fairly good values of the B-coefficients are predicted for the current sequence (as will be described in Sec. 2.4 below). The corresponding G-values are known through Eqs. (7). Using the Radau spacings, the integrator steps through the sequence. At each step the position (and velocity if necessary) is obtained from Eqs. (9), (10) using the B-values. Then the force is calculated at the predicted position, and each new force value determines an improved G-value as in Eqs. (4). Every B-value that depends on this G-value is immediately upgraded using Eqs. (5). Essentially, the B-values determine the predicted positions, the forces at these positions determine the G-values, and these determine better B-values. After stepping through a sequence once, the integrator has considerably improved B- and G-values. A second pass rofines these to high accuracy. The procedure becomes very clear when one follows through the listing of the 15th-order integrator in Sec. 4.

On initial starting, the initial B-values are all zero, and the predicted positions are inaccurate. Nonetheless, the process converges to high accuracy with enough iterations. It is worthwhile to take 6 iterations to start the 15th-order integrator and 10 iterations to start in 27th order.

2.4 Prediction of the B-values for the Next Sequence

Convergence is speeded greatly by using past information, and after the first sequence only 2 passes are required. Let B_1, ... , B_4 be accurate values from the previous sequence, and let $Q = T(new)/T(old) = T'/T$ be the ratio of the sequence sizes. An analytic continuation of the curve for F from one sequence to the next (with a change in origin and this change in scale) requires that

$$
\begin{aligned}
B_1(new) &= Q\ (B_1 + 2B_2 + 3B_3 + 4B_4)\ , \\
B_2(new) &= \qquad Q^2(B_2 + 3B_3 + 6B_4)\ , \\
B_3(new) &= \qquad\qquad Q^3(B_3 + 4B_4)\ , \\
B_4(new) &= \qquad\qquad\qquad Q^4\ B_4\ .
\end{aligned}
\tag{13}
$$

These are excellent values with which to start the new sequence, especially with the refinement discussed in the next section. Equations (13) can easily be extended to any order by noting the pattern of binomial coefficients in the columns. Thus, if there were a B_5 term, the constants in the added column then are 1, 5, 10, 10, 5 from bottom to top.

2.5 Refinements

In the technical report, Everhart (1974b), we advocating absorbing the constants 1/6, 1/12, 1/20, etc, that appear in Eqs. (9) and (11) into the B-values. This can be done, and it does speed the computation some 10-20%, but it has the disadvantage of changing all the above equations and making the program difficult to check. We have discarded this idea.

Two other improvements are incorporated: First, we have found in some quite recent tests that if the equations are written out without loops then the integration is 25% to 30% faster! This is entirely practical through 15th order, as listed in the Sec. 4.

A second improvement has been incorporated. If one saves the values of B(new) predicted by Eqs. (13) and compares them later with the final B-values of the sequence, the difference is slowly-varying, and can be applied in advance as a correction. This simple change in the algorithm cuts the global error by a factor of 2 to 10 and is quite worthwhile.

2.6 Gauss-Radau, Gauss-Lobatto, and Gauss-Legendre Spacings

The use of Gaussian spacings for the substeps enhances the integration order for Class I, IIS, and II equations alike. For quadratures one ordinarily uses Gauss-Legendre spacings, but for solving differential equations with an implicit algorithm there is reason to use the Gauss-Radau or Gauss-Lobatto spacings. They use F_1 at $h_1=0$, and F_1 is not recalculated in the iteration.

Radau and Lobatto spacings may be found to 30 significant digits in Tables 12 and 11 of Stroud and Secrest (1966) to extremely high orders. These span the range -1 to +1, and they must be rescaled to the range 0 to +1 for the present application. The spacings so rescaled may be found in Table I of our 1974 paper for orders through 15th. There is a typographical error in h_2 for 15th order. The correct number is in the listing in Sec. 4. The reason Radau spacings, which give odd orders, are preferred to Lobatto spacings is that the order is one higher for a given number of terms. Thus Eq.(2) with terms through 4th order in time gives 9th order with Radau spacings and 8th order with Lobatto spacings.

2.7 Sequence Size Control

To control T one can monitor the last term in the y-expansion, here $B_4 T^2/30$. This term is $B_4'T'^2/30$ for the next sequence (primed values), which will be controlled to have magnitude 10^{-L}. Now $B_4' = (T'/T)^4 B_4$, as in the last of Eqs. (13), so the condition on T' is

$$10^{-L} = B_4'T'^2/30 = (T'/T)^4 B_4 T'^2/30 = H_4 T'^6, \tag{14}$$

whence $H = B_4/(30T^4)$, or rather the largest such term in absolute value in any of the equations. Upon solving for T' one finds $T' = (10^{-L}/H)^{1/6}$ is appropriate for the next sequence in a 9th-order integration.

In the case of a 15th-order integration of a system of simultaneous Class IIS or Class II equations, the result of an analogous calculation is

$$T' = (10^{-L}/H)^{1/9}, \quad (15\text{th-order case}) \tag{15}$$

where H is taken from the largest value of

$$H = B_7/(72T^7) \quad . \tag{16}$$

In 15th order values of L will run from 6 or 7 for low-accuracy calculations to 10 or more for high-accuracy problems. The global error for the entire integration is much smaller than 10^{-L}. See Sec. 3.2 below.

We can give no strong theoretical reason why the above calculation should be a good sequence size control. However, practical experience over a 10-year time span with invariably accurate results suggest that the above sequence size control is appropriate and usable.

The initial trial sequence size may be furnished by the programmer. If he makes no choice then the value +0.1 is chosen for forward integration and -0.1 for backward integration. This always gets one started. Choosing too large a starting value of T is not a serious problem. If the value of T(new) called for by the sequence size control is less than T(start), then the program itself chooses a more appropriate starting value and restarts.

The integrator can be set to accept a constant step size, and this is appropriate in some cases, as in Secs. 3.3 and 3.4 below.

3. RECENT PRACTICAL TESTS

Among the best of other integrators is DVDQ, a variable-order Adams (multistep) method by Krogh (1973) that solves equations of Class I, IIS, and II directly. There are DIFSYS and DIFSY2 by Bulirsch and Stoer (1966), extrapolation methods that solve Class I and IIS, respectively. Another is RKN8(9), a Runge-Kutta-Nystrom integrator of 8th order by Fehlberg (1972) that solves Class IIS. The first of these is accurate though not always fastest, the extrapolation methods have a low overhead per function call and are fast, and the RKN8(9) integrator is easy to program and fast for comparitively low accuracy problems. In some celestial mechanics problems, the customary integration is by a multistep method of high order. In an earlier time predictors were hand-calculated by differencing, but with machines it is often advantageous to predict with Lagrangian formulas.

Comparitive tests of RADAU and other integrators, by House, Weiss, and Weigandt (1978), have studied numerical integration of stellar orbits, while Papp, Innamen, and Patrick (1977, 1980) studied numerical orbit computations in galaxy models. These authors found RADAU to be a good compromise between speed and efficiency for their problems. It was chosen by Shefer (1982), who used it in 11th order for his study of perturbed orbits. In a review article Batrakov (1982) speaks of RADAU as being among the best of modern integrators, particularly for the Encke method. After extensive comparitive tests Carusi, Kresak, Perozzi, and Valsecchi (1985) chose RADAU in 19th order for their study of the long term evolution of all the short-period comets.

3.1 A particular 3-body Problem

A periodic orbit in the restricted three-body problem was described by Newton (1959). In the earth-moon system, idealized to circular orbits, a small body follows a complicated path with 3 loops passing very near the earth and the moon. Step size must be changed frequently on this orbit. The equations are:

$$y_1'' = 2y_2' + y_1 - u'(y_1 + u)/r_1^3 - u(y_1 - u')/r_2^3 \,,$$

$$y_2'' = -2y_1' + y_2 - u'y_2/r_1^3 - u\, y_2/r_2^3 \qquad . \tag{17}$$

Here $r_1 = ((y_1 + u)^2 + y_2^2)^{\frac{1}{2}}$, $r_2 = ((y_1 - u')^2 + y_2^2)^{\frac{1}{2}}$,

and $u = 1/82.45$, $u' = 1 - u$.

The initial conditions are:

$$y_1 = 1.2, \quad y_1' = 0 \,,$$

$$y_2 = 0, \quad y_2' = -1.04935\ 75098\ 30319\ 90731\ 04104\ 34... \,, \tag{18}$$

and $t_{final} = 6.19216\ 93313\ 19639\ 70699\ 23217\ ...$.

A CDC 6600 computer was used with RADAU in orders 7, 11, and 15 in single precision (60-bit word) and orders 19, 23, and 27 in double precision (120-bit word). In every case RADAU used less function calls than the other integrators. However, for the same absolute error, DVDQ was close at intermediate acccuracies. DIFSYS used twice as many function calls because it solved this system as 4 first order equations. However, its low overhead per function call made it comparable in timing. Gallaher and Perlin (1966) gave the long numbers in the problem to 21 digits. The last 5 or 6 digits were found by RADAU in 27th order.

3.2 An Elliptical Orbit Problem

Using the IBM-PC with a 64-bit double word one integrates 8 times around an ellipse of eccentricity 0.6 looking at the error of closure. Here 11th and 15th order were about comparable in timing for accuracies from 5 to about 12 digits.

Quite recently we tested this problem again using a Cray-1 computer in double precision (128-bit word). Our integrator could be set for 15th, 19th, 23rd, or 27th order (version RA27). The limiting accuracy was in the 24th or 25th decimal digit because of roundoff. The high orders reached this accuracy easily, but, surprizingly, so did 15th order at L=14. (15th order used more function calls.) This 15th order case is most interesting. The last term in the series for each sequence was held to 10^{-14} or less, but the global error was 10^{-24} in hundreds of sequences. The error in one sequence must have been very much smaller than this latter figure. More than anything else, this fact shows the extraordinary properties of the truncated series developed from Gaussian spacings.

In these highly accurate calculations it is advantagous to iterate 3 times through each sequence.

This test is useful, but if one must integrate such an eccentric two-dimensional orbit, he should use Levi-Civita regularization (1903), see Bettis and Szebehely (1972). This removes the singularities from the differential equations and speeds this problem by a factor of eight.

3.3 The Outer Planets Problem

We tested the outer planet integration problem of Eckert, Brouwer, and Clemence (1951). They used a 12th-order multistep Lagrangian integrator in the predict-evaluate mode with 40-day steps. Their result was the positions of the 5 outer planets to 9 decimals for the years 1653-2060. Their calculation required 120 hours on a very early large computer. We used RADAU in 15th order, taking 320-day sequences (7 substeps/sequence) and integrating from a starting point in 1941 back to 1653 in 3.7 seconds on a CDC 7600 computer in single precision (60-bit word). The agreement for Jupiter with the 1951 paper was exact to the number of digits published. The other planets sometimes differed by 1 or 2 in the last digit. It should be noted that the multistep method would be just as fast on a modern computer. The advantage of RADAU is the it is self-starting, whereas the multistep methods are difficult to start.

Recently this same problem was done to the same accuracy with an IBM Personal Computer (equipped with the 8087 co-processor) in double-precision FORTRAN (64-bit word). Using RADAU in 15th order, the time was 8.7 minutes. It is practical to do such problems on this home computer.

3.4 A Class I Problem

Most of the problems in celestial mechanics involve 2nd-order differential equations, but the 1st-order equation is important in many fields. We tried the equation

$$y' = t(1 - y) + (1 - t) \exp(-t), \qquad y(0) = 1, \tag{19}$$

described by Krogh (1973), for which the solution is

$$y = 1 - \exp(-t) + \exp(-t^2/2). \tag{20}$$

The test is to integrate out to t=10. This equation is difficult to integrate because instability usually sets in as the integration proceeds. Indeed, RADAU became unstable and could not do this when allowed to pick its own sequence sizes, but when a constant sequence size of 0.2 or smaller was specified, then the error was in the 16th digit. For this test RADAU was in 15th order with a 64-bit double word. This is a 'stiff' equation. Often systems of stiff equations have both very large and very small eigenvalues. Though not developed for these, RADAU can handle some cases. The equations of celestial mechanics are not stiff.

This problem is somewhat unusual in that the independent variable
t appears explicitly on the right. This causes no difficulty with the
integrator, but does not often happen in celestial mechanics problems.

4. LISTING OF RADAU INTEGRATOR

It is not easy to program a complicated integrator, such as RADAU, even
when given a full mathematical description of the algorithm. A FORTRAN
listing of the subroutine is therefore given here. There was a choice:
(1) Giving a version called RA7 which uses loops to calculate the
series and where one may choose 7th, 11th, or 15th order. (2) Giving
a 15th-order version called RA15 where the series are written out
explicitly. Choice (2) was made: because it is some 30% faster, because
15th-order is often suitable, and because it is easier to compare the
written-out series with the mathematical description.
   It will be evident how to handle any number NV of simultaneous
differential equations. In the listing will be seen loops with index K
running from 1 to NV which bring all equations along together.
   It should not be difficult to write (say) an 11th-order version by
modification of the 15th-order listing in the appendix. Dimension H for
6 and put the Radau spacings for order 11 in the DATA statement. The
dimensioning of other quantities is a little excessive, but will do no
harm. Each series in written out only through terms with indices (5,K).
The index 7 in most loops is changed to 5, and 8's are changed to 6's.
The quantity PW becomes 1./7. Besides RA7 described already, another
version available on request is RA27. This will integrate in 15th, 19th,
23rd, or 27th order. This requires a computer with a 120- or 128-bit
double word and will integrate to 24 decimal-digit accuracy.

4.1 FORTRAN Listing of RA15, the 15th-order version of RADAU

```
      SUBROUTINE RA15(X,V,TF,XL,LL,NV,NCLASS,NOR)
C  Integrator by E. Everhart, Physics Department, University of Denver
C  This 15th-order version is called RA15. Order NOR is 15.
C  y'=F(y,t)     is    NCLASS=1,          y"=F(y,t) is NCLASS= -2,
C  y"=F(y',y,t) is    NCLASS=2
C  TF is t(final) - t(initial). (Negative when integrating backward.)
C  NV = the number of simultaneous differential equations.
C  Change dimensioning if NV is greater than 18.
C  LL controls accuracy. Thus SS=10.**(-LL) controls the size of
C  the last term in a series. Try LL=8 and work up or down from there.
C  However, if LL.LT.0, then XL is the constant sequence size used.
C  A non-zero XL sets the size of the first sequence regardless of
C  LL's sign.  Zero's and Oh's look alike on this printer. Use care!
C  X and V enter as the starting position-velocity vector (values of y
C  and y' at t=0) and they output as the final position-velocity vector.
C  Integration is in double precision. A 64-bit double-word is assumed.
C  In some computers IMPLICIT REAL*8 should be IMPLICIT DOUBLE PRECISION
      IMPLICIT REAL*8 (A-H,O-Z)
      REAL*4 TVAL,PW
```

```
C   The vectors X and V are dimensioned for unity below, because they
C   appear in the call. Storage for them is set out in the main program.
        DIMENSION X(1),V(1),F1(18),FJ(18),C(21),D(21),R(21),Y(18),Z(18),
       A    B(7,18),G(7,18),E(7,18),BD(7,18),H(8),W(7),U(7),NW(8)
        LOGICAL NPQ,NSF,NPER,NCL,NES
        DATA NW/0,0,1,3,6,10,15,21/
        DATA ZERO, HALF, ONE,SR/0.0D0, 0.5D0, 1.0D0,1.4D0/
C   These H values are the Gauss-Radau spacings, scaled to the range 0
C   to 1, for integrating to order 15.    H(1) = 0.D0 always.
        DATA H/        0.D0, .05626256053692215D0, .18024069173689236D0,
       A.35262471711316964D0, .54715362633055538D0, .73421017721541053D0,
       B.88532094683909577D0, .97752061356128750D0/
C   The sum of these H-values should be 3.7333333333333333
        NPER=.FALSE.
        NSF=.FALSE.
        NCL=NCLASS.EQ.1
        NPQ=NCLASS.LT.2
C   y'=F(y,t),NCL=.TRUE.   y"=F(y,t),NCL=.FALSE.   y"=F(y',y,t),NCL=.FALSE.
C   NCLASS=1, NPQ=.TRUE.   NCLASS=-2,NPQ=.TRUE.   NCLASS= 2,   NPQ=.FALSE.
C   NSF is .FALSE. on starting sequence, otherwise .TRUE.
C   NPER is .TRUE. only on last sequence of the integration.
C   NES is .TRUE. only if LL is negative. Then the sequence size is XL.
        DIR=ONE
        IF(TF.LT.ZERO) DIR=-ONE
        NES=LL.LT.0
        XL=DABS(XL)*DIR
        PW=1./9.
C   Evaluate the constants in the W-, U-, C-, D-, and R-vectors
        DO 14 N=2,8
        WW=N+N*N
        IF(NCL) WW=N
        W(N-1)=ONE/WW
        WW=N
    14  U(N-1)=ONE/WW
        DO 22 K=1,NV
        IF(NCL) V(K)=ZERO
        DO 22 L=1,7
        BD(L,K)=ZERO
    22  B(L,K)=ZERO
        W1=HALF
        IF(NCL) W1=ONE
        C(1)=-H(2)
        D(1)=H(2)
        R(1)=ONE/(H(3)-H(2))
        LA=1
        LC=1
        DO 73 K=3,7
        LB=LA
        LA=LC+1
        LC=NW(K+1)
        C(LA)=-H(K)*C(LB)
```

```
          C(LC)=C(LA-1)-H(K)
          D(LA)=H(2)*D(LB)
          D(LC)=-C(LC)
          R(LA)=ONE/(H(K+1)-H(2))
          R(LC)=ONE/(H(K+1)-H(K))
          IF(K.EQ.3) GO TO 73
          DO 72 L=4,K
          LD=LA+L-3
          LE=LB+L-4
          C(LD)=C(LE)-H(K)*C(LE+1)
          D(LD)=D(LE)+H(L-1)*D(LE+1)
    72    R(LD)=ONE/(H(K+1)-H(L-1))
    73    CONTINUE
          SS=10.**(-LL)
C    The statements above are used only once in an integration to set up
C    the constants. They use less than a second of execution time. Next
C    set in an estimate to TP based on experience. Same sign as DIR.
          TP=0.1DO*DIR
          IF(XL.NE.ZERO) TP=XL
          IF(TP/TF.GT.HALF) TP=HALF*TF
          NCOUNT=0
          WRITE (*,3)
    3    FORMAT(/' No. of calls, Every 10th seq.X(1),X(2),T,TM,TF')
C    An * is the symbol for writing on the monitor. The printer is unit 4.
C    Line 4000 is the starting place of the first sequence.
 4000    NS=0
          NF=0
          NI=6
          TM=ZERO
          CALL FORCE (X, V, ZERO, F1)
          NF=NF+1
C    Line 722 begins every sequence after the first. First find new
C    G-values from the predicted B-values, following Eqs. (7) in text.
  722    DO 58 K=1,NV
          G(1,K)=B(1,K)+D(1)*B(2,K)+
      X    D(2)*B(3,K)+D(4)*B(4,K)+D( 7)*B(5,K)+D(11)*B(6,K)+D(16)*B(7,K)
          G(2,K)=          B(2,K)+
      X    D(3)*B(3,K)+D(5)*B(4,K)+D( 8)*B(5,K)+D(12)*B(6,K)+D(17)*B(7,K)
          G(3,K)=B(3,K)+D(6)*B(4,K)+D( 9)*B(5,K)+D(13)*B(6,K)+D(18)*B(7,K)
          G(4,K)=          B(4,K)+D(10)*B(5,K)+D(14)*B(6,K)+D(19)*B(7,K)
          G(5,K)=                    B(5,K)+D(15)*B(6,K)+D(20)*B(7,K)
          G(6,K)=                              B(6,K)+D(21)*B(7,K)
    58    G(7,K)=                                        B(7,K)
          T=TP
          T2=T*T
          IF(NCL) T2=T
          TVAL=DABS(T)
C    Writing to the screen during the integration lets one monitor the
C    progress. Values are shown at every 10th sequence.
          IF(NS/10*10.EQ.NS) WRITE(*,7) NF,NS,X(1),X(2),T,TM,TF
    7    FORMAT(1X,2I6,5F12.5)
```

```
C  Loop 175 is 6 iterations on first sequence and 2 iterations therafter
         DO 175 M=1,NI
C  Loop 174 is for each substep within a sequence.
         DO 174 J=2,8
         JD=J-1
         JDM=J-2
         S=H(J)
         Q=S
         IF(NCL) Q=ONE
C  Here Y is used for the value of y at substep n. We use Eq. (9).
C  These collapsed series are broken into two parts because an otherwise
C  excellent  compiler could not handle the complicated expression.
         DO 130 K=1,NV
         A=W(3)*B(3,K)+S*(W(4)*B(4,K)+S*(W(5)*B(5,K)+S*(W(6)*B(6,K)+
      V     S*W(7)*B(7,K))))
         Y(K)=X(K)+Q*(T*V(K)+T2*S*(F1(K)*W1+S*(W(1)*B(1,K)+S*(W(2)*B(2,K)
      X    +S*A))))
         IF(NPQ) GO TO 130
C  Next are calculated the velocity predictors if needed for general
C  Class II. Here Z is used as the value of y' at substep n. (Eq. (10))
         A=U(3)*B(3,K)+S*(U(4)*B(4,K)+S*(U(5)*B(5,K)+S*(U(6)*B(6,K)+
      T     S*U(7)*B(7,K))))
         Z(K)=V(K)+S*T*(F1(K)+S*(U(1)*B(1,K)+S*(U(2)*B(2,K)+S*A)))
  130    CONTINUE
C  Find forces at each substep.
         CALL FORCE(Y,Z,TM+S*T,FJ)
         NF=NF+1
         DO 171 K=1,NV
C  Find G-values from the force FJ found at the current substep. This
C  section, including the many-branched GOTO, uses Eqs. (4) of text.
         TEMP=G(JD,K)
         GK=(FJ(K)-F1(K))/S
         GO TO (102,102,103,104,105,106,107,108),J
  102    G(1,K)=       GK
         GO TO 160
  103    G(2,K)=       (GK-G(1,K))*R(1)
         GO TO 160
  104    G(3,K)=       ((GK-G(1,K))*R(2)-G(2,K))*R(3)
         GO TO 160
  105    G(4,K)=       (((GK-G(1,K))*R(4)-G(2,K))*R(5)-G(3,K))*R(6)
         GO TO 160
  106    G(5,K)=       ((((GK-G(1,K))*R(7)-G(2,K))*R(8)-G(3,K))*R(9)-
      X        G(4,K))*R(10)
         GO TO 160
  107    G(6,K)=       (((((GK-G(1,K))*R(11)-G(2,K))*R(12)-G(3,K))*R(13)-
      X        G(4,K))*R(14)-G(5,K))*R(15)
         GO TO 160
  108    G(7,K)=((((((GK-G(1,K))*R(16)-G(2,K))*R(17)-G(3,K))*R(18)-
      X        G(4,K))*R(19)-G(5,K))*R(20)-G(6,K))*R(21)
C  TEMP is now the improvement on G(JD,K) over its former value.
C  Now we upgrade the B-value using this difference in the one term.
```

```
C  This section is based on Eqs. (5).
  160    TEMP=G(JD,K)-TEMP
         B(JD,K)=B(JD,K)+TEMP
         GO TO (171,171,203,204,205,206,207,208),J
  203    B(1,K)=B(1,K)+C(1)*TEMP
         GO TO 171
  204    B(1,K)=B(1,K)+C(2)*TEMP
         B(2,K)=B(2,K)+C(3)*TEMP
         GO TO 171
  205    B(1,K)=B(1,K)+C(4)*TEMP
         B(2,K)=B(2,K)+C(5)*TEMP
         B(3,K)=B(3,K)+C(6)*TEMP
         GO TO 171
  206    B(1,K)=B(1,K)+C(7)*TEMP
         B(2,K)=B(2,K)+C(8)*TEMP
         B(3,K)=B(3,K)+C(9)*TEMP
         B(4,K)=B(4,K)+C(10)*TEMP
         GO TO 171
  207    B(1,K)=B(1,K)+C(11)*TEMP
         B(2,K)=B(2,K)+C(12)*TEMP
         B(3,K)=B(3,K)+C(13)*TEMP
         B(4,K)=B(4,K)+C(14)*TEMP
         B(5,K)=B(5,K)+C(15)*TEMP
         GO TO 171
  208    B(1,K)=B(1,K)+C(16)*TEMP
         B(2,K)=B(2,K)+C(17)*TEMP
         B(3,K)=B(3,K)+C(18)*TEMP
         B(4,K)=B(4,K)+C(19)*TEMP
         B(5,K)=B(5,K)+C(20)*TEMP
         B(6,K)=B(6,K)+C(21)*TEMP
  171    CONTINUE
  174    CONTINUE
         IF(NES.OR.M.LT.NI) GO TO 175
C  Integration of sequence is over. Next is sequence size control.
         HV=ZERO
         DO 635 K=1,NV
  635    HV=DMAX1(HV,DABS(B(7,K)))
         HV=HV*W(7)/TVAL**7
  175    CONTINUE
         IF (NSF) GO TO 180
         IF(.NOT.NES) TP=(SS/HV)**PW*DIR
         IF(NES) TP=XL
         IF(NES) GO TO 170
         IF(TP/T.GT.ONE) GO TO 170
    8    FORMAT (2X,2I2,2D18.10)
         TP=.8D0*TP
         NCOUNT=NCOUNT+1
         IF(NCOUNT.GT.10) RETURN
         IF(NCOUNT.GT.1) WRITE (4,8) NOR,NCOUNT,T,TP
C  Restart with TP=0.8*T if new TP is smaller than original T on 1st seq
         GO TO 4000
```

```
 170   NSF=.TRUE.
C Loop 35 finds new X and V values at end of sequence. Eqs. (11),(12).
 180   DO 35 K=1,NV
       X(K)=X(K)+V(K)*T+T2*(F1(K)*W1+B(1,K)*W(1)+B(2,K)*W(2)+B(3,K)*W(3)
      X      +B(4,K)*W(4)+B(5,K)*W(5)+B(6,K)*W(6)+B(7,K)*W(7))
       IF(NCL) GO TO 35
       V(K)=V(K)+T*(F1(K)+B(1,K)*U(1)+B(2,K)*U(2)+B(3,K)*U(3)
      V      +B(4,K)*U(4)+B(5,K)*U(5)+B(6,K)*U(6)+B(7,K)*U(7))
  35   CONTINUE
       TM=TM+T
       NS=NS+1
C  Return if done.
       IF(.NOT.NPER) GO TO 78
       WRITE(*,7) NF,NS,X(1),X(2),T,TM,TF
       WRITE(4,7) NF,NS
       RETURN
C  Control on size of next sequence and adjust last sequence to exactly
C  cover the integration span. NPER=.TRUE. set on last sequence.
 78    CALL FORCE (X,V,TM,F1)
       NF=NF+1
       IF(NES) GO TO 341
       TP=DIR*(SS/HV)**PW
       IF(TP/T.GT.SR) TP=T*SR
  341  IF(NES) TP=XL
       IF(DIR*(TM+TP).LT.DIR*TF-1.D-8) GO TO 77
       TP=TF-TM
       NPER=.TRUE.
C  Now predict B-values for next step using Eqs. (13). Values from the
C  preceding sequence were saved in the E-matrix. The correction BD
C  is applied in loop 39 as described in Sec. 2.5.
  77   Q=TP/T
       DO 39 K=1,NV
       IF(NS.EQ.1) GO TO 31
       DO 20 J=1,7
  20   BD(J,K)=B(J,K)-E(J,K)
  31   E(1,K)=      Q*(B(1,K)+ 2.D0*B(2,K)+ 3.D0*B(3,K)+
      X             4.D0*B(4,K)+ 5.D0*B(5,K)+ 6.D0*B(6,K)+ 7.D0*B(7,K))
       E(2,K)=                 Q**2*(B(2,K)+ 3.D0*B(3,K)+
      Y             6.D0*B(4,K)+10.D0*B(5,K)+15.D0*B(6,K)+21.D0*B(7,K))
       E(3,K)=                        Q**3*(B(3,K)+
      Z             4.D0*B(4,K)+10.D0*B(5,K)+20.D0*B(6,K)+35.D0*B(7,K))
       E(4,K)=  Q**4*(B(4,K)+ 5.D0*B(5,K)+15.D0*B(6,K)+35.D0*B(7,K))
       E(5,K)=                Q**5*(B(5,K)+ 6.D0*B(6,K)+21.D0*B(7,K))
       E(6,K)=                       Q**6*(B(6,K)+ 7.D0*B(7,K))
       E(7,K)=                              Q**7*B(7,K)
       DO 39 L=1,7
  39   B(L,K)=E(L,K)+BD(L,K)
C  Two iterations for every sequence. (Use 3 for 23rd and 27th order.)
       NI=2
       GO TO 722
       END
```

4.2 The main program

Whereas the integrating subroutine RA15 of the previous section retains
the same form for different problems, the main program is specific to
the particular problem. Besides providing the parameters in the call
to RA15, it must give inital conditions, final value of the independent
variable, and display or use the final values. A listing of JSUNP.FOR
that controls the outer planet program will be sent on request. This
gives initial positions and velocities of the 5 planets at the start.
When combined with the programs in Sec. 4.1 and 4.3 all parts will be
in hand for doing the outer planet problem of Sec. 3.3

4.3 The FORCE subroutine

This example of a force subroutine is rather complicated, but instructive and useful.

```
      SUBROUTINE FORCE(X,V,TM,F)
C  The FORCE subroutine for the 5 outer planet integration.
      IMPLICIT REAL*8 (A-H,O-Z)
C The above Implicit statement assumes an 8-byte double word (64 bits).
C X, V. and F are dimensioned unity because they appear in the call.
      DIMENSION X(1),V(1),F(1),RM(5),PM(5),R(5),RH(5,5)
C  The reciprocal masses of the 5 planets, units of reciprocal sun.
      DATA RM/1047.355D0, 3501.6D0, 22869.D0, 19314.D0, 360000.D0/
      DATA SC,SCZ,KSA/0.D0, 0.D0, 0/
      IF(KSA.EQ.1) GO TO 5
      KSA=1
      SCZ=-(1.720209895D-2**2)*(800.D0**2)
      SC=-1.8938494521574133D2
C  SCZ is the Gaussian constant for an 800-day time unit, and SC is the
C  same except the mass of the sun is augmented by masses of inner
C  planets, Mercury through Mars.
C  X, V, and F are dimensioned for 15 in calling programs. Indices 1,2,3
C  are for x,y,z for Jupiter, 4,5,6 are for x,y,z Saturn, 7,8,9 are for
C  x,y,z  Uranus, 10,11,12 for x,y,z Neptune, and 13,14,15 for Pluto.
      DO 4 I=1,5
    4 PM(I)=-SCZ/RM(I)
    5 DO 10 N=1,5
      J=(N-1)*3+1
      R(N)=1.D0/DSQRT(X(J)**2+X(J+1)**2+X(J+2)**2)**3
      IF(N.EQ.5)  GO TO 10
      NA=N+1
      DO 9 L=NA,5
      K=(L-1)*3+1
      RH(N,L)=1.D0/DSQRT((X(J)-X(K))**2+(X(J+1)-X(K+1))**2+(X(J+2)
     A      -X(K+2))**2)**3
    9 RH(L,N)=RH(N,L)
C  indices K and L run 1-15, indices N and L for the planets run 1-5.
C  The mass factors are in PM, the distance from the sun of each planet
C  contribute to R, and the planet-to-planet distances contribute to RH.
   10 CONTINUE
```

```
      DO 20 N=1,5
      J=(N-1)*3+1
      SCM=(SC-PM(N))*R(N)
      F(J  )=SCM*X(J  )
      F(J+1)=SCM*X(J+1)
      F(J+2)=SCM*X(J+2)
C  The F-values above are for the sun-planet forces/unit mass.
      DO 20 L=1,5
      IF(L.EQ.N) GO TO 20
      K=(L-1)*3+1
      F(J  )=F(J  )+PM(L)*((X(K  )-X(J  ))*RH(N,L)-X(K  )*R(L))
      F(J+1)=F(J+1)+PM(L)*((X(K+1)-X(J+1))*RH(N,L)-X(K+1)*R(L))
      F(J+2)=F(J+2)+PM(L)*((X(K+2)-X(J+2))*RH(N,L)-X(K+2)*R(L))
C  The mutual planetary perturbation forces/unit mass are added on. The
C  first part of the second term is due to the planet-to-planet force,
C  and the second part is the indirect term because the sun at the
C  origin is not at the center of mass of the system.
   20 CONTINUE
      RETURN
      END
```

ACKNOWLEDGEMENTS

I thank the National Science Foundation, who sponsored the work of
visiting scientists at the National Center for Atmospheric Research,
where the original testing of the integrator was carried out. To
Dr. W. H. Goodyear, Dr. Paul R. Beaudet. Dr. Paul N. Swarztrauber,
Dr. J. M. A. Danby, and Dr. Fred Fernald I owe a note of appreciation
for their help. The use and testing of the integrator by Dr. B. G.
Marsden, Dr. A. Carusi, Dr. G. B. Valsecchi, Dr. E. Perozzi, Mr. Shio
Oikawa, and many others has been encouraging.

REFERENCES

Batrakov, Yu. N. :1982, 'Methods of Computation of the Perturbed Motion
     of Small Bodies in the Solar System' in Sun and Planetary System,
     (Proceedings of the VI European Meeting in Astronomy), W. Fricke and
     G. Teleki, editors, D. Reidel Publishing Co., Dordrecht. pp415-419.
Bettis, D. G. and Szebehely, V. :1972, 'Treatment of Close Approaches in
     the Numerical Integration of the Gravitational Problem of N-bodies'
     Gravitational N-Body Problem, M. Lecar, editor, D. Reidel Publishing
     Co., Dordrecht, pp388-405. See p395.
Bulirsch, R. and Stoer, J. :1966. 'Numerical Treatment of Ordinary Dif-
     ferential Equations by Extrapolatiom Methods' Num. Math., 8, pp1-13.
Butcher, J. C. :1964, 'Integration Processes Based on Radau Quadrature
     Formulas', Math. Comp. 18, pp233-344.
Carusi, A., Kresak, L., Perozzi, E., and Valsecchi, G. B. :1985 'The
     Long-Term Evolution Project' in Dynamics of Comets: Their Origin
     and Evolution, A. Carusi and G. B. Valsecchi, editors, D. Reidel Pub-
     lishing Co.,Dordrecht. (IAU Colloq. 83, Rome, June 1984) This volume.

Eckert, W. J., Brouwer, D., and Clemence, G. M. :1951 'Coordinates of
    the Five Outer Planets, 1653-2060' Astron. Papers Am. Ephemeris, 12.
Everhart, E. :1974a, 'Implicit Single-Sequence Methods for Integrating
    Orbits', Celest. Mech. 10, pp35-55.
Everhart, E. :1974b 'An Efficient Integrator of Very High Order and
    Accuracy' Denver Res. Inst. Tech. Report, 1 July 1974 (unpublished).
Everhart E. and Marsden, B. G. :1983, 'New Original and Future Cometary
    Orbits' Astron. J., 88, pp135-137.
Fehlberg, E. :1972, 'Classical Eighth- and Lower Order Runge-Kutta-
    Nystrom Formulas with Stepsize Control for Special Second-Order
    Differential Equations', NASA Tech. Rept., NASA TR R-381.
Gallaher, L. J. and Perlin, I. E. :1966, 'A comparison of Several
    Methods of Numerical Integration of Nonlinear Differential Equa-
    tions'. Presented at the SIAM meeting, Univ. of Iowa, March. 1966
    (unpublished). See Krogh, 1973.
House, F., Weiss, G., and Weigandt, R. :1978, 'Numerical Integration of
    Stellar Orbits', Celest. Mech., 18, pp311-318.
Krogh, F. T. :1973, 'On Testing a Subroutine for Numerical Integration
    of Ordinary Differential Equations', J. Assoc. Comput. Mach., 20,
    pp545-562.
Levi-Civita, T. :1903, 'Traiettorie singulari ed urti nel problema
    ristretto del tre corp', Annali di Mat. 9, pp1-32.
Marsden, B. G., Sekanina, Z., and Everhart, E. :1978, 'New Osculating
    Orbits for 110 Comets and Analysis of Original Orbits for 200
    Comets', Astron. J., 83, pp64-71.
Newton, R. R. :1959, 'Periodic Orbits of a Planetoid', Smithson.
    Contrib. Astrophys. 3, No. 7, pp69-78.
Papp, K. A., Innanen, K. A., and Patrick, A. T. :1977, 1980, 'A Compari-
    son of Five Algolrithms for Numerical Orbit Computation in Galaxy
    Models', Celest. Mech., 18, pp277-286, and 21, 337-349.
Shefer, V. A. :1982, 'Variational Equations of the Perturbed Two Body
    Problem in Regularized Form', Institute of Theoretical Astronomy,
    Leningrad. Publication No. 37. An 11th-order version of RADAU is used.
Stroud, A. H. and Secrest, D. :1966, Gaussian Quadrature Formulas
    Prentice Hall, Inc. Englewood Cliffs, N.J., See Table 12 (Radau
    spacings).

QUESTION: (A. Milani)   At the University of Pisa we have been using an
    integrator similar to yours of the implicit Runge-Kutta type which
    uses information from a previous step to start the iteration.
    However, we use the Gauss-Legendre spacings because we get order 2n
    with n sub-steps. Why do you think the Gauss-Radau spacing is better?

ANSWER: (E.Everhart) When one uses Gauss-Radau spacings there is a
    force found at the starting point of a sequence, and this is not re-
    peated when one iterates. The trial expression of Eq. (9) is already
    correct through the quadratic terms, and the iteration should con-
    verge quicker. None-the-less, a very fine integerator can surely be
    built on Gauss-Legendre spacings and I am delighted to hear of yours.
    Please send me a listing and description when you can.

THE LONG-TERM EVOLUTION PROJECT

A. Carusi[1], L. Kresák[2], E. Perozzi[1], G.B. Valsecchi[1]

[1]IAS-Reparto di Planetologia, C.N.R.
Viale Università 11, 00185 Rome, Italy

[2]Astronomical Institute, S.A.V.
84228 Bratislava, Czechoslovakia

ABSTRACT. The Long-Term Evolution Project (LTEP), realized in collaboration by the IAS-Reparto di Planetologia (Rome, Italy) and the Astronomical Institute of SAV (Bratislava, Czechoslovakia), has been developed with the aim of giving a general insight into the dynamical evolution of short-period comets. The motion of all the known short-period comets has been investigated over a long time span (over 800 years) taking care, as far as possible, to eliminate the sources of possible discrepancies within the computations. An internally consistent data-set and an atlas of orbital evolutions are the first outputs of this project. The main characteristics of the LTEP are discussed, together with some general remarks on its importance for cometary studies, its limitations and the future developments.

1. INTRODUCTION

The orbital evolution of short-period comets is characterized by a wide range of possible regimes of motion that a single comet can pass through, or that can be observed comparing the histories of different objects. Peculiar dynamical events are rather frequent: strong gravitational interactions, mainly with Jupiter, leading sometime to a temporary energetic binding to the planet, and resonant motion about low-order resonances with Jupiter and Saturn. Our observational evidence of these processes is very limited and heterogeneous. The discovery probability of a comet depends to a great extent on its orbital elements (in particular on the perihelion distance and revolution period), and on its variable absolute brightness. The osculating orbits of individual comets are of very different accuracy, mainly depending on the number of apparitions linked up by the computation. Every strong

A. Carusi and G. B. Valsecchi (eds.), Dynamics of Comets: Their Origin and Evolution, 203–214.
© *1985 by D. Reidel Publishing Company.*

perturbation outside this interval tends to degrade appreciably the reliability of the extrapolation beyond it. Moreover, the nongravitational effects, not only differing from one comet to another but also changing with time in each individual case, represent a severe constraint independent of the accuracy of the starting orbit.

These general remarks on the dynamical properties of short-period comets outline the difficulties that arise when approaching the study of
the evolution of their orbits; although many authors have dealt with
this problem, their results do not provide a complete and homogeneous
coverage of the cometary population as a whole.

The wide choice between the special perturbation methods available
to investigate cometary motion results in fact in a widespread variety
of accuracy and stability characteristics of the integrations carried
out by different authors. This, together with the numerical model
adopted (e.g. number of perturbing bodies included in the computations,
modelling of the nongravitational forces) can play an important role for
the critical effect that strong gravitational interactions can have on
the reliability of the results (see, e.g., Carusi et al., 1981). Furthermore, the number of objects taken into consideration and the time
span covered by the integrations are, again, variable parameters from
author to author.

The Long-Term Evolution Project (LTEP) has been carried out in order
to provide a more general approach to the study of the dynamical evolution of short-period comets: all the known sample has been integrated
for a long time span (over 800 years), taking care to eliminate, as far
as possible, the main sources of discrepancies within the computations.
The resulting data-base will then allow the development of different
kinds of studies: focusing on the evolution of peculiar objects as well
as investigating the behaviour of the population of periodic comets as
a whole.

2. OUTLINE

The aims of generality of the LTEP point out some of its basic characteristics:
    - The comet sample includes all the known short-period comets: their
number, updated to 31 December 1983, is 126.
    - The source of the starting data is, as far as possible, the same
for all the comets: osculating elements are taken from the 4th edition
of the "Catalogue of Cometary Orbits" (Marsden, 1982), unless new comets
were discovered after its publication or more accurate elements became
available; in this case data were extracted from the corresponding IAU
or Minor Planet Circulars. The starting elements almost invariably cor-

respond to the last apparition of the comet included in the orbit computation; predicted elements were always discarded.

- In order to satisfy the requirements of computing all the orbital
evolutions using the same numerical integration technique, while ensur-
ing a high rate of accuracy to them, many tests and comparisons have
been carried out using different integrating methods. These will be
described in detail in the next section. The best compromise between
accuracy and computational speed has been reached by the single-step
RADAU integrator (Everhart; 1974a, 1974b, 1985) solving barycentric
equations of motion.

- The perturbations of all the planets on the comet motion have always been included. In order to speed up the computations and to keep,
as much as possible, the same gravitational environment during the integration of different objects, only the motion of the comet has been computed; positions and velocities of the Sun and the planets have been
read from the JPL DE-102 Long Ephemeris (Standish et al., 1982).

- Each comet has been followed for a time span of 821 years covering
the arc from 1585 to 2406 AD, as representative of a reasonable fraction
of the active lifetime of a short-period comet. It appeared to be long
enough to recognize the basic dynamical behaviours of each object, without rendering completely unrealistic the results for the propagation of
the errors due to the numerical approach (round-off errors, truncation
errors, numerical instabilities after repeated planetary close encounters).

- Two evolutions for each comet have been computed: one backwards,
from the starting date to JD 2300000.5 (1585 February 1.0), the other
forwards, from the starting date to JD 2600000.5 (2406 June 17.0), adding up to a total of exactly 300,000 days.

- The forces acting on a comet were assumed to be of purely gravitational origin; nongravitational effects have always been neglected,
mainly because the related parameters are not yet known for all the
short-period comets and are variable with time. Their inclusion is
planned for further developments of the project.

- A quality class has been defined in order to give an estimate of
the reliability of every integration. It is determined by the accuracy
of the starting elements and by the importance of the nongravitational
forces acting on the comet.

The whole LTEP required about 25 CPU hours, using an UNIVAC 1100/82
machine, and a total of about 70 hours of I/O were necessary to handle
the input and output data and to generate auxiliary and back up files.
The total amount of time covered by the integrations is about 100,000
years; the mean length of the time step was 43.6 days but, since an
automatic step-size control was provided, it fell down to a fraction
of an hour in case of extremely close approaches to Jupiter, or grew up

to one year for comets moving temporarily well outside the planetary
region. The longest integration was that of P/Encke (15,128 time steps),
the shortest that of P/Wilk (2,074 time steps), with an average of 6,877
time steps per evolution (backwards and forwards).

All the information about every cometary evolution has been retained;
the output files, stored on tape, contain the Julian Date and the six
heliocentric ecliptic coordinates of the comet at each integration step.
These can be considered as the basic data files; from them, auxiliary
files are generated for specific purposes.

3. THE NUMERICAL METHOD

The advantages of suitable testing procedures for the selection of
a numerical integration technique have been outlined by many authors
(e.g. Krogh, 1973; Lapidus and Seinfeld, 1971). In our case the information needed was mainly concerned with computational speed requirements, in order to guarantee the feasibility of the whole project, and
with the accuracy that the integrator could ensure over the wide range
of orbital characteristics of the short-period comets sample. The procedure described by Everhart (1974a) was chosen because of its good
accuracy/speed ratio. The test was the "closure error test", which
consists in performing forward-backward integrations parametrized on
the numerical approach in question (the core integrator, the force formulation, the accuracy requirements, etc.). For each test-integration,
the relative closure error Δr, defined as the normalized difference between the given starting point and the computed one, and the number of
function calls N_f, i.e. the number of force evaluations needed to carry
out the whole integration, are computed. Characteristic curves can then
be isolated on the corresponding (Δr, N_f) plane, which allow both to
analyze the behaviour of individual methods and to make comparisons between them.

This procedure was applied to some 100 years test-integrations of cometary orbits; they were chosen among the most representative of the
sample, in order to have an estimate of the efficiency of each numerical
approach over different possible regimes of motion. Some of the results
obtained are shown in fig. 1: the moderately perturbed, but quite dis-
similar, orbits of comets P/Encke and P/Halley, and the more chaotic
orbit of P/de Vico-Swift have been integrated several times, requesting
an increasing accuracy to the two best integrators at our disposal: the
multi-step DVDQ, developed by Krogh (1970), and the single-step RADAU,
by Everhart (1974a, 1974b, 1985).

Connecting the points in the (Δr, N_f) plane, corresponding to the same
orbit and obtained using the same integrator, a family of curves is

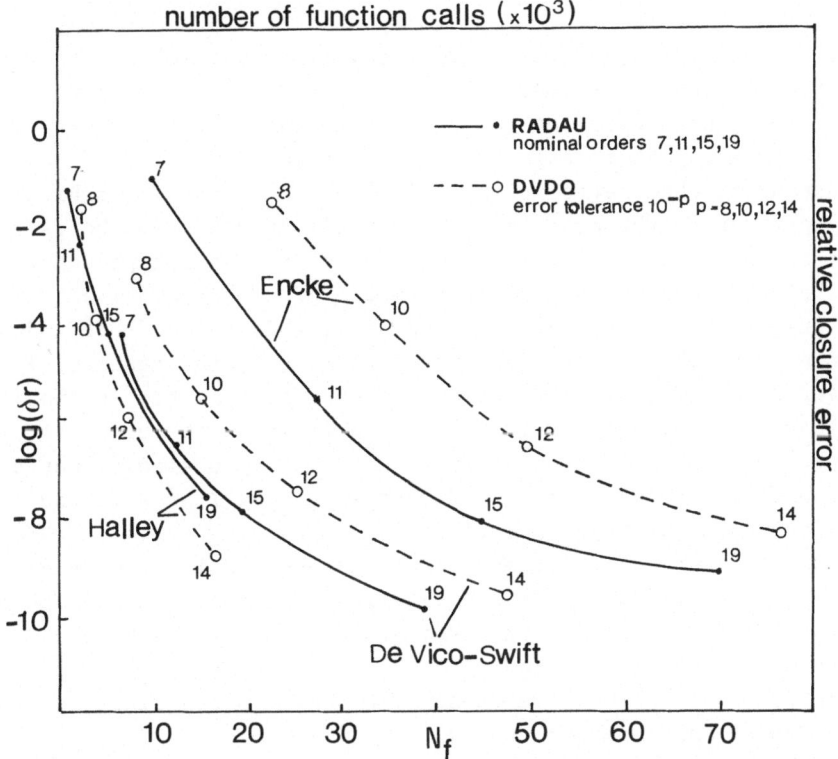

Figure 1.   Some of the results obtained applying the closure error test
to 100 years integrations of short-period comet orbits.   For each of the
test objects, curves have been fitted through the points obtained using
the same integrator, but requesting an increasing accuracy in the computations.   This was done changing the parameter that controls the local
error tolerance of DVDQ, and varying the nominal order of the integrations with RADAU.

drawn.   They define some basic characteristics of the methods tested
through their slope (related to the order of the integrator, i.e. the
gain in precision when halving the step-size), and their position within
the plane (the more the line is displaced towards the lower-left corner
of the plane, the faster and more accurate the integration has been).
The distance between the lines corresponding to the same object allows
a quantitative estimate of the difference in computational speed and accuracy of the two integrators; note the considerable improvements that
can be achieved by a proper choice.   The crossing of the lines related
to comet P/Halley, reverting the results obtained for the other two objects, shows the different sensitivity of the integrators when treating

different types of orbits.

Several test-integrations were performed, focused mainly on the numerical method and on the formulation of the equations of motion used.
They led to the final choice of the double precision, 19th order version
of RADAU, integrating barycentric equations of motion.

Notwithstanding these efforts in optimizing, as far as possible, the
numerical approach to the problem, some general comments on the reliability of our integrations are needed. As to the testing procedure
used: the "closure error" is just an estimate of the internal accuracy
of an integrator (Everhart, 1974a; Zadunaisky, 1979); round-off and
truncation errors will anyway affect the accuracy of the results, i.e.
their convergence to an exact solution. Even more dramatic is the loss
of significance that can arise from the uncertainty with which the osculating period of a short-period comet is known: 5 to 6 significant digits of the best determined orbits contrast with only 1 or 2 in the case
of some of the one-apparition comets.

The propagation of both of these errors is not easily predictable as
it depends upon the stability characteristics of the numerical method,
and upon the orbit itself. The reliability of the results becomes critical especially when strong gravitational interactions occur (see e.g.
Carusi et al., 1985), which is a rather frequent event during the dynamical evolution of short-period comets.

It is necessary to point out that our integrations should be regarded
as a representative sample of evolutions of observable short-period comets passing, near their midpoints, close to the osculating orbits of
all known objects of this type.

4. THE ATLAS OF COMETARY EVOLUTIONS

The basic output of the LTEP has been collected into a book: "Long-
Term Evolution of Short-Period Comets" (Carusi et al., 1984), in which
the dynamical history of each comet is presented in a compact form,
with the aid of plots and tables. Each comet is described in a General
Section, whose format is the same throughout the book (as an example,
the one relative to comet P/Oterma is reported in fig. 2). A short text
is followed by a table showing the evolution of the orbital elements at
equispaced intervals of 50,000 days; the core of the section consists of
four plots showing the behaviour of the Tisserand invariant of the comet
with respect to Jupiter, of its inclination, perihelion and aphelion
distances in time, thus allowing an overall view of the main events
characterizing each evolution. In the case of comet P/Oterma, for example, the rather appreciable variation in absolute value of its Tisserand
invariant arount 1770 implies the occurrence of a strong perturbation by

P/Oterma

The orbital evolution of P/Oterma is quite unique in two respects. At the beginning its perihelion is far outside the orbit of Jupiter, and the Tisserand invariant relative to Jupiter is extraordinarily high. It is only a deep encounter with Saturn in the XVIIIth century which permits the comet to be captured by Jupiter into a temporary satellite orbit, to make three revolutions in an orbit of extraordinarily small aphelion distance and, after another symmetrical close encounter and stay in a satellite orbit, to be ejected into an orbit similar to the previous one. This evolution is assisted by permanently low eccentricity and inclination.

42 revolutions. 3 apparitions. Quality class 5.

Epoch	T	a	e	ω	Ω	I
1956.0606	1956.06105	3.95999	0.14449	354.872	155.109	3.992
1585.0201	1586.05116	8.54140	0.25940	194.385	82.782	4.206
1721.1225	1732.09061	8.02791	0.21650	189.229	80.212	4.163
1858.1117	1861.01223	7.30148	0.21030	235.409	38.038	3.095
1995.1010	2002.12217	7.27569	0.24400	55.745	330.875	1.933
2132.0901	2135.12183	6.98534	0.16419	120.951	282.430	1.652
2269.0725	2278.05037	7.62596	0.20376	180.688	259.562	1.868
2406.0617	2403.07087	7.56274	0.20343	182.673	252.106	1.912

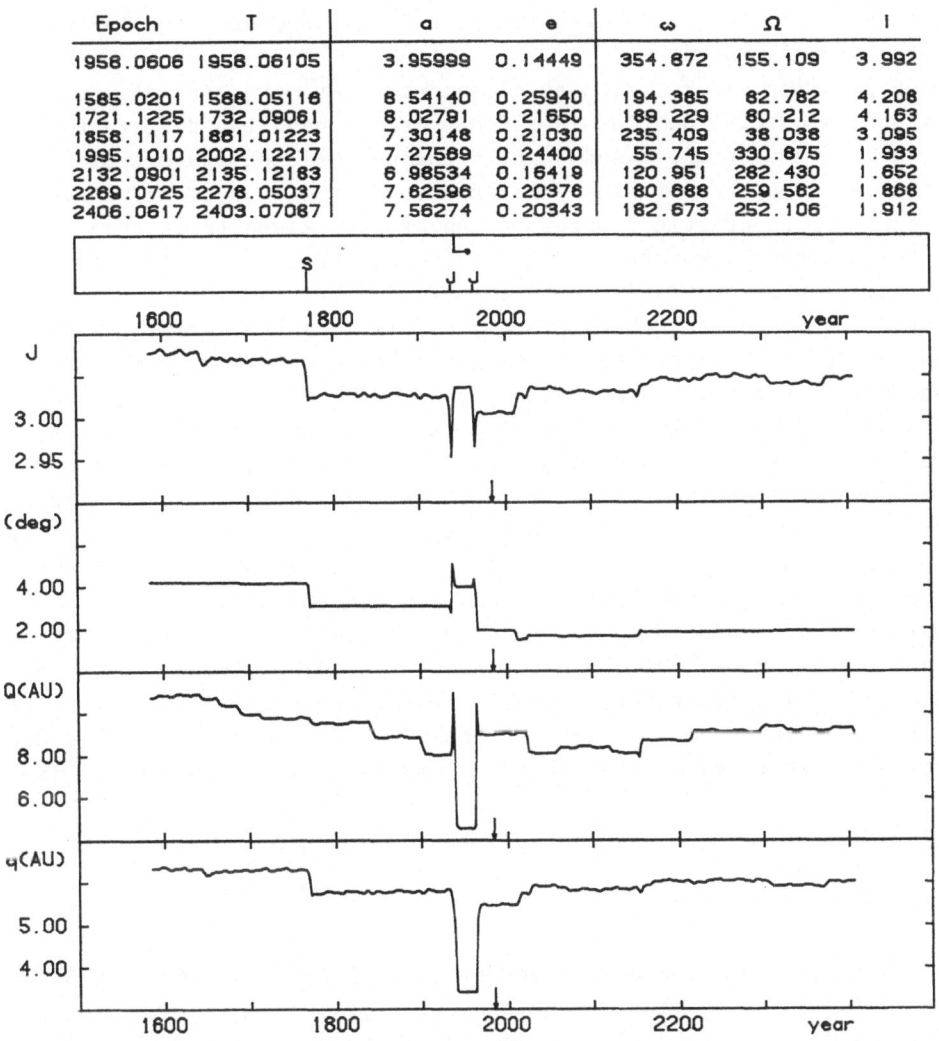

Figure 2.   An example of the General Section: the orbital evolution of P/Oterma.   For more explanations see text.

a planet other than Jupiter (in fact it was shown to be Saturn). Besides, the strong reduction of the perihelion distance of the comet, due
to a close encounter with Jupiter in 1937, is clearly responsible of its
discovery, indicated in the narrow window over the main plots. Here additional general information is given: the time of the first observation
is marked by a vertical bar; beginning at the lower end of it a horizontal line covers the time span between the first and last observations.
A dot indicates the starting date of the integration, corresponding to
the elements listed in the first line of the table. At the bottom of
the window, letters J, S, and U indicate the passage of the comet within
0.5 AU from Jupiter, Saturn, and Uranus, respectively. No encounters
with Neptune or Pluto have been found, and the ones with the terrestrial
planets are not listed.

When peculiar events, such as close encounters or resonant motion,
are found, a Special Section is added. It can have two different for-
mats, as shown in figs. 3 and 4 for P/Shajn-Schaldach and P/Kopff.

If a very deep and/or slow complex encounter with a planet takes place, resulting in drastic changes of the orbital elements of the comet,
its motion within a sphere of 2 AU radius around the planet is traced
(fig. 3). Tables containing both the heliocentric and planetocentric
elements are added, together with plots following the variations of the
distance from the planet, of the planetocentric binding energy and the
trajectory of the comet during the encounter.

The second type of special section is presented when a comet exhibits
libration cycles around some resonance with Jupiter (fig. 4): plots
showing the jovicentric trajectory are then included, together with a
polar diagram where the radius vector is the osculating mean motion of
the comet. Further information, such as the number of apparitions or
the quality class of the orbit can be found in the text.

The atlas is produced in separate sheets, in order to allow both an
internal rearrangement of the ordering of the comets (e.g. by alphabet,
period, perihelion distance, number of apparitions), and an easy updating of its content, as the LTEP continues to produce when new periodic
comets are discovered or the orbital elements of known comets are improved.

5. CONCLUSIONS

The aims and the general characteristics of the Long-Term Evolution
Project - an investigation on the orbital evolution of all the known
short-period comets for a long time span - have been reviewed. Although
some constraints on the reliability of the integrations are due to the
uncertainty with which the osculating elements of the comets are known,

P/Shajn-Schaldach

Encounter with Jupiter

1939 – 1948

61	3	

Heliocentric elements

t	f	q	Q	a	e	i	ω	Ω
1939.065	292.07	4.2962	5.1840	4.7401	0.0936	10.81	179.83	208.89
1948.839	260.47	2.2346	5.2901	3.7624	0.4061	6.14	215.08	187.47

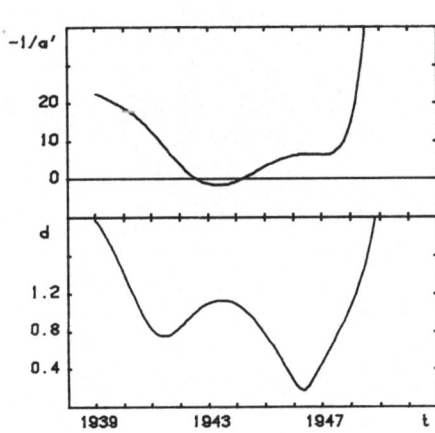

Jovicentric elements

t	d	a'
1939.065	1.9533	-0.045
1941.434	0.7468	-0.116
1946.335	0.1800	-0.150
1948.839	1.9815	-0.013

t	e'	i'
1939.065	38.52	27.11
1941.434	7.54	103.21
1946.335	2.20	59.20
1948.839	45.77	84.35

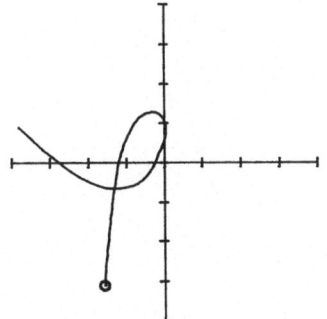
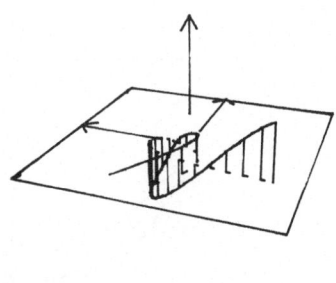

Figure 3.   An example of the Special Section – Planetary Encounters: the 1939-1948 close encounter of P/Shajn-Schaldach with Jupiter. The period of a temporary satellite capture is represented by the negative part of the -1/a plot; the distance d and the semimajor axes a, a' are in AU. The two plots at the bottom of the section depict the orbital pattern of the comet: the one on the left shows the x-y projection of the comet trajectory in the frame centred on Jupiter and rotating with its radius vector.   This plot extends from -2 to +2 AU in both directions, and a small circles indicates the beginning of the path.   A three dimensional perspective view of the same path is presented on the right.

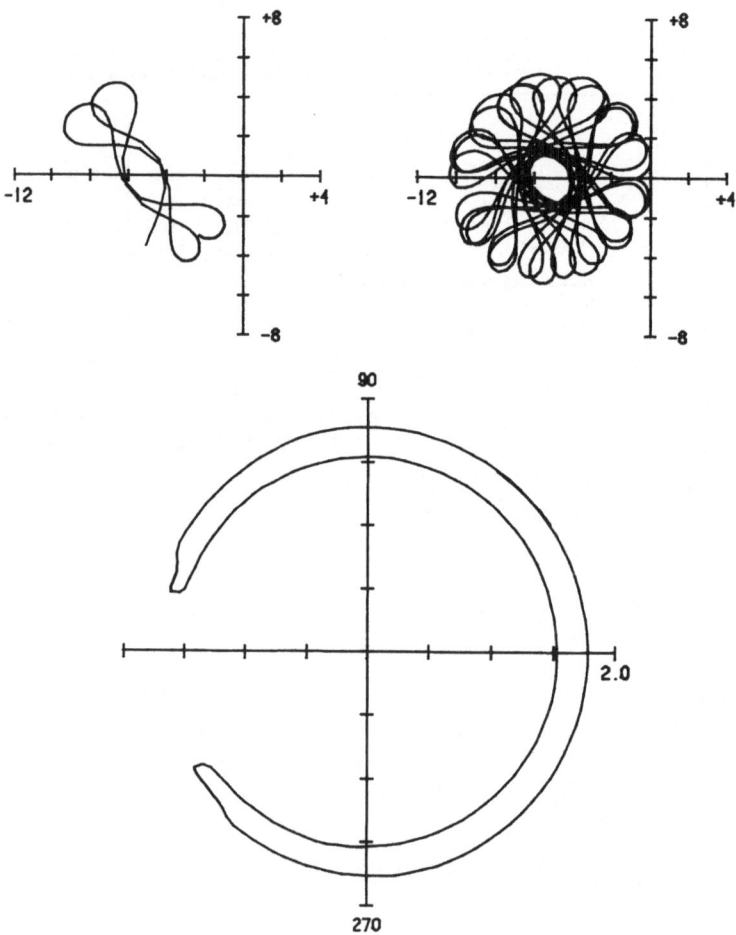

P/Kopff

Libration about the 1/2

resonance with Jupiter

2077 – 2250

Figure 4.   An example of the Special Section – Resonances: the libration
of P/Kopff around the 1/2 resonance with Jupiter.   Usually only one libration cycle is shown: the upper left plot contains the jovicentric
trajectory of the comet for the first revolutions, in order to show its
shape, while the right-hand plot follows the orbital path over the whole
time span indicated.   Below is the horseshoe-shaped libration pattern
with respect to Jupiter, during one libration cycle.   The position angle
is the libration argument, counted counter-clockwise; the radius vector
measures the mean motion of the comet, in tenth of a degree per day.

and to the numerical approach used, the LTEP has created a homogeneous data-base on the dynamics of short-period comets. It allowed the publication of a book in which the history of each member of this population is shortly described, and the presentation of some of the results obtained after a first survey of the output data (Carusi et al., 1985a).

The project is planned to continue in the future to ensure both a regular updating and a refinement of the computations. Possible future developments of the LTEP will include the influence of nongravitational forces on the motion of the comets; taking into account the presence of satellite systems or of planetary oblateness during close encounters; a uniform treatment of asteroidal objects moving in comet-like orbits, and extension of the time span covered by integrations for individual interesting objects.

REFERENCES

Carusi, A., Kresák, L., Valsecchi, G.B.: 1981, Astron. Astrophys., 99, 262-269.

Carusi, A., Kresák, L., Perozzi, E., Valsecchi, G.B.: 1984, IAS-Internal Report 12.

Carusi, A., Kresák, L., Perozzi, E., Valsecchi, G.B.: 1985a, this volume.

Carusi, A., Perozzi, E., Pittich, E., Valsecchi, G.B.: 1985b, this volume.

Everhart, E.: 1974a, Celest. Mech., 10, 35-55.

Everhart, E.: 1974b, Denver Res. Inst. Techn. Report, 1 July 1974.

Everhart, E.: 1985, this volume.

Krogh, F.T.: 1970, JPL Technology Utilization Document n. CP-2308.

Krogh, F.T.: 1973, J. Ass. Comp. Mach., 20, 545-562.

Lapidus, L., Seinfeld, J.H.: 1971, Numerical Solution of Ordinary Differential equations, Academic Press, New York and London.

Marsden, B.G.: 1982, Catalogue of Cometary Orbits, IAU Central Bureau for Astron. Telegrams, Cambridge, Mass.

Standish, E.M., Keesey, M.S.W., Newhall, X X: 1976, JPL Techn. Report n. 32-1603.

Zadunaisky, P.E.: 1979, Celest. Mech., 20, 209-230.

DISCUSSION

H. Scholl: How did you handle the problem of close encounters between comets and Jupiter? Comment: it does not appear very fair to me to compare DVDQ and RADAU, since DVDQ is not specially designed for solving

Newton's equations of motion but is designed for integrating ordinary
differential equations in general, whereas RADAU is specially made for
solving Newton's equations of motion.

E. Perozzi: The numerical procedure used was to integrate always solar
system barycentric equations of motion; no change in the force modelling
has been performed during the integrations, even when close encounters
occurred. This choice came from the following considerations:

   - a regularization of the equations of motion is difficult where, as
in our case, only the motion of the object is integrated. First of all
the change of independent variable, needed in a regularization procedure, is inconsistent with the format of the JPL Ephemeris (which uses
Julian Dates) requesting an additional integration in order to go back
to the usual time at each integration step. Furthermore, the structure
of the integrator used (RADAU 19) implies a substep sampling of the perturbation within every step when the distance of the integrated object
from the perturbing body, needed to regularize the problem, is still not
computed;

   - Encke's formulation of the equations of motion did not show the expected advantages with respect to the barycentric formulation, especially in computational speed. Additional refinements of the integration
method are, nevertheless, possible in connection with the numerical instabilities arising when strong gravitational interactions occur.

   For what concern the comment on the comparison between DVDQ and
RADAU, it must be stressed that, Although RADAU has been developed by E.
Everhart, who works in cometary dynamics, it has been presented as a
general method for solving ordinary differential equations in an unfortunately still unpublished paper (Everhart, 1973; see also this volume).
In that work the author discussed the efficiency of RADAU in a number of
test cases, not only concerning celestial mechanics: the appendix listing of the program is the one used in our work. Furthermore, it is
worthwhile to remark that, even if strong structural differences do, in
fact, exist between the two integrators (DVDQ is a quite sophysticated
routine, while RADAU is more straightforward), our testing procedure was
specifically directed toward an optimal choice for solving a specific
dynamical problem: the integration of short-period comet orbits. RADAU
and DVDQ appeared, in the end, the most efficient integrators at our
disposal, in the sense that they could really compete in accuracy and
computational speed. Our final result, however, has no aim of generality, as the efficiency of the single integrator will depend on the problem to be solved.

P.E. Zadunaisky: The closure test may be useless if symmetries are involved in the numerical integrator.

E. Perozzi: The closure test was useful in comparing some different methods. Everhart's method does not involve symmetries.

ERRORS IN NUMERICAL INTEGRATIONS AND CHAOTIC MOTIONS.

A. Milani and A.M. Nobili
Universita' di Pisa
Dipartimento di Matematica
Piazza dei Cavalieri 2
I-56100 Pisa Italy

ABSTRACT. The methods to estimate the integration errors, including the effects of truncation, rounding off, and instability of the solutions, are discussed. Polynomial error accumulation depends upon numerical method, stepsize, orbital period and also eccentricity; it is also machine dependent. Comets correspond to the most difficult case of exponentially diverging orbits; however they can be very close to resonant ordered regions.

Numerical integration is an essential tool in the study of cometary orbits; however the results of the integration are not always reliable. Integration error is a complex phenomenon; it is not a purely numerical effect, but the result of a complex interaction between the approximations introduced in the computation and the physical instability of the real orbit.

In this paper we review the possible causes of integration error, and try to give explicit estimates of their size, both for the local error (i.e. within one integration step) and for the accumulated error. For the sake of this discussion, we shall distinguish among the possible causes of error the truncation of the discretisation formula, the rounding-off of the numbers in the computer arithmetic unit, the errors due to the use of implicit formulae, the physical model errors and the errors in the initial conditions.

Truncation errors are conceptually the best understood; nevertheless practically useful formulae to estimate the order-of - magnitude of both the local and the propagated truncation errors really applicable to the orbits of celestial bodies are not available in the standard literature. In section 1 we try to fill this gap.

A. Carusi and G. B. Valsecchi (eds.), Dynamics of Comets: Their Origin and Evolution, 215–226.
© *1985 by D. Reidel Publishing Company.*

Rounding-off is difficult to study, because there is no way to understand the process underlying the accumulation of this kind of errors without a deep understanding of the way the arithmetic unit of the computer works and of the way a high level language code is translated into machine code. In section 2 we review briefly some recent improvements in the understanding of the machine - dependent round - off errors.

Implicit formulae are always used together with a control of their convergence, therefore the contribution to the integration error from their imperfect convergence is always minor and anyway well under the control of the programmer. Sometimes it is difficult to explain why the errors arising from this source are so small, as was discussed in this meeting by E. Everhart: we had the same experience with our implicit Runge-Kutta method.

The physical model errors will be discussed elsewhere in these proceedings. We are left with the error in the initial conditions: strictly speaking it is not an error arising within the process of computing the orbit; however the way it accumulates as time goes by is relevant for our discussion of the integration errors, for the very simple reason that every error results in the displacement of the computed orbit on a nearby orbit whose initial conditions were different. Thus the instability of the real, physical orbit does introduce a numerical instability as well, and the "numerical" error cannot accumulate slower that the rate of growth of the separation of two nearby orbits. This kind of instability is, unfortunately for us, specially relevant for cometary orbits, as it is discussed in section 2.

1.   THE TRUNCATION ERROR

The local truncation error is the difference between the actual orbital motion between time t and time t + h and the solution of the discretized problem actually solved in the computational algorithm for the corresponding step; it is often estimated with a formula based only on the product hn (n is the mean motion); however every such formula is valid for circular orbits only. A more general formula valid for nonzero eccentricities is given here.

Let us assume the integration is performed with a Stormer predictor:

$$\underline{x}_{k+1} = 2\underline{x}_k - \underline{x}_{k-1} + h^2 \sum_{j=0}^{m} b_j \, \nabla^j \underline{\ddot{x}}_k \tag{1}$$

where ∇ is the backward difference operator $\nabla f(t) = f(t) - f(t-h)$. Since formula (1) is obtained by truncation of the summation to order m, the local truncation error is:

$$R_{m+1} = b_{m+1} h^2 \nabla^{m+1} \ddot{\underline{x}}_k + \dots$$
(2)

and by the Lagrange formula:

$$\nabla^j_{f(t)} = h^j f^{(j)}(t^*), \qquad t - jh < t^* < t$$
(3)

$$R_{m+1} = b_{m+1} h^{m+3} \underline{x}^{(m+3)}(t) + \dots$$
(4)

If the exact solution were a circular orbit with mean motion n, then for m even (m=2y):

$$\underline{x}^{(m+3)} = (-1)^y n^{m+2} \dot{\underline{x}}$$
(5)

and for m odd (m=2y+1):

$$\underline{x}^{(m+3)} = (-1)^y n^{m+3} \underline{x}$$
(6)

and the magnitude of the local error is:

$$\left| R_{m+1} \right| = b_{m+1} (hn)^{m+3} a$$
(7)

(a the semimajor axis). However we are mainly interested in the propagation of this local error; this phenomenon can be studied in different ways (Kinoshita, 1968; Henrici, 1962) but the most natural way for astronomers is to use Gauss perturbation equations: the local error (4) can be interpreted as the effect of a constant perturbing force F :

$$F = b_{m+1} h^{m+1} \underline{x}^{(m+3)}(t^*) + \dots$$
(8)

The resulting perturbations on the orbital elements will depend on the direction as well as the magnitude of the force F: the main effect, the one growing quadratically with the time, will be the perturbation in longitude arising from a secular perturbation in the semimajor axis. All the other perturbative effects will result in much smaller accumulated errors; the short term perturbations will produce errors of the order of

$$F/n^2 \simeq b_{m+1} (hn)^{m+1} a$$
(9)

The Gauss equation for the semimajor axis is:

$$\dot{a} = 2 \langle F, \dot{\underline{x}} \rangle / n \, a + \text{terms of order} \geqslant 1 \text{ in e.}$$
(10)

For m even (m=2y) we will have for a circular orbit:

$$\dot{a}(t) = 2b_{m+1}(-1)^{y}(hn)^{m+1} \langle \underline{\dot{x}}(t), \underline{\dot{x}}(t^{*}) \rangle / na + \ldots \tag{11}$$

where t^{*} is a few steps before t, by (3); the angle between $\underline{\dot{x}}(t)$ and $\underline{\dot{x}}(t^{*})$ is thus of the order of mhn/2 and:

$$\langle \underline{\dot{x}}(t), \underline{\dot{x}}(t^{*}) \rangle = n^{2}a^{2}(1 + 0(n^{2}h^{2}m^{2}/4)) \tag{12}$$

We can assume the 0(...) term in (12) to be of higher order, and by substituting in (11):

$$\dot{a}/a = 2b_{m+1}(-1)^{y}(hn)^{m+1}n + \ldots \tag{13}$$

the corresponding coefficient of the quadratic error accumulation in longitude is then:

$$\dot{n}/2 = -1.5\,b_{m+1}(-1)^{y}(hn)^{m+1}n^{2} + \ldots \tag{14}$$

It is worth remarking that a different formula holds for m odd (m=2y+1): because of (6) the direction of the "truncation perturbing acceleration" F is more radial than along track, and from a formula analogous to (12):

$$\langle \underline{\dot{x}}(t), \underline{x}(t^{*}) \rangle = na^{2}\,0(\,n\,h\,m/2) \tag{15}$$

we find that (13), (14) are modified by a factor 0(nhm/2): the secular effect appears to be of higher order; however in most integrations nh is not much smaller than 2/m. The results given by formulae (9) and (14) do not change if another integration method is used; only the constant b_{m+1} does change.

It is intuitively obvious that the error estimates given by (9), (13) and (14) give a grossly underestimated error for eccentric orbits; however how fast the error grows with eccentricity is somewhat surprising. There is a way to compute an estimate of the truncation error in the integration of an eccentric orbit just by recomputing formulae (5), (6). For an eccentric orbit the exact solution, in an appropriate coordinate system xyz, can be expanded in a Fourier series in the mean anomaly M with coefficients formed by Bessel functions (Wintner, 1941; Kovalevski, 1963):

$$x/a = -3e/2 + \sum_{p=1}^{+\infty} 1/p\left\{ J_{p-1}(pe) - J_{p+1}(pe) \right\} \cos pM$$

$$y/a = (1-e)\sum_{p=1}^{+\infty} 1/p\left\{ J_{p-1}(pe) + J_{p+1}(pe) \right\} \sin pM \tag{16}$$

These series have the D'Alembert property, that is the lowest order term of the coefficient of the cosine (or sine) of pM is $O(e^{p-1})$; however, when the derivatives of order m+3 are computed, e^{p-1} is multiplied by an high power of p:

$$R_{m+1} = b_{m+1} a \ (hn)^{m+3} \sum_{p=1}^{+\infty} g(e,p) \ trig \ (pM) \qquad (17)$$

with trig (pM) a trygonometric vector function of length 1; the coefficients g is:

$$g(e,p) = \frac{e^{p-1} \ p^{p+m+2}}{2^{p-1} \ p!} \ + O(e^{p+1}). \qquad (18)$$

The truncation error (17) is then the sum of different harmonic components; at the perihelion all the error harmonics are in phase, and the size of the local error can be estimated by the sum of the series

$$S_1 = \sum_{p=1}^{+\infty} g(e,p) \sim Z_1 = \sum_{p=1}^{+\infty} \frac{e^{p-1} \ p^{p+m+2}}{2^{p-1} \ p!} \qquad (19)$$

The accumulated along-track error however is not the result of the error at a specific point but rather of the average error in energy; for the purpose of an estimate of the along-track quadratic error one should consider the different error harmonics as independently acting, thus use the root mean square sum of their size:

$$S_2^2 = \sum_{p=1}^{+\infty} g^2(e,p) \sim Z_2^2 = \sum_{p=1}^{+\infty} \left(\frac{e^{p-1} \ p^{p+m+2}}{2^{p-1} \ p!} \right)^2 \qquad (20)$$

and substitute the resulting "average" energy error in the same formulae used to obtain (14):

$$\dot{n}/2 = -1.5 \ b_{m+1} (-1)^Y (hn)^{m+1} Z_2 \ n^2 + \ldots . \qquad (21)$$

In table 1 the prediction given by formula (21) is compared with the results of a test performed by Cohen et al. (1973) with m-12, h=40 days, n = mean motion of Jupiter. The error is an along track acceleration in arcsec/year^2.

The comparison shows that our formula gives the right value of the along track error for low eccentricities, and a good order - of - magnitude estimate for moderate values of e. Of course the use of Z_2 instead of S_2 results in inaccuracies for large e because the neglected higher order (in e) terms are not much smaller. The Cohen et al. (1973) test were performed as a preparation for a long integration of planetary orbits; the effect on cometary orbits of the same phenomenon is striking. In table 2 we have listed the predicted increase of the error with respect to a circular orbit as a function of e, both for the local error

TABLE 1

e	error predicted by (21)	error found in the test integration
0	-0.4×10^{-12}	
.01	-1.8×10^{-12}	-1.6×10^{-12}
.02	-5.3×10^{-12}	-5.7×10^{-12}
.03	-12.7×10^{-12}	-12.1×10^{-12}
.04	-26.5×10^{-12}	-24.3×10^{-12}
.05	$-52. \times 10^{-12}$	$-43. \times 10^{-12}$
.06	$-97. \times 10^{-12}$	$-68. \times 10^{-12}$

at perihelion and for the accumulated along-track quadratic error. It can be appreciated that the error grows very fast with e, much faster than the cubic growth hypothesized by Cohen et al.; for cometary orbits, this implies that the use of a fixed stepsize algorithm is not recommended and the use of a short stepsize is not a good solution.

TABLE 2

e	increase of local error at perihelion	increase in accumulated error
.05	3×10^{2}	1×10^{2}
.1	5×10^{3}	2×10^{3}
.15	6×10^{4}	2×10^{4}
.2	6×10^{5}	2×10^{5}
.25	6×10^{6}	2×10^{6}
.3	6×10^{7}	2×10^{7}
.35	7×10^{8}	2×10^{8}
.4	9×10^{9}	2×10^{9}

Of course a solution to the problem of the increase of the error with eccentricity is the use of a time element s such that ds/dt is proportional to 1/r (r=distance from the Sun); then the time element is essentially the eccentric anomaly E and formulae like (16) are substituted by $x=a(\cos E-e)$, $y=a(1-e^{2})^{1/2}\sin E$ that do not contain the higher harmonics. With a time element proportional to the eccentric anomaly, the local and the accumulated along track error are still given by (9) and (14) respectively, for every eccentricity e.

2. THE ROUNDING-OFF ERROR

The rounding-off error is usually discussed with reference to the treatment by Brouwer (1937). However that celebrated paper was written before the era of the electronic computers, and thus was based on assumptions which are not necessarily applicable to the process of orbit computation as it is usually performed today.

The hypotheses under which Brouwer theorem applies are as follows: A) the error (or at least the significant part of it) is done in summing up the values of the previous second derivatives, or their differences, multiplied by h^2, to the previous step to get the new value, as in (1). The round-off errors done within the computation of the acceleration at each step are of lesser importance, because h^2 is small. B) the local round off error can be modelled as a random variable, uniformly distributed between $-d/2$ and $d/2$, where d is the "machine precision", i.e. the value of the last bit, or 1 in the last recorded digit; in particular its mean, or expected, value is exactly zero. C) the orbit is in itself orbitally stable, e.g. because the perturbations are negligible. D) other errors are uninfluent, e.g. the truncation error is smaller (in the manual variable order computation algorithms used at the time, this was checked at each step anyway).

Under these assumptions, Brouwer proved that the error in the orbital elements will be distributed as a gaussian random variable, with zero mean and root mean square value growing with the number N of integration steps as $N^{3/2}$ for the anomaly (i.e. along track) and as $N^{1/2}$ for the other five (i.e. cross track). The problem is now to assess how valid are the assumptions A, B, C and D: In particular B requires that the rounding off is performed by computing the sum in (1) with all the significant digits conserved, and then rounded: this is not at all the way the arithmetic operations are performed in modern computers, or better: not at all the way in which the compilers instruct the arithmetic unit of the CPU to operate. In reality, numbers are usually truncated, just forgetting the significant digits that are on the right of the maximum allowable mantissa length. As a result, the "rounf-off" error has an average not equal to zero but to $-d/2$; the latter formula being exactly true only for fixed point arithmetic and provided the negative numbers are represented in complement: the sign of the coordinates modifies the expected error if the negative numbers are represented as modulus plus sign. As a result, the same "random walk" argument used by Brouwer gives an expected secular error in semimajor axis; the expected error along track grows as N^2, for the fixed point, modulus plus sign arithmetic (Fabri and Penco, 1984). Different results are obtained for different arithmetics; floating point arithmetics generates a "quantum

effect" by which the relative error depends upon the absolute value of
the semimajor axis.

Fabri and Penco have proved rigorously (at least for circular
orbits) the existence of errors growing faster than Brouwer's rule by
questioning only assumption B: the round-off error is still modelled in
a statistical way, but with a realistic distribution (of course the
computer errors are not random at all, being uniquely determined by the
algorithm and by the initial conditions: the statistic refers to integrations with different initial conditions). However the other assumptions seem questionable as well. We will discuss assumption C in
section 3. Assumptions A and D more or less mean that when different
error sources are present, the local error will propagate according to
the propagation trend of the larger local error: e.g. if a rounding-off
error propagating as $N^{3/2}$ (because assumption B is valid; e.g. by the
use of guard-digits) is larger than the quadratically-propagating
truncation error, the latter is "masked" and is not allowed to grow at
its own pace. There are no rigorous arguments to support this way of
thinking. We conclude that, unless special attention is paid not only to
the integration algorithm but even to the computer firmware, the integration error is bound to grow at least quadratically with time.

3. ERRORS IN CHAOS

We have at last to question the assumption upon which most of the
discussions of the numerical errors, including the two previous
sections, are based: that the errors done in propagating an unperturbed
Keplerian orbit are representative of the errors that would result from
the integration, with the same method, of a perturbed orbit.

As it is known since the times of Poincare', the perturbed motion
does not belong to an integrable system; this means that some orbits
will lie on manifolds of quasi-periodic solutions (Arnold,1963), but in
between these there will be chaotic regions where homoclinic orbits
generate hyperbolic sets (Smale, 1967). The technical definition of an
hyperbolic set does not matter here; the essential property of a chaotic
region containing an hyperbolic set can be described with the device
first used by Henon and Heiles (1964) for their numerical investigations
of non-integrable dynamical systems: if two orbits with initial conditions very close together are propagated (e.g. numerically), the distance between points corresponding to the same time will grow exponentially
with time, and the logarithm of the ratio between the initial distance
and the distance after some fixed time T will be a measure of the
instability. Indeed this same logarithm divided by T is related to the
Liapounov characteristic exponent (Benettin et al., 1980).

What about the errors in the numerical orbit propagation when the real orbit lies in such a chaotic region? A very simple, and often overlooked, result says that no numerical method can be more stable than the exact solution the method is used to compute. Because does not matter how small is the local integration error: since it is anyway nonzero, after the first step the numerical method will really integrate an orbit starting from different initial conditions: if the latter diverges exponentially from the exact solution, so will the numerical solution (if it is "convergent", i.e. unless it really solves an other equation).

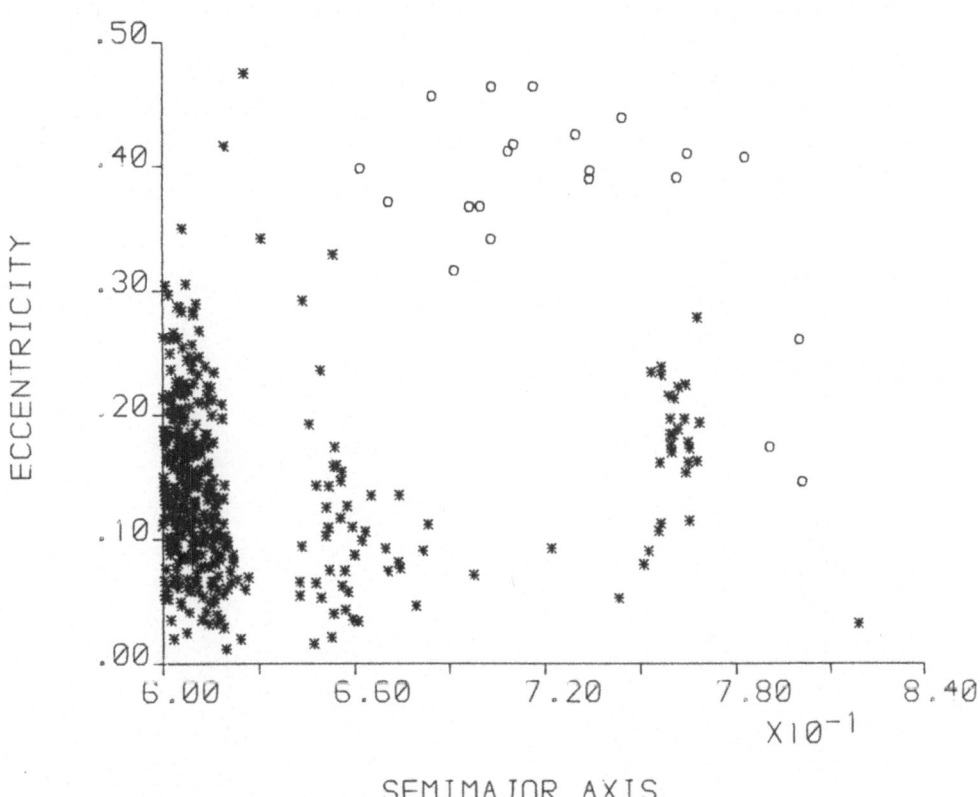

SEMIMAJOR AXIS

Figure 1: Numbered asteroids (stars) from the TRIAD file and periodic comets (circles) from the catalogue of perihelion passages by Marsden and Roemer, 1982. Semimajor axis (as a fraction of Jupiter's) and eccentricity are plotted for the region 3.2AU<a<4.3AU, 0<e<0.48. The region that appears void in this plot, between the Hildas and the 2/1 gap and for moderate eccentricities is indeed occupied occasionally by comets on temporary "transfer" orbits.

Cometary orbits have the special feature of (almost) always lying in chaotic regions. This is in reality an observational selection effect: a comet can exibit a spectacular coma only provided that its orbit "recently" underwent a change resulting in a large decrease of its perihelion distance; this change results from some strong perturbation, and the strong perturbation in turn generátes chaotic behaviour. Because of the very complex structure of the resonances with the major planets, the regions of the phase space where both chaotic behaviour and abrupt orbital elements changes can occur are intermingled with "ordered" regions of dominant quasi-periodic behaviour. This is best illustrated by the boundary region between the outer asteroid belt and the belt of the comets of the Jupiter family (Figure 1).

In between there is a "grey" belt of orbits that are neither obviously cometary nor asteroidal: the best criterion to predict wether a given set of initial conditions will give rise to an ejection from the region plotted in Figure 1 (hence is "cometary" even if in a transient almost quiescent state) is to compute the divergence ratio as described above (Milani and Nobili, 1984b).

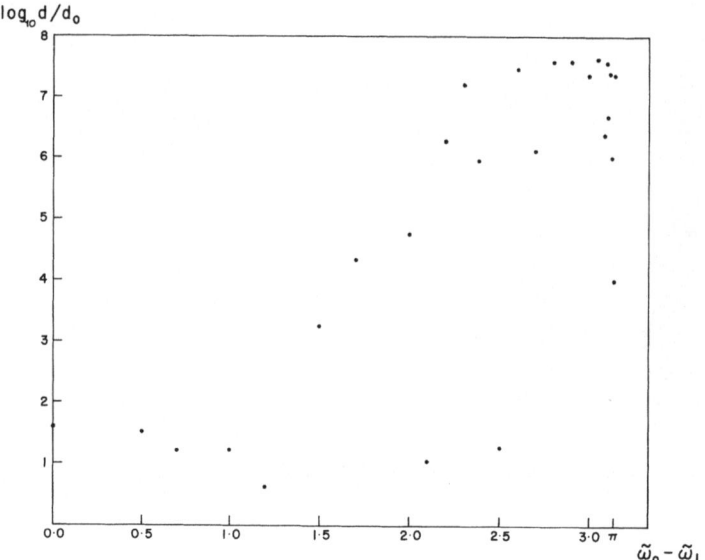

Figure 2: The divergence, i.e. the ratio between the very small initial distance d_0 and the distance d after 50 synodic periods, is plotted for the orbit of a comet with initial conditions a=0.68 a_J , e=0.226 (i.e. in the depleted region of figure 1) as a function of the initial angle between the comet's and Jupiter's perihelion, in radians. Not only the values change dramatically, but no smooth curve can be fitted to the points of this plot: the chaotic behaviour can appear and disappear with very small changes in the initial conditions.

But even the chaotic behaviour is not a stable prediction: because in between a chaotic region there are islands of ordered behaviour, where resonant asteroidal orbits can survive the perturbations by the planets; this can be seen by computing the divergence ratio for many nearby orbits (Figure 2) or by plotting the orbital elements and looking for resonant behaviour (Milani and Nobili, 1984a).

We are not stating that the orbits of comets are impossible to compute. For a fixed span of time, by complying with a list of cautions that are suggested by the discussion presented in this paper, it is possible to compute the orbit of a comet to a reasonable accuracy, unless very close approaches to a planet do occur. However the very long-term evolution of the single cometary orbits is not accessible to our computations, and this state of affairs being due to the very nature of the problem it is not likely to change soon; the qualitative behaviour of cometary orbits is on the contrary amenable to study, in a statistical sense, and numerical integrations are an essential tool for this purpose.

REFERENCES

Arnold, V.I. (1963) Usp. Mat. Nauk. 18, 91.
Benettin, G., Galgani, L., Giorgilli, A. and Strelcyn, J.M. (1980) Meccanica, March 1980, 9.
Brouwer, D. (1937) Astr. J. 46, 149.
Fabri, E. and Penco, U. (1984) "Propagation of the Round-off Errors in Numerical Integrations", in preparation.
Heinrici, P. (1962) Discrete Variable Methods in Ordinary Differential Equations, John Wily & Sons, New York-London.
Henon, M. and Helies, C. (1964) Astron. J. 69, 73.
Kinoshita, H. (1968) Pubbl. Astr. Soc. Japan 20, 1.
Kovalevsky,J. (1963) Introduction a la mecanique celeste, Armand Colin, Paris.
Milani, A. and Nobili, A.M. (1984a) Celestial Mechanics, in press.
Milani, A. and Nobili, A.M. (1984b) Astron. Astrophys., in press.
Smale, S. (1967) Bull. A.M.S. 73, 747.
Wintner, A. (1941) The Analytical Foundations of Celestial Mechanics, Princeton Univ. Press.
Cohen, C.J., Hubbard, E.C. and Oesterwinter, C. (1973) Astr. Pap. Am. Ephem. 22, pt.1.

Discussion

Marsden : The inadequacy of the Brouwer error-accumulation model was not
a problem with automatic computers until after 1960. One could always
program the early computers to act in the same way as mechanical desk
calculators with regard to rounding. The problem arose with the
introduction of purely binary computers and of high-level computer
languages. I recall that around 1961 an assistant at Yale, using one of
the new computers, found a rather dramatic decrease in the semimajor
axis of an orbit. This was on a friday afternoon. Brouwer then spent the
whole weekend integrating the two-body problem over several revolutions,
using a desk calculator,but truncating rather than rounding. By monday
morning he was convinced that his 1937 paper did not apply in the case
of truncation.

Milani : However this was not published.

Valsecchi : What model did you use for your integration of asteroid
orbits? What is your expectation about the nature and number of
protective mechanisms using more complex models?

Milani : The elliptic restricted planar 3-body model. When the third
dimension is taken into account, the protection mechanism based on the
inclination can play a role; this has been shown by Froeschlé and Scholl
(Astron. Astrophys. 1979). There is at least one asteroid protected in 3
different ways, one based on the inclination: it is 721 Tabora.

Zadumansky : For testing the accuracy of numerical experiments your
methods are good. However Lyapounov's theory of stability may give
valuable qualitative indications.

Milani : That is true as a matter of principle. However if you happen to
find an unpredicted resonance, the theoretical predictions have to be
changed.

ONE OF THE PROBLEMS OF LONG-TERM INTEGRATIONS OF COMETARY ORBITS

A. Carusi[1], E. Perozzi[1], E.M. Pittich[2], G.B. Valsecchi[1]

[1]IAS-Reparto di Planetologia, C.N.R.
Viale Università 11, 00185 Rome, Italy

[2]Astronomical Institute, S.A.V.
84228 Bratislava, Czechoslovakia

ABSTRACT. One of the problems of the numerical integrations of cometary orbits is that of their numerical stability. For those bodies which undergo close encounters with the giant planets the problem has a specific feature. Apart from the numerical instability of integrations in the mathematical sense, there is an additional source of instability due to the inaccuracy of initial data, i.e. the orbital elements. An example of this case is presented in this paper by the numerical integration of the motion of comet P/Shajn-Schaldach over an interval of 368.4 years, within which six close encounters with Jupiter occurred. The inaccuracy of the starting orbital elements of this comet is modelled by changes in the last digit of each element of the central orbit, determined by the set of the best orbital elements. In the process of integration, eight model orbits experience differential perturbations with respect to the central orbit. The numerical instability, caused by the inaccuracy of starting orbital elements and represented by the dispersion of these orbits, tends to increase abruptly beyond each encounter with Jupiter. It is shown that, with the attainable accuracy of the osculating elements representing the observations, one or two approaches to within 1 AU from Jupiter can make the orbit entirely indeterminate.

1. INTRODUCTION

The study of cometary evolution is not possible without the knowledge of individual cometary motions during longer time spans. Comets moving in the solar system undergo perturbations by the gravitational attraction of the planets, and in some cases by nongravitational forces, too. Their motion in full range corresponds to the motion in a n-body system. There is generally only one way for the solution of such a dynamical

227

A. Carusi and G. B. Valsecchi (eds.), Dynamics of Comets: Their Origin and Evolution, 227–235.
© 1985 by D. Reidel Publishing Company.

problem: the numerical integration of differential equations of motion.

The solution of these equations by means of difference approximations or polynomial function representations is subject to numerical
truncating errors. Several authors have compared the relative efficiencies of integrating differential equations by means of the classical
single-step or multi-step methods and the recurrent power series methods.
Generally they have investigated these methods from the point of view of
cumulation of local truncating error versus total computer time. They
analyzed the accuracy of the integration in difficult problems, such as
close encounters (Bettis and Szebehely, 1972; Everhart, 1974a,b; Roy et
al., 1972; Schubart and Stumpff, 1966; Sitarski, 1979; Szebehely and Bettis, 1972). For a detailed analysis of the effects of infinitesimal
changes in starting elements and of integrator's characteristics the
reader may refer to Oikawa and Everhart (1979).

In the dynamical evolution of some comets there is a series of
close encounters with giant planets. These events are the sources of
other inaccuracies in the determination of cometary orbits. Close encounters not only require convenient integration methods, but they can
cause indeterminacy of the orbit. This is a consequence of the inaccuracy of the initial orbital elements from which the numerical integration
begins.

2. SELECTION OF THE MODEL ORBITS

The number of close planetary ecounters after which the effect of the inaccuracy of starting orbital elements becomes serious is generally unknown, and to investigate this problem we must start from a convenient
example. For this purpose, it is necessary to select a comet undergoing
repeated close encounters - one whose motion is also not affected by nongravitational forces - and follow it over a relatively short time span,
in which the cumulation of local truncating errors can be neglected.
Otherwise, it is not possible to distinguish the contribution of the numerical effects due to the integration and those caused by the inaccuracy
of the initial orbital data.

One of the most suitable comets for our purpose is P/Shajn-Schal-
dach. The comet has a well determined orbit (Marsden, 1977, 1978b), the
nongravitational effects on it prior to 1945 should have been negligible,
since its perihelion distance at every return was larger than 4.1 AU, and
there were repeated close encounters with Jupiter at relatively short
time intervals (Pittich, 1981). The high value of the Tisserand invariant, 2.93, makes these encounters very effective.

The inaccuracy of the starting orbital elements for the numerical

integration can be modelled by small changes of the known elements of
the comet. To this purpose, the tenth digit of each element of P/Shajn-
Schaldach (Marsden, 1977) was modified by an increment d = 0, ±0.25,
±0.50, ±0.75, ±1.00 (see Table I). The set of the nine potential orbits
of P/Shajn-Schaldach is defined by these modified elements. The orbits
are designated by their corresponding d-value: 000, ±025, ±050, ±075,
±100; for example, in the +025 case, a q-value of 2.233905338 becomes
2.23390533825, and a similar increment is made to all the elements.

TABLE I.   Starting orbital elements

Epoch	1949 Nov. 20.0 ET	
T	1949 Nov. 26.9192662 + d ET	
ω	215°3052345 + d	
Ω	167°3928552 + d	1950.0
i	6°1520408 + d	
q	2.233905338 + d AU	
e	0.404977568 + d	

d = 0.00, ±0.25, ±0.50, ±0.75, ±1.00
     in the last significant digit

The evolution of these orbits was calculated from 1949 November 20.0
(JD 2433240.5) back in time until 1585 February 1.0 (JD 2433240.5), with
the same procedure used in Carusi et al. (1984). The main characteristics of the integrations are: the use of an equatorial reference frame
centred in the barycentre of the solar system; the integration of the
barycentric equations of motion with the subroutine RADAU (Everhart,
1974b; for a listing and general comments on the subroutine see also
Everhart, this volume) to the 19th order; the positions of the Sun and
planets taken from the JPL DE-102 Long Ephemeris (Newhall et al., 1983).
The evolution data of P/Shajn-Schaldach model orbits used in this paper
will be published elsewhere (Carusi et al., 1985).

3. ORBITAL INSTABILITY AFTER CLOSE ENCOUNTERS

During the considered interval of 368.4 years, P/Shajn-Schaldach passed
very close to Jupiter six times in the orbits 000 and -025, and five
times in the other model orbits (see Table II). The minimum jovicentric
distance was in three cases (orbit 000) and in two cases (all other model
orbits) not greater than 0.22 AU.
   The differences between the individual elements of the central orbit

and those of each other model orbit within the considered interval of time are shown in Figures 1 and 2.

TABLE II. Jovicentric distances of P/Shajn-Schaldach (AU) at close encounters with Jupiter for various model orbits.

Date	−100	−075	−050	−025	orbit 000	+025	+050	+075	+100
1590 02 15				.7142					
1592 02 05					.1606				
1651 09 03	.7268	.7098	.6930	.6763	.6597	.6432	.6269	.6108	.5948
1780 09 06	.5490	.5489	.5489	.5488	.5487	.5486	.5486	.5485	.5484
1875 09 29	.2125	.2125	.2125	.2125	.2125	.2125	.2125	.2125	.2125
1941 06 15	.7374	.7473	.7473	.7473	.7473	.7473	.7473	.7473	.7473
1946 04 10	.1828	.1828	.1828	.1828	.1828	.1828	.1828	.1828	.1828

Since a good starting orbit was available, the small changes of the elements did not practically affect the dynamical evolution of this co-met during the close encounters from 1946 until 1780 in a backward inte-gration, but later the situation has become quite different. There are increasing differences in the model orbits prior to the next encounter in 1651. However, it is still possible to define the orbit of this comet in the limited space occupied by all the model orbits. This bundle of orbits is quite stable during the interval without other close planetary encounters. The cross-section of the bundle in its steady state is strongly dependent upon the accuracy of the starting elements; the sharp-er the actual orbit is defined, the smaller is the cross-section of the bundle at any moment.

There is one interesting feature of the evolutions of these ficti-tious objects: after the encounter of 1651 the orbits 000 and −025 are much more sensitive than the others. This fact follows from the position of the comet relative to Jupiter at the time of next encounter. Similar cases are described by Carusi et al. (1981).

The encounters of the 000 orbit in 1592 and the −025 orbit in 1590 indicate how much the model orbits diverge beyond that point; the spread of the bundle after (backward in time) these encounters is no longer com-pensated by some convergece at next encounter.

4. INACCURACY OF ACTUAL ELEMENTS

The maximum change of one unit in the tenth significant digit (10^{-7} day

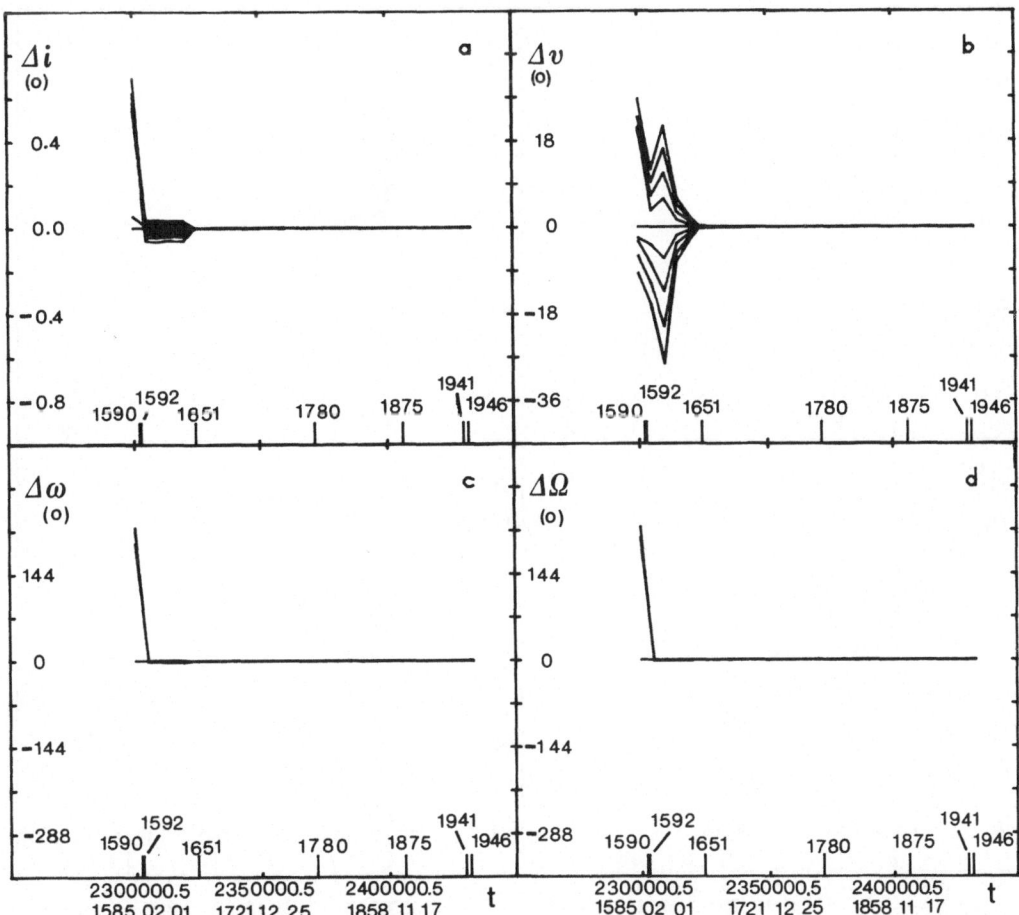

Figure 1. The differences between the evolved orbital elements and the elements of the standard 000 orbit are plotted versus time; a: inclination; b: true anomaly; c: argument of perihelion; d: longitude of node.

in T, 10^{-7} deg in angular elements, 10^{-9} AU in q and 10^{-9} in e) was intentionally set unrealistically low. Even though the starting elements are based on linking up of two apparitions spaced by three revolutions, their probable errors are four to five orders of magnitude greater than d, and they are mutually strongly correlated (in particular, $\Delta\omega \sim -\Delta\Omega$ and $\Delta q \sim \Delta(1-e)$). Our choice is equivalent to rounding-off all elements upwards and downwards; proceeding in time, the differences tend to increase as the orbits diverge by differential perturbations. Only after some time they reach realistic values. At that time they already bear signatures of the sensitivity of individual elements to differential perturbations; from there on, the divergence can be compared with that due to the actual uncertainty of the orbit.

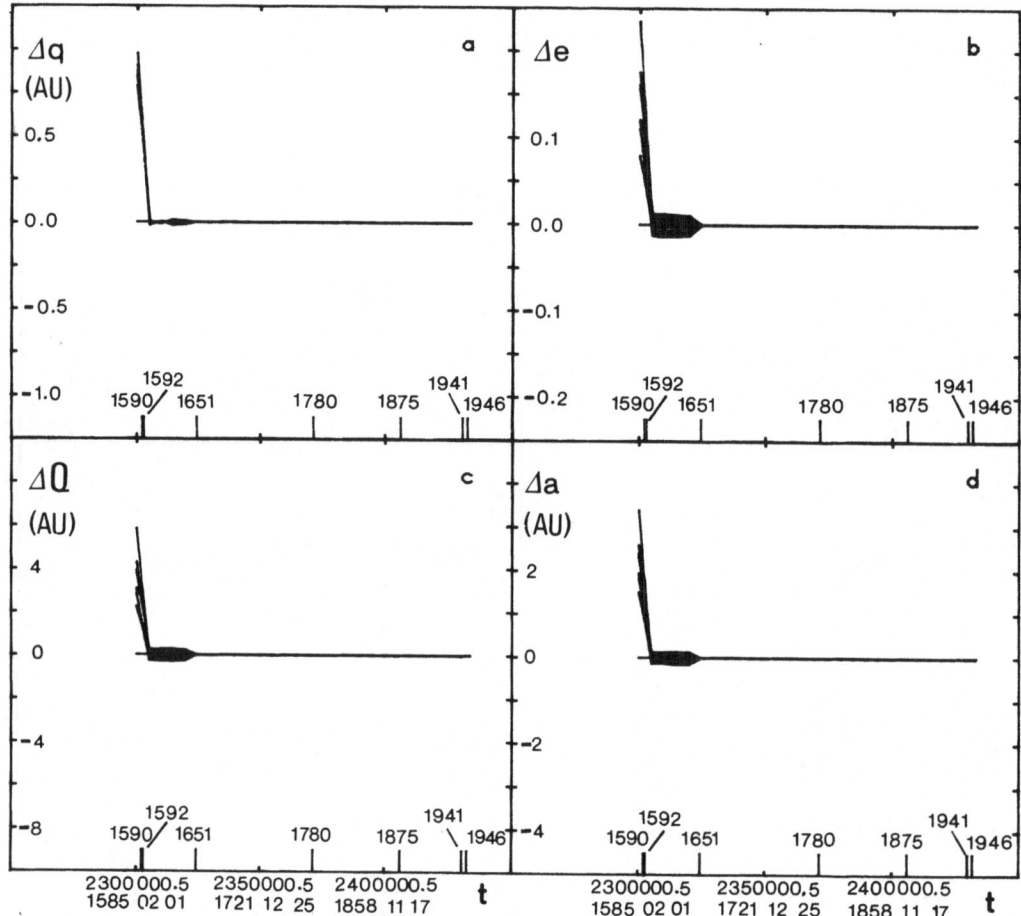

Figure 2.   The same as Fig. 1; a: perihelion distance; b: eccentricity; c: aphelion distance; d: semimajor axis.

The differences between the central orbit and the others in all osculating elements at four epochs placed approximately half-way between the close encounters with Jupiter are listed in Table III.   They are expressed as the means and standard deviations of the quantities obtained dividing the final differences with respect to orbit 000 by the arbitrary starting change in the corresponding element and adding the base 10 logarithms of the ratios so obtained.

The values of Table III can be compared with the corresponding values for three available element sets (Table IV).   They are:

a) Marsden's (1978a) orbit for 1949 November 20, based on 68 observations between 1949 August 28 and 1972 January 20, and Marsden's (1971) orbit for 1949 November 20, based on 64 observations between 1949 September 18 and 1949 December 20.   Here the differences should be characteristic for the improvement by linking up two apparitions against a final one-apparition orbit of average accuracy;

b) Marsden's (1978b) orbit for 1971 October 16, based on 68 observations between 1949 August 28 and 1972 January 20, and Nakano's (1984) orbit for 1971 October 16, based on 37 observations between 1971 September 15 and 1978 October 31. Here the differences should be characteristic for two independent linkages of pairs of apparitions, one proceeding forwards and one backwards, and meeting at the same osculating date;

c) Nakano's (1984) mean errors for the 1979 orbit (mean residual ±1.05" in the apparent position). Here the differences should account for a high-quality least-square solution, as internal errors not taking into account the correlations between individual elements.

TABLE III. Means and standard deviations of the differences in the osculating elements for each model orbit

Orbit	2420000.5		2390000.5		2350000.5		2310000.5	
	m	s	m	s	m	s	m	s
−100	2.12±1.56		3.27±1.13		5.56±1.13		7.29±1.36	
−075	1.59±0.86		3.32±1.33		5.55±1.13		7.17±1.36	
−050	2.85±1.13		3.35±1.37		5.26±1.13		7.00±1.36	
−025	2.83±1.14		3.18±0.91		4.96±1.13		6.70±1.36	
+025	2.99±1.01		3.42±1.11		4.96±1.13		6.72±1.36	
+050	2.38±1.11		3.00±0.74		5.26±1.13		7.02±1.36	
+075	2.11±1.26		3.61±1.03		5.44±1.13		7.19±1.36	
+100	2.93±1.12		3.61±1.06		5.56±1.13		7.32±1.36	

TABLE IV. Mean values and standard deviations from differences of osculating elements of observed orbits

Orbit	mean	st. deviation
a	4.76	0.68
b	4.55	0.66
c	4.42	0.64

5. CONCLUSIONS

The analysis of the model and actual orbits of P/Shajn-Schaldach allows
some general conclusions. Determination of orbits of periodic comets
experiencing close encounters with the giant planets is practically impossible in a long time interval. The orbits become poorly determined
and have to be replaced by bundles of changing cross-section within the
time intervals between successive encounters. The inaccuracy of orbital
elements tends to increase abruptly after each encounter with a giant
planet.

The actual uncertainty of the orbit of P/Shajn-Schaldach is comparable to the dispersion of our model orbits before the encounter with
Jupiter of 1780, as can be deduced by the values given in Tables III and
IV. This means that even a well determined orbit like that of P/Shajn-
Schaldach becomes entirely indeterminate after one or two passages within
1 AU from Jupiter. Even for an unrealistic accuracy of the starting orbital elements - by four to five orders of magnitude higher than their
observational uncertainty - a considerable divergence takes place after
five or six encounters with Jupiter.

Of course, these results apply to a particular comet, with a high
value of the Tisserand invariant, and therefore a low relative speed at
encounters with Jupiter. Nevertheless, the results seem to indicate that
in other cases the complete indeterminacy of the orbits can simply take
place in somewhat longer time spans.

ACKNOWLEDGMENTS. The authors are grateful to L. Kresák for critically
reading an earlier version of this paper, and to the referees E. Everhart
and A. Manara for useful suggestions.

REFERENCES

Bettis, D.G., and Szebehely, V.: 1972, in "Gravitational N-Body Problem"
    (M. Lecar ed.), IAU Coll. 10, pp. 388-405.
Carusi, A., Kresák, L., and Valsecchi, G.B.: 1981, Astron. Astrophys. 99,
    262-269.
Carusi, A., Kresák, L., Perozzi, E., and Valsecchi, G.B.: 1984, IAS In-
    ternal Report n. 12.
Carusi, A., Perozzi, E., Pittich, E.M., and Valsecchi, G.B.: 1985, in
    preparation.
Everhart, E.: 1974a, Celestial Mechanics 10, 35-55.
Everhart, E.: 1974b, Denver Research Institute Technical Report.
Marsden, B.G.: 1971, Quart. J. 12, 268.

Marsden, B.G.: 1977, private communication.

Marsden, B.G.: 1978a, Quart. J. 19, 52.

Marsden, B.G.: 1978b, Quart. J. 19, 54.

Nakano, S.: 1984, Nakano Circ. 450.

Newhall, X X, Standish, E.M., and Williams, J.G.: 1983, Astron. Astrophys. 125, 150-167.

Oikawa, S., and Everhart, E.: 1979, Astron. J. 84, 134-139.

Pittich, E.M.: 1981, Bull. Astron. Inst. Czechosl. 32, 340-345.

Roy, A.E., Moran, P.E., and Black, W.: 1972, Celestial Mechanics 6, 468-482.

Schubart, J., and Stumpff, P.: 1966, Veroff. Astron. Rechen-Inst., Heidelberg, n. 18, Karlsruhe.

Sitarski, G.: 1979, Acta Astronomica 29, 401-411.

Szebehely, V., and Bettis, D.G.: 1972, in "Gravitational N-Body Problem" (M. Lecar ed.), IAU Coll. 10, pp. 136-147.

THE ROLE OF THE RESEARCHES OF E.I. KAZIMIRCHAK-POLONSKAYA ON THE DYNAMICAL EVOLUTION OF SHORT-PERIOD COMETS

N.A. Belyaev
Institute of Theoretical Astronomy
Leningrad
USSR

In 1961 Kazimirchak-Polonskaya (1961a, 1961b) published comprehensive reviews of all investigations on the dynamics of cometary orbits as well as on close approaches
of the short-period comets with Jupiter for the time span
covering 1770-1960 and for the first time put forward the
basic problems of cometary astronomy from the standpoint
of celestial mechanics. Chebotarev (1971) pointed out that
"these reviews, supplemented with extensive references,
can serve as a valuable manual for all researchers of cometary motions". In 1967 Kazimirchak-Polonskaya (1967a)
developed the advanced problems in a definitive form. The
corresponding member of the USSR Academy of Science
M.F. Subbotin (at that time the Head of the Institute)
characterized the above mentioned works as "the general
plan for cometary studies in the important branch of cometary astronomy".

This plan incorporated, in particular, construction
of computer program complexes and the development of highly accurate numerical theories of cometary motion for the
entire period of observations taking account of all planetary perturbations and non-gravitational effects as well
as a study of the secular evolution of cometary orbits including the large transformations of these which take place in the spheres of action of Jupiter and the other outer
planets.

The first comprehensive study on the evolution of cometary orbits of numerous group of real comets, taking
account of all planetary perturbations over a time span of
400 yr (1660-2060), containing also both the development
of the capture theory from a completely new perspective
and the presentation of a new hypothesis of cometary origin, was published by Kazimirchak-Polonskaya in 1967 (1967b).

Pages 453-457 of the above mentioned work are devoted
to a critical review of the classical theories of capture
by Jupiter, carried out by Laplace, H. Newton, Callandreau,

A. Carusi and G. B. Valsecchi (eds.), Dynamics of Comets: Their Origin and Evolution, 237–241.
© *1985 by D. Reidel Publishing Company.*

Tisserand. The author evaluates their achievements and reveals the methodological shortcomings which inevitably resulted in discrepancies between the conclusions of these
authors and the observational data. Overcoming these difficulties by her accurate research technique, Kazimirchak-
Polonskaya has shown that owing to a new presentation and
solution of the problem, contradictions of the theory with
observations can be completely eliminated.

In short, the essence of this work is as follows. Kazimirchak-Polonskaya selected a large series of real comets either belonging to various planetary families or
having transplutonian orbits: P/Lexell (1770 I), P/Hersch-
el-Rigollet (1788 II), P/Olbers (1815), P/de Vico-Swift
(1844 I), P/d'Arrest (1851 II), P/Stephan-Oterma (1867 I),
P/Neujmin 3 (1929 III), P/Whipple (1933 IV), P/Oterma
(1942 VII), P/Kearns-Kwee (1963 VIII), etc. The detailed
comprehensive studies of the motion of these comets taking
account of perturbations, as a rule, from eight planets
(Venus-Pluto) over the time span of 400 yr (1660-2060) illustrated by numerous tables, led the author to the following conclusions:

1. "The outer planets (Jupiter-Neptune) with their
vast spheres of action are the powerful transformers of
cometary orbits, essentially determining their evolution,
changing their spatial orientation, shape and dimensions,
transferring comets from one family into another, and in
some cases either removing comets to the periphery of the
planetary system and even outside its boundaries or capturing them from the nearly parabolic orbits" (p.453).
Strong perturbations by Jupiter, particulary following
close approaches, are very often favourable either for discovery of new comets or rediscovery of the lost ones;
sometimes on the contrary they may be a cause for the loss
of comets both temporary and permanent (1967b, pp.453 and
459).

2. The suggested hypothesis of cometary origin represents a combination of a diffusion theory of K.A. Shtejns
(Shtejns, 1960, 1961, 1962, 1964) with the capture theory
of Kazimirchak-Polonskaya.

Integrating the equations of diffusion with due consideration of cometary disintegration as a function of
perihelion distance Shtejns arrives at a formulation of
the laws of diffusion, two of these being essential for
our purposes: (a) comets with smaller values of the semi-
major axes have also smaller inclinations; (b) greater
perihelion distances correspond to smaller eccentricities.

At this point Shtejns' investigations, which he carried out employing statistical methods for nearly parabolic and long period comets with semi-major axes $a > 40$ a.
u., are completed.

The fourth stage of the comets' orbital evolution at

a < 50 a.u. - the capture by the major planets - represents the principal subject matter of Kazimirchak-Polonskaya's studies, performed by the precise numerical methods of celestial mechanics.

Thus, in accordance with her scheme (1967b, pp.458-
460), due to diffusion at the periphery of the planetary
system, there arises a concentration of perihelia of numerous comets, invisible from the Earth with quasi-circular
orbits having small inclinations. The slow motion of these
comets at great heliocentric distances and the vast spheres of activity of the outer planets (particularly of Neptune with radius 0.580 a.u.), favour rather prolonged encounters of comets with these planets as well as great transformations of their orbits. In this way favourable conditions are created for capture of these invisible comets
into the region of the Solar system planets.

Capture according to a new interpretation of Kazimirchak-Polonskaya is a complex process, developing over centuries, millennia and even million of years*. It can start
from the periphery of the planetary system, when Neptune
is capturing a comet (with perihelion in the vicinity of
its orbit) into its sphere of influence and represents
an evolutionary process - a successive translational transfer from one planetary family into the other until it reaches that of Jupiter. But the capture can also proceed
catastrophically either from nearly parabolic orbits or by
means of transferring of a comet from one planetary family
into another which is at a considerable distance from it.
Sometimes a comet (as, for instance, P/Oterma) can be transferred due to perturbing force of Jupiter from one planetary family into the other and then returned back into the
original family. More often the capture represents a complex many-staged process where all stages of an evolutionary
and catastrophic development are either combined or inter-
changed. At the final stage (relative to a terrestrial observer) the cometary orbit approaches the Earth's orbit,
contributing to the discovery of the comet.

In the case of a very deep penetration of the comet

---

* Let it be briefly noted, that in the classic presentation of the problem the capture was considered:
1/ by Jupiter only,
2/ from a parabolic orbit (owing to the interstellar
origin of comets),
3/ as a rule, in the problem of two bodies (outside the
Jupiter sphere of activity: Sun-comet; inside the sphere: Jupiter-comet),
4/ as one single passage through the sphere of activity.
As a consequence of the four points indicated above the
capture turned out to be an exceptionally rare occurence.

into the Jupiter's sphere of influence there arise in a
heliocentric frame of reference, short-term, osculating,
strongly elongated ellipses instantaneous parabolae, and
hyperbolae (for instance, in P/Brooks 2 orbit's transformations in 1886). In exceptional cases, there appear even
short-period osculating ellipses with retrograde motion
and very small perihelion distances (see, for example,
Table 18, illustrating the transformation of P/Lexell's
orbit deep within the Jupiter's sphere of influence in
1779) but on leaving Jupiter's sphere of influence, these
osculating conic sections are swiftly transformed into the
short-period ellipses with direct motion.

Therefore, the short-period comets from Jupiter's family do not leave its sphere along elliptical orbits with
retrograde motion. And if, in exceptional cases, hyperbolic orbits are preserved on leaving this sphere of influence, then their perihelia are situated either on the Jupiter's orbit or in its close vicinity, and that is why
such comets can not be observed from the Earth. Thus, in
Kazimirchak-Polonskaya's capture theory all contradictions
of the classic theories with observational data have been
overcome.

Such is, in brief, a general picture of the capture
and origin of comets outlined by Kazimirchak-Polonskaya
in her paper (1967b).

The results of this work were successfully reported at
the meeting of the IAU Commission 20 during the General
Assembly Session of the International Astronomical Union,
in Prague (1967). After Dr.Kazimirchak-Polonskaya's report,
President of the IAU Commission 20, Prof.G.A.Chebotarev
(at that time the Head of the Institute of Theoretical
Astronomy) proposed that a decision be taken to carry out
a special symposium on the motion, evolution of orbits
and origin of comets, in Leningrad. The proposal was una-
nimously approved and the Symposium was organized and carried out by the Institute of Theoretical Astronomy of the
USSR Academy of Science (Leningrad) in 1970. Up to the
present time it remains the only IAU Symposium on cometary
problems.

A wide horizon and a profound knowledge of the history of evolution of cometary dynamics and cosmogony as well
as her new results in 1967 on the secular evolution of cometary orbits and the capture of comets; her further investigations in this field (Kazimirchak-Polonskaya, 1971)
along with the works dealing with motion and the secular
evolution of meteor showers, accompanied by large transformations of their orbits while passing through the Jupiter
sphere of influence; all this enabled Kazimirchak-Polonskaya to elaborate the entire program of IAU Symposium 45 and
together with G.A. Chebotarev take an active part in its
organization.

G.A. Chebotarev in (1972, p.4) evaluates the results
of Kazimirchak-Polonskaya's works in the following way:
"By studying the motions of a number of comets over the
interval 1660-2060 Kazimirchak-Polonskaya has obtained for
the first time a real picture of the evolution of cometary
orbits". In his paper (1971, p.640) Chebotarev stresses
that "Kazimirchak-Polonskaya's studies have for the first
time provided a convincing basis for the hypothesis of capture of the short-period comets by the major planets and
made necessary a critical revision of hypotheses of cometary origin. Her works represent an important contribution
both into the modern theoretical astronomy and dynamic cosmogony. By a decree of the Presidium of the USSR Academy
of Science of 1969, Jan. 24, E.I.Kazimirchak-Polonskaya
was awarded a prize named after an outstanding comet's researcher A.F.Bredikhin for a series of her works concerned
with the theory of short-period comet motions and the evolution of their orbits".

REFERENCES

Chebotarev, G.A. (1971). Byull. ITA. 12, 639.
Chebotarev, G.A. (1972). IAU Symposium No.45,"The Motion,
Evolution of Orbits, and Origin of Comets", 1.
Kazimirchak-Polonskaya, E.I. (1961a). Trudy ITA. 7,3.
Kazimirchak-Polonskaya, E.I. (1961b). Trudy ITA. 7, 19.
Kazimirchak-Polonskaya, E.I. (1967a). Trudy ITA. 12, 3.
Kazimirchak-Polonskaya, E.I. (1967b). Astron.Zh. 44, 439.
Kazimirchak-Polonskaya, E.I. (1971). Byull.ITA. 12, 796.
Shtejns, K.A. (1960). Uch. Zap. Latv. Gos. Univ. 38, 69.
Shtejns, K.A. (1961). Astron. Zh. 38, 107, 304.
Shtejns, K.A. (1962). Astron. Zh. 39, 915.
Shtejns, K.A. (1964). Uch. Zap. Latv. Gos. Univ. 68, 39.

REVIEW OF STUDIES ON CAPTURE OF COMETS BY NEPTUNE AND ITS ROLE IN THE
DYNAMIC EVOLUTION OF COMETARY ORBITS

E.I. Kazimirchack-Polonskaya

Institute of Theoretical Astronomy
Leningrad
U.S.S.R.

ABSTRACT. A historical review of investigations on encounters of minor
bodies with outer planets is presented. The dominating role of Jupiter
in the capture of comets and in determining their secular evolution is
emphasized. Participation in the above process of Saturn, Uranus and,
especially, Neptune is investigated. In the present paper are considered
five fictitious comets, penetrating into the depths of Neptune's sphere
of action, where they undergo large orbital transformations and capture
by Neptune. Two near-circular orbits are transformed into elliptical
ones, their perihelia being moved inside the planetary system; one hyperbolic orbit is transformed into a transplutonian one, another into a typical short-period orbit of the Neptune's family with a perihelion accessible to terrestrial observers. One fictitious comet is captured by Neptune as a stable satellite.

1. REVIEW OF INVESTIGATIONS ON ENCOUNTERS OF MINOR BODIES WITH OUTER
PLANETS (1967-1984)

Modern investigations on the secular evolution of cometary orbits are
mainly concerned with large transformations of orbits within Jovian
sphere of action and with the dominant role of Jupiter in cometary cosmogony.

Since 1967 the problem has attracted many researchers in the field in
USSR and abroad and has been extensively and thoroughly developed. The
following papers by Kazimirchak-Polonskaya (1967a,b; 1971, 1972, 1976,
1978a, 1982a,b), Kazimirchack-Polonskaya and Shaporev (1976), Belyaev
(1967, 1973a,b), Belyaev and Khanina (1972), Belyaev and Shaporev (1974),
Belyaev and Merzlyakova (1976) deal with large orbital transformations in
Jupiter's sphere of activity as well as with capture and secular orbital

243

A. Carusi and G. B. Valsecchi (eds.), Dynamics of Comets: Their Origin and Evolution, 243–258.
© 1985 by D. Reidel Publishing Company.

evolution (400 yr) of several dozens of the real short-period comets (Lexell, Brooks 2, Wolf, Kearns-Kwee, Schwassmann-Wachmann 3, Oterma, Gehrels 3 etc. being included). On the basis of the data obtained Kazimirchack-Polonskaya (1967, 1972, 1978a) has formulated basic regularities in the multi-staged capture process of comets as well as those occurring in large transformations and during dynamic evolution of cometary orbits on a cosmogonic time scale.

Shtejns studied diffusion of long-period comets taking into account large stellar perturbations of their orbits (1961a,b; 1962) and the dynamic evolution of these (1964). Chebotarev (1972) gave a review of previous investigations on the problems under consideration.

Marsden in his pioneer work (1967) when studying motions of 100 periodic comets over the time span 1725-1965 comes to a conclusion that there exist various types of resonances in the motions of comets and Jupiter, as well as gaps in the Jovian family of comets.

Everhart (1972, 1973, 1976, 1982) on the basis of his large-scale studies of fictitious comets reached very interesting conclusions, especially in paper (1976) concerning the origin of various types of comets.

Kresák (1972, 1974) has initiated a new series of numerous studies dealing with investigations of encounters of fictitious minor bodies with Jupiter. Subsequently this attracted attention of Italian astronomers Carusi, Pozzi, Perozzi and Valsecchi and has been extensively elaborated by Kresák and the above mentioned authors (Carusi and Pozzi, 1978a,b; Carusi, Pozzi, Valsecchi, 1979; Carusi and Valsecchi, 1980; Carusi, Kresák and Valsecchi, 1981; Carusi and Valsecchi, 1981; Carusi, Kresák and Valsecchi, 1982; Carusi and Valsecchi, 1982a). In the first two papers Carusi and Pozzi presented their method for studying the close encounters with Jupiter and applied it for study of 3,000 fictitious minor bodies. In the rest of the above cited papers attention is given primarily to the temporary satellite captures as well as the encounters with Jupiter of those objects whose orbits are nearly tangent to that of the planet. Along with fictitious objects the authors have studied encounters with Jupiter and transformations of orbits also of real comets. Thus Carusi and Valsecchi (1982b) studied the dynamic evolution of 22 short-period comets having low velocity encounters with Jupiter.

Even earlier Rickman (1974) investigated the capture of fictitious comets; Buckley (1977) and Rickman (1979) studied motions and encounters with Jupiter of six new comets discovered during 1974-1977.

Rickman and Karm (1982), and Karm and Rickman (1982), performed an analysis of approaches to Jupiter of all short-period comets prior to their discovery. Vaghi and Rickman (1982) studied orbital evolution of 9 comets which had moved for some time in resonance orbits close to commensurability 2:1.

In this connection the author would also like to draw attention to an interesting investigation by Fernández (1980).

Such are in brief the principal works on close cometary encounters with Jupiter, large transformations and dynamic evolution of orbits both of real and fictitious objects.

None the less to get a more complete picture of the dynamic evolution of cometary orbits it was necessary to reveal, if only in general outline, the dynamics of the Saturn, Uranus and Neptune's spheres of action and to clarify the influence of these planets on the evolution of cometary orbits on a cosmogonic time scale.

The first studies in this field were carried out by the author (Kazimirchak-Polonskaya, 1972). However, insufficiency of material on the motions and observations of the real comets, belonging to the most distant planetary families, forced the author to turn first to fictitious objects. Then, considering their dynamical evolution over very long time intervals, we linked these fictitious objects with the real short-period comets of the Jupiter's family.

Using our method and complex of programs for electronic computers (1972, 1982a), we studied the secular orbital evolution both of fictitious and real comets, belonging to the Saturn and Uranus' families (Kazimirchak-Polonskaya, 1972, 1974, 1976, 1978b) as well as the large orbital transformations in the spheres of activity of these planets. During these investigations the author succeeded to reveal a capture of fictitious comet N-3 (1972) and N-4 (1974, 1976) into the Neptune's satellite orbit, the post-encounter orbit of N-4 being located between those of the two natural satellites of Neptune, namely Triton and Nereid.

Of a great interest (in the author's view) are penetrations of fictitious comets into the depths of the Neptune's sphere of action (1976, 1978b). In her paper (1978b) the author revealed the role of Neptune in the dynamic evolution of cometary orbits on a cosmogonic time scale and treated the problem of cometary origin as well.

The problem of cometary capture by Saturn, Uranus and Neptune has also attracted attention of some astronomers abroad. Thus, Everhart (1976) in his interesting generalizing conclusions notes that some of the short period comets could have been formed inside the orbit of Neptune. He maintains also that orbits of long-period comets of small inclination and perihelia near Jupiter's orbit could have evolved into orbits typical of short-period comets. One phase of their evolution is represented by near-circular orbits located just outside the Jupiter's orbit. The results of our studies have shown that this Everhart's conclusion remains valid if we extend it over unobservable long-period comets of small inclination and perihelia close to the Saturn's, Uranus' and especially Neptune's orbit.

Exceptionally valuable contributions of a very large cosmogonic scale into the above discussed problems of cometary encounters with the four

outer planets have been made in papers: Carusi and Valsecchi (1982c),
Carusi, Perozzi and Valsecchi (1983). The first paper dealt with 1,000
fictitious minor bodies and 4,000 of their approaches to outer planets
which resulted in 46 TSCs (Temporary Satellite Capture) by Jupiter, 16
by Saturn, 4 by Uranus and 4 by Neptune. In a more recent paper (1983)
the authors have investigated the effects of low velocity encounters between fictitious minor bodies and the four outer planets as well as their
consequences. Using definition of the TSC, given by Rickman and Malmort
(1982), the authors on the one hand have established the capture zones
for each outer planet and on the other hand determined, as a consequence,
numerous TSCs which the authors divided into two types respectively. The
other consequences of encounters (according to the authors) are: the exchange of perihelion with aphelion of the minor body orbit and a change
of the minor body semi-major axis from one greater than that of the planet to the smaller one, or vice-versa. These changes due to large perturbations amounted to 117 in Jupiter, 57 in Saturn, 11 in Uranus and 17 in
Neptune's families. Influence of all these processes on the dynamic evolution of the minor body orbits on a secular scale is discussed.

An outstanding completion of all these efforts is a unique Catalogue
by Carusi, Kresák, Perozzi and Valsecchi (1984) containing data of the
secular evolution of 126 comets with a period less than 200 years em-
bracing the time span of 821 yr (1585-2406).

2. CAPTURE OF COMETS BY NEPTUNE AND ITS ROLE IN THE DYNAMIC EVOLUTION OF COMETARY ORBITS

From the author's point of view (1972, 1976, 1978b) of a particular interest are penetrations of comets into the depths of the Neptune's sphere
of activity. Owing to the extent of this sphere, the slow motion of comets and the great heliocentric distance, such comets can dwell in the
Neptune's sphere of activity for about 10-15-20... yr. Below are given
several of the most interesting closest encounters of fictitious comets
with Neptune and large transformations of their orbits.

The author (1972, 1976, 1978b) has treated two types of comets: 1)
those entering Neptune's sphere of action on near-circular orbits, which
can arise owing to diffusion of the long-period comets, and 2) those penetrating this sphere on interstellar hyperbolic orbits.

1. Thus, comet N-2 (1972, Fig. 15) is entering the Neptune sphere of
action along a near-circular orbit (e=0.029) with a revolution period of
169 yr and perihelion near the orbit of Neptune. The comet penetrates
deeply this sphere to a minimum distance $\Delta N_{min}=0.00093$ AU and leaves it
after a large transformation on an elliptic orbit (e=0.170) with revolution period 132 yr, the perihelion near Uranus orbit and the aphelion

located in close vicinity to the Neptune's orbit; the apsidal line makes
a forward turn of 161°.

Even more large orbital transformation, that of comet N-5, is represented in Fig. 14 (1976) and in Fig. 2 (1978b, in this paper comet N-5
received the designation "Neptunian 2"). We reproduce also here this
characteristic transformation of the orbit (see Fig. 1). Comet N-5 is

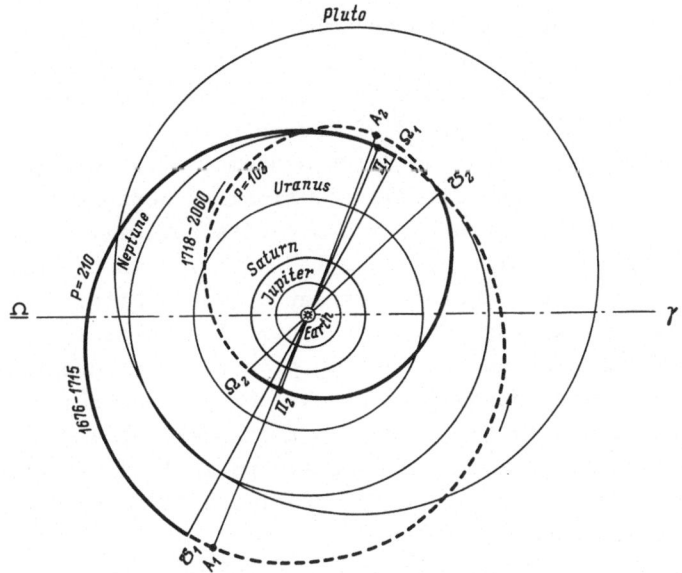

Figure 1. Capture of the N-5 comet by Neptune and transformation of its
orbit within Neptune's sphere of action. The part of cometary orbit located under the ecliptic is always designated by a dotted line.

moving at the periphery of the planetary system on an orbit having eccentricity e=0.157, its perihelion on the Neptune's orbit, its aphelion far
beyond this orbit, the inclination being 8.7° and the revolution period
210 yr. The comet penetrates Neptune's sphere of action even more deeply
than comet N-2, to ΔN_{min}=0.00042 AU, and having done several revolutions
around the planet leaves it on an orbit which is even more elliptic than
before (e=0.434) with inclination 22.1°, revolution period 103 yr, perihelion between the orbits of Uranus and Saturn and aphelion near the or
bit of Neptune. The whole process of the evolution and catastrophic
transformation of the comet N-5 orbit as well as those of each orbital
element are shown over the time interval 1676-2058 in Table 2 of paper
(1978b).

These two large transformations of the orbits of comets N-2 and N-5
in the depths of the Neptunian sphere of action have the following features in common: the ellipticity of an orbit is increased, its revolu-

tion period is reduced, the new aphelion is located in the vicinity of an earlier perihelion near the orbit of Neptune, whereas the perihelion of the transformed orbit has moved inwards to the planetary system in a heliocentric direction, but both comets remain invisible from the Earth on this stage of their orbital evolution. This process may be considered as the first stage of cometary capture by Neptune.

The further dynamic evolution of the cometary orbit of N-5 (and similarly that of comet N-2) may proceed quite differently. For instance, after a long time (several millennia) the comet N-5 may repeatedly undergo the closest approaches with Neptune and the planet can either transfer it back to an orbit close to the original one (see for realization of such possibility Fig. 13 of the paper of 1972) or remove it beyond the planetary and even the Solar system limits (we have integrated several variants like the cited), or, finally, Neptune can increase the ellipticity of the comet orbit and move its perihelion still farther inside the planetary system, retaining its aphelion in the vicinity of Neptune's orbit. However, another variant of the orbital evolution of the comet is also quite feasible: over a very long time interval, under the influence of the continuously acting planetary perturbations, the cometary perihelion can approach to Uranus' orbit, and the comet may at some time penetrate deeply into the planet's sphere of action. An alternative situation may arise: the cometary perihelion may approach Saturn's orbit and then at a known period of time a close encounter between comet and Saturn will occur.

Then a new stage of the capture of comet N-5 by Saturn will come into effect, the comet leaving its sphere of action in a transformed orbit, whose aphelion will be close to the previous perihelion; the comet will belong to Saturn's family and its perihelion will advance to Jupiter's orbit and will be located either on the inner or outer side of it. At this stage of evolution, the final step of the capture of the comet by Jupiter into its family will occur; the cometary aphelion will be located near the orbit of Jupiter and the perihelion will appear in the region of visibility and thus the comet will be discovered by a terrestrial observer.

Such is a rough scheme of the dynamical evolution of cometary orbits in the form of multi-stage capture of the first type of comet.

At this stage we have closely advanced to the real short-period comets. Namely in this way a number of newly discovered comets were captured from the Saturnian to Jovian families: P/Oterma in 1937-1938, P/Gehrels 3 in 1970 etc. Likewise the whole series of comets (P/Schwassmann-Wachmann 3 in 1882, P/Brooks 2 in 1886, P/West-Kohoutek-Ikemura in 1972 etc.) were captured into Jupiter family from the boundary of the two planetary families Saturn-Uranus.

2. Kazimirchak-Polonskaya (1976, 1978b) outlined the two types of

the cometary capture by Neptune from hyperbolic orbits, whose perihelia
(influenced by stellar perturbations or due to other reasons) have pene-
trated into the periphery of the planetary system and are located between the orbits of Neptune and Uranus. A catastrophic transformation
of the orbit of comet N-6 in the Neptune sphere of action is depicted in
paper (1976) and Fig. 15. In paper (1978b) this comet was designated
"Neptunian 1"; its dynamic evolution over the time span 1710-2060 is represented there in Fig. 1 and Tab. 1. The comet N-6 (see Fig. 2) enters

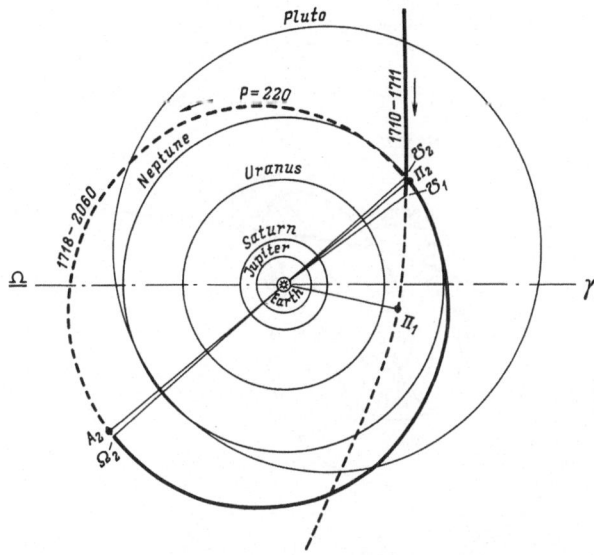

Figure 2.   Capture by Neptune of the interstellar comet N-6 and its
transformation into an unobservable periodic comet.

the Neptune sphere of action on a hyperbolic orbit with retrograde motion
(indicated by an arrow in Fig. 2) and with orbital inclination 135° prior
to perihelion passage Π_1. Its penetration into the depths of the Neptune
sphere of action attaining ΔN_{min}=0.00046 AU, produces as usual a motion
of the comet along a temporary satellite orbit. After a catastrophic
transformation of the N-6 orbit in the course of the first stage of capture by Neptune, the comet leaves its sphere of action on a near-circular orbit (e=0.181) with a direct motion (i=16.0°), the period of revolution reaching nearly 220 yr, the aphelion being in the transplutonian
area and the perihelion in close vicinity to Neptune's orbit (q=29.8 AU);
this betokens a future (though a very remote one) inevitable penetration
of the comet into Neptune's sphere of action and new large orbital transformations.
    Comparing the orbit of comet N-6 after it had left this sphere with

that of comet N-5 before it entered the sphere, one can see that these
are of the same type. Therefore, further stages of the N-6 orbit as
well as the new stages of its capture by Neptune and the other outer
planets take a similar course as outlined above for the comet N-5.

Comet N-7 represents an exceptionally interesting object (see Fig. 3).
Capture of the comets N-7 and the large transformation of its orbit in
the Neptune sphere of action are depicted in Fig. 16 (1976) and in Fig.
3 in paper (1978b) where the comet was designated as "Neptunian 3". It
penetrates the Neptune's sphere of action, after perihelion passage on

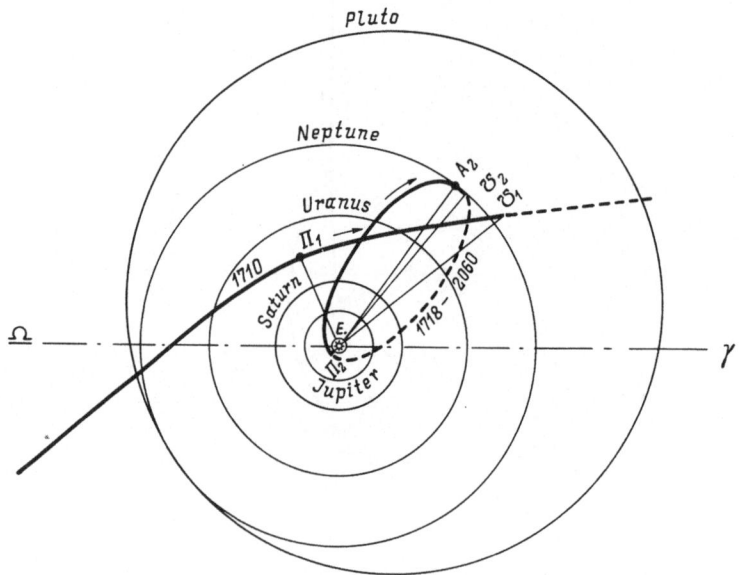

Figure 3.   Capture by Neptune of the N-7 comet from a hyperbolic orbit
and its transformation into a short-period retrograde comet of Neptune's
family.

a hyperbolic orbit with retrograde motion (i=159°), and attains ΔN_{min} =
0.00035 AU. It implies that the comet dashes through perineptunian at
a distance of about one Neptune's diameter from its surface. This case
resembles the closest approach of P/Brooks 2 to Jupiter in 1886. In the
depths of the Neptune's sphere of action within the powerful gravity
field of the planet, a one-stage capture of the comet N-7 by Neptune and
catastrophic transformation of its orbit are accomplished: the interstellar comet leaves the Neptune's sphere of action as a typical short-period comet of Neptune's family. Its orbit represents a stable (that is
until a new close encounter with Neptune, Jupiter or some other outer
planet), strongly elongated ellipse with a retrograde motion (i=136.5°),
the period of revolution being about 62 yr, aphelion near the orbit of

Neptune and perihelion distance 1.2 AU, and consequently perihelion is situated in close vicinity to Earth's orbit. At this stage of evolution the comet becomes accessible to terrestrial observers.

The orbital elements of the comet N-7 on leaving the Neptune sphere of action are very much like those of real comets of Neptune's family (see Table 1).

TABLE 1.   Comparative characteristics of orbits of N-7 and other short-period comets of Neptune's family. q and Q = perihelion and aphelion distances. e,q,P,i: Marsden (1982); a,Q: calculated by Kazimirchak-Polonskaya

Comet	a (AU)	e	q (AU)	Q (AU)	P (yr)	i (°)
N-7 (1985)	15.7	0.92	1.23	30.2	62.3	136
P/Pons-Gambart (1827 II)	14.9	0.95	0.81	29.0	57.5	136
P/Westphal (1852 IV)	15.5	0.92	1.25	29.8	61.2	41
P/Dubiago (1921 I)	15.7	0.93	1.11	30.3	62.3	22

The dynamic evolution of the comet over the period 1710-2056 is presented in Table 4 of the paper (1978b).

On the basis of the given figures and tables one can draw the following conclusions (Kazimirchak-Polonskaya, 1978b, pp. 411-413).

1. Neptune, due to its greatest heliocentric distance and its vast sphere of action participates actively in the capture of comets and like Jupiter is a powerful transformer of cometary orbits, substantially determining their dynamic evolution on a cosmogonic scale.

2. Capture of comets from the periphery of the planetary or Solar system (or even from the interstellar space), and their introducing into a region of visibility, represents a very complicated multi-stage process taking place over centuries, millennia and even millions of years in which all the outer planets participate (particularly Jupiter and Neptune), owing to the powerful dynamic transformations which occur deep within their spheres of action.

3. It is important, however, to point out the essential differences in the dynamics of the sphere of action of Jupiter and those of the other outer planets. Jupiter with its powerful mass is exerting a strong perturbing action on the cometary orbit long before it enters its sphere of action and for a long time after leaving it. Neptune having a significantly lesser mass is devoid of such a strong transforming influence

even within its sphere of action until the comet reaches $\Delta N < 0.1$ AU.
But when a comet is sweeping past at a short distance from the Neptune's
surface, its gravity influence produces large or even catastrophic transformations of the cometary orbit.

   Analogous observations (although with some modifications) apply also
to dynamic transformations and the evolution of cometary orbits when comets penetrate into the depths of the Saturn's and Uranus' spheres of
action.

   4. On the basis of all her investigations the author established
that beyond the Neptune's orbit and between the orbits of Neptune and
Uranus as well as between those of Uranus and Saturn, Saturn and Jupiter
there exist vast belts (or reservoirs) containing perihelia of the numerous unobservable comets.

   Majority of these comets in the multi-stage capture by Neptune-Jupiter
are gradually moving to the Sun and in vicinity of their perihelia they
get into the region of visibility and owing to this situation the comets
are discovered as new short-period comets. Existence of these belts is
substantiated by studies of Whipple (1972) and Carusi, Perozzi and Valsecchi (1983).

3.   CAPTURE OF A COMET ON THE NEPTUNE'S SATELLITE ORBIT

Finally we present in Fig. 4 the unusual orbit of comet N-4 captured by

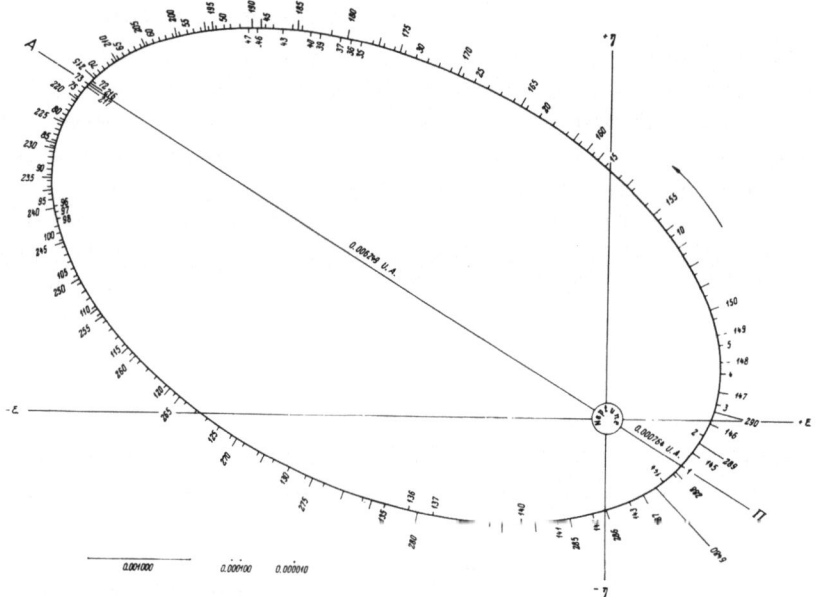

Figure 4.   Comet-Satellite N-4 captured by Neptune.

Neptune as a stable satellite having an orbit intermediate between those of Neptune satellites, Triton and Nereid (see Table 2). The detailed catastrophic transformation of this extraordinary comet in the depths of Neptune's sphere of action is examined in the author's paper (1974) and presented also in (1976, Fig. 17, Tab. 2).

TABLE 2.   Characteristics of orbits of Neptune's satellites: Triton, Nereid and N-4

Satellite	Mean distance from Neptune (10^3 km)	Δ_{min} (10^3 km)	Δ_{max} (10^3 km)	Sidereal period (d:hr:min:sec)	e^1	i^2
Triton	354	354	354	5:21:02:39	0.00	132.79°
N-4	524	91	957	11:05:31:12	0.73	60.87°
Nereid	5570	1337	9803	359:09:36	0.76	4.97°

1: eccentricity of mean orbit
2: inclination relative to ecliptic

REFERENCES

Belyaev, N.A.: 1967, Sov. Astron. Zh., 44, 461–470.
Belyaev, N.A.: 1973a, Problemy Kosmicheskoj Fiziki, 8, 112–117.
Belyaev, N.A.: 1973b, Trudy Astron. Obs., Kazań, 39, 102–109.
Belyaev, N.A., Khanina, F.B.: 1972, The Motion, Evolution of Orbits and Origin of Comets (G.A. Chebotarev, E.I. Kazimirchak-Polonskaya and B.G. Marseden eds.), Reidel, Dordrecht, 167–172.
Belyaev, N.A., Shaporev, S.D.: 1975, Problemy Kosmicheskoj Fiziki, 10, 9–16.
Belyaev, N.A., Merzlyakova, M.A.: 1976, Byull. Inst. Teor. Astron., 14, 206–209.
Buckley, R.J.: 1977, J. Brit. Astron. Assoc., 87, 444–456.
Carusi, A., Pozzi, F.: 1978a, Moon and Planets, 19, 65–70.
Carusi, A., Pozzi, F.: 1978b, Moon and Planets, 19, 71–87.
Carusi, A., Pozzi, F., Valsecchi, G.B.: 1979, Dynamics of the Solar System (B.L. Duncombe ed.), Reidel, Dordrecht, 185–189.
Carusi, A., Valsecchi, G.B.: 1980, Moon and Planets, 22, 113–124.
Carusi, A., Kresák, L., Valsecchi, G.B.: 1981, Astron. Astrophys., 99, 262–269.

Carusi, A., Valsecchi, G.B.: 1981, Astron. Astrophys., 94, 226–228.

Carusi, A., Kresák, Valsecchi, G.B.: 1982, Bull. Astron. Inst. Czechosl., 33, 141–150.

Carusi, A., Valsecchi, G.B.: 1982a, Sun and Planetary System (W. Fricke, G. Teleki eds.), Reidel, Dordrecht, 379–384.

Carusi, A., Valsecchi, G.B.: 1982b, Comparative Study of the Planets (A. Coradini, M. Fulchignoni eds.), Reidel, Dordrecht, 131–148.

Carusi, A., Valsecchi, G.B.: 1982c, Sun and Planetary System (W. Fricke, G. Teleki eds.), Reidel, Dordrecht, 385–388.

Carusi, A., Perozzi, E., Valsecchi, G.B.: 1983, Dynamical Trapping and Evolution in the Solar System (V.V. Markellos, Y. Kozai eds.), Reidel, Dordrecht, 377–395.

Carusi, A., Kresák, L., Perozzi, E., Valsecchi, G.B.: 1984, IAS-Internal Report n. 12, Rome.

Chebotarev, G.A.: 1972, The Motion, Evolution of Orbits and Origin of Comets (G.A. Chebotarev, E.I. Kazimirchak-Polonskaya, B.G. Marsden eds.), Reidel, Dordrecht, 1–5.

Everhart, E.: 1973, The Motion, Evolution of Orbits and Origin of Comets (G.A. Chebotarev, E.I. Kazimirchak-Polonskaya, B.G. Marsden eds.), Reidel, Dordrecht, 360–363.

Everhart, E.: 1973, Astron. J., 78, 316–328.

Everhart, E.: 1976, The Study of Comets (B. Donn, M. Mumma, W. Jackson, M. A'Hearn, R. Harrington eds.), NASA SP-393, 445–464.

Everhart, E.: 1982, Comets (L.L. Wilkening ed.), Univ. Arizona Press, Tucson, 659–664.

Fernández, J.A.: 1980, Mon. Not. R. astr. Soc., 192, 481–491.

Karm, J., Rickman, H.: 1982, Bull. Astron. Inst. Czechosl., 33, 359–370.

Kazimirchak-Polonskaya, E.I.: 1967a, Sov. Astron. Zh., 44, 439–460.

Kazimirchak-Polonskaya, E.I.: 1967b, Trudy Inst. Teor. Astron., 12,63–85.

Kazimirchak-Polonskaya, E.I.: 1971, Byull. Inst. Teor. Astron., 9, 796–812.

Kazimirchak-Polonskaya, E.I.: 1972, The Motion, Evolution of Orbits and Origin of Comets (G.A. Chebotarev, E.I. Kazimirchak-Polonskaya, B.G. Marsden eds.), Reidel, Dordrecht, 373–397.

Kazimirchak-Polonskaya, E.I.: 1974, Asteroids, Comets, Meteoric Matter (C. Cristescu, W.J. Klepczynski, B. Milet eds.), Academ. Republ. Social. România, 205–221.

Kazimirchak-Polonskaya, E.I.: 1976, The Study of Comets (B. Donn, M. Mumma, W. Jackson, M. A'Hearn, R. Harrington eds.), NASA SP-393, 490–536.

Kazimirchak-Polonskaya, E.I., Shaporev, S.D.: 1976, Sov. Astron. Zh., 53, 1306–1314.

Kazimirchak-Polonskaya, E.I.: 1978a, Problemy issledovaniya Vselennoj, 7, (Akad. Nauk SSSR), 340–383.

Kazimirchak-Polonskaya, E.I.: 1978b, Problemy issledovaniya Vselennoj,
    7, (Akad. Nauk SSSR), 384-417.
Kazimirchak-Polonskaya, E.I., Chernykh, N.S.: 1980, Sov. Astron. Zh.,
    57, 378-386.
Kazimirchak-Polonskaya, E.I.: 1982a, Trudy Inst. Teor. Astron., 18,3-77.
Kazimirchak-Polonskaya, E.I.: 1982b, Trudy Inst. Teor. Astron., 18, 91-
    106.
Kazimirchak-Polonskaya, E.I.: 1982c, Byull. Inst. Teor. Astron., 15, 213-
    216.
Kresák, L.: 1972, The Motion, Evolution of Orbits and Origin of Comets
    (G.A. Chebotarev, E.I. Kazimirchak-Polonskaya, B.G. Marsden eds.),
    Reidel, Dordrecht, 503-514.
Kresák, L.: 1974, Asteroids, Comets, Meteoric Matter (C. Cristescu, W.J.
    Klepczynski, B. Milet eds.), Academ. Republ. Social. România, 193-203.
Marsden, B.G.: 1967, Science, 155, 1207-1213.
Marsden, B.G.: 1982, Catalogue of Cometary Orbits, Cambridge, Mass., USA.
Rickman, H.: 1974, Asteroids, Comets, Meteoric Matter (C. Cristescu, W.J.
    Klepczynski, B. Milet eds.), Academ. Republ. Social România, 187-191.
Rickman, H.: 1979, Dynamics of the Solar System (R.L. Duncombe ed.),
    Reidel, Dordrecht, 293-298.
Rickman, H., Karm, J.: 1982, Sun and Planetary System (W. Fricke, G. Te-
    leki eds.), Reidel, Dordrecht, 389-390.
Rickman, H., Malmort, A.M.: 1982, Sun and Planetary System (W. Fricke,
    G. Teleki eds.), Reidel, Dordrecht, 395-396.
Shtejns, K.A.: 1961a, Sov. Astron. Zh., 38, 107-114.
Shtejns, K.A.: 1961b, Sov. Astron. Zh., 38, 304-309.
Shtejns, K.A.: 1962, Sov. Astron. Zh., 39, 915-920.
Shtejns, K.A.: 1964, Uch. Zap. Latv. Gos. Univ., 68, 39-64.
Vaghi, S., Rickman, H.: 1982, Sun and Planetary System (W. Fricke, G.
    Teleki eds.), Reidel, Dordrecht, 391-394.
Whipple, F.L.: 1972, The Motion, Evolution of Orbits and Origin of Co-
    mets (G.A. Chebotarev, E.I. Kazimirchak-Polonskaya, B.G. Marsden eds),
    Reidel, Dordrecht, 401-408.

CONCERNING THE PAPERS BY R. DVORAK AND C. MARCHAL ON THE CAPTURE OF A COMET ON THE NEPTUNE'S SATELLITE ORBIT

The results of the investigations by Kazimirchak-Polonskaya (1974) were
disputed by R. Dvorak (1976, Astron. Astrophys., 49, 293-298) and R.
Dvorak and C. Marchal (1978, Astron. Astrophys., 69, 373-374).

Before passing to comments on these two papers it should be emphasized that the capture of a comet in the Neptune's sphere of action on the

Satellite orbit is an exceptionally rare occurrence. The author (1974)
intended to determine such initial system of elements of a fictitious
comet at the epoch of its entry into the Neptune's sphere of action
which would allow realization of such an exclusive case. The system of
N-4 elements at the epoch T_o=1726, March 28.0 UT satisfies the required
condition (Table 1). Numerous experiments carried out by the author
have shown that even the slightest changes although of a single element
of the system prevent the capture of a comet on the satellite orbit of
Neptune. Computations were performed with consideration of perturbations by eight planets (Venus-Pluto) using a specially devised method
and complex of programs with increased precision (18 significant digits)
constructed by the author as well as her stock of coordinates of major
planets for 400 yr. The method and complex of programs are briefly giv-
en in (1972) and a more detailed presentation may be found in (1982).
All results on N-4 orbital transformations in Neptune's sphere of action
and those pertaining to its transition on the satellite orbit are presented in Tables I-IV and Fig. 1 (1974) but unfortunately due to purely
technical reasons with the restricted number of significant digits. In
1974 the Institute of Theoretical Astronomy has moved to a new place of
residence and at that time a fraction of the computer archives of the
published papers was destroyed and among those – computations pertaining
to the paper (1974). Therefore the author is not in a position to re-
establish the initial system of elements in the original form. R. Dvorak
(1976) made up his mind to verify my results, proceeding from the calculated by him heliocentric elements of N-4 for the epoch t_o=1738 Sept.
25.14 UT, i.e. for the moment of the comet transition into a stable satellite orbit around Neptune. In his studies of motion of the comet and
when calculating coordinates of the major planets, Dvorak employed the
n-body program by Shubart and Stumpff with precision to 13 significant
digits introducing, according to his judgement required additions and
changes conformably to the problem under consideration. Besides Dvorak
utilized also the other unpublished materials. Having performed a very
labour-consuming and initiative work Dvorak improved the initial system
of elements for t_o, computed on its basis the neptunocentric coordinates
of the comet $\bar{\xi}$, $\bar{\eta}$, $\bar{\zeta}$ for duration of 11 days, compared these with those
of Kazimirchak-Polonskaya ξ, η, ζ (Table 3) and noted that the compared
coordinates were in good agreement. Proceeding from the elements obtained for the epoch t_o, he integrated the equations of the comet motion
back to the epoch t_1=1738 July 2.6 UT and compared his results with those
of the author (Table 4). He found large discrepancies between the results.

From our point of view these discrepancies are due to the following
causes: 1. Discrepancies between the results in the sixth decimal point
(AU) (Table 3) for the period of 11 days: $\xi-\bar{\xi}$=+45, $\eta-\bar{\eta}$=+56, $\zeta-\bar{\zeta}$=-42 are

tōo big, which is a direct consequence of the disagreement between the initial systems of elements of N-4 for the epoch t_o of Dvorak and the author. 2. Computations of both researchers were performed proceeding from differing initial conditions, using different methods and programs of various precision as well as coordinates of major planets obtained in different ways. No wonder that such verification of the author's results in a very complicated and difficult problem was unsuccessful which was admitted by Dvorak himself (1976, p. 297).

However, distrusting "the surprising" results of the author and seeking to find errors in her calculations, R. Dvorak in cooperation with C. Marchal undertook a new investigation (1978) of a more general theoretical nature. As its basis they assumed a priori statement that a duration of a capture of a fictitious comet on the planetary satellite orbit and a duration of an escape of the natural satellite of this planet from its stable orbit are equal. It is obvious both from the title of the paper: "Duration of escape or capture of a satellite", and from its content.

Correctness of such fundamental, non-evident statement should first have been proved. The authors have not done that and could not have done that since this statement is at variance with real facts for the two above mentioned processes: the process of capture of a fictitious comet and that of an escape of the natural satellite of the planet, are substantially different.

In fact, all prominent researchers of large transformations of cometary orbits in Jupiter's sphere of action (N. Belyaev, A. Carusi, A. Dubiago, E. Everhart, L. Kresák, M. Kamienski, B. Marsden, H. Rickman, G. Valsecchi and others) well know that these grand transformations due to Jupiter's attraction are swiftly realized and have duration of several years (rarely decades) and more often several months. This regularity was proved by numerous examples in exceptionally interesting studies by A. Carusi, E. Perozzi, G.B. Valsecchi (1983) as well as by a grand scale Catalogue (A. Carusi, L. Kresák, E. Perozzi, G.B. Valsecchi, 1984). When modelling transformations of cometary orbits the analogous velocity occurs in Neptune's sphere, specifically in transformation of N-4 orbit.

On the contrary, it is known that an escape of a natural satellite of the major planet from its stable orbit (if it is at all possible) is an exceptionally slow process and it is realized during many millennia and even millions of years.

Such misconception of the substance of a capture process of the comet as well as an erroneous formulation of the problem based on the wrong a priori statement have inevitably led R. Dvorak and C. Marchal to a fantastically unlikely conclusion that Kazimirchak-Polonskaya (1974) made an appalling error, having determined a duration of the capture of N-4 "of about 100 days instead of 135 centuries" (1978, p. 374).

By the way Dvorak and Marchal have made another mistake: the capture of comet N-4, i.e. a grand transformation of its orbit in the Neptune's aphere of action and transition of the comet on the satellite orbit of the planet had been realized not over a period of 100 days as stated by these authors, but over a period of about 12 years (from 1726 to 1738), as shown in the paper by Kazimirchak-Polonskaya (1974, Tables I-III).

It should be noted, in conclusion, that a high accuracy of the method and complex of programs by the present author have been repeatedly verified when studying the motion and comparing the calculations with observations of a series of short-period comets. Thus, for instance, in the paper by Kazimirchak-Polonskaya (1982) was presented a numerical theory of P/Wolf (1884 III) motion over 100 years (1884-1984) taking into account perturbations caused by 9 planets (Mercury-Pluto), nongravitational effects in all elements, as well as influence of minor planets and the Galilean satellites of Jupiter, Ganymede and Callisto, during a deep penetration of the comet into Jupiter's sphere of activity in 1922. This numerical theory made it possible a good representation of 1500 observations of the comet over 100 yr period. Using the elements and ephemeris of the author, the observer G.B. Gibson of the Mount Palomar Observatory was able to rediscover the comet as an object of 20^m with following differences: 1983 Aug. 123479, $\Delta\alpha\cos\delta = +0.2''$, $\Delta\delta = -0.0''$ (Kazimirchak-Polonskaya, 1983, Kometn. Tsirc., Kiev, 316, 1-2) long before its passage through perihelion in 1983.

SECTION V

DYNAMICS OF COMETS

STATISTICAL AND NUMERICAL STUDIES OF THE ORBITAL EVOLUTION OF SHORT-PE-
RIOD COMETS

A. Carusi and G.B. Valsecchi

IAS-Reparto di Planetologia, C.N.R.
Viale Università 11, 00185 Rome, Italy

ABSTRACT.   The evolutionary paths connecting cometary reservoirs in the
outer solar system and the population of periodic comets involve stellar
and planetary perturbations. Close planetary encounters play a special
role, making the dynamical evolution slow or fast depending on the orbital elements of the comet at various stages of the process. In this
paper numerical work done in the last decades on this subject, using
both real and fictitious objects, is reviewed.

1.   INTRODUCTION

It is well known that the lifetime of comets in the inner solar system
is limited to very much less than the age of the system itself both physically, because of progressive gas and dust loss from the nucleus, and
dynamically, due to the instability of their motion against ejection on
hyperbolic orbits; in fact, these arguments hold for all comets, no matter if of long or short period, and the conventional explanation for the
mere fact that we do observe comets is that reservoirs sufficient for
the replenishment of both cometary populations exist in the outer solar
system (Oort cloud, Uranus-Neptune planetesimals).

 The formation of the Oort cloud is reviewed in this book by Fernandez, and its dynamical evolution is discussed by Weissman. Doubts have
been recently cast on its survival over the age of the solar system (Napier and Staniucha, 1982) because encounters with giant molecular clouds
would strip comets from the cloud repeatedly and with great efficiency,
but Bailey (1983a) has shown that an appropriate revision of the extension and especially the structure of the cloud could ensure the necessary
stability (more on this can be found, in this book, in the above mentioned papers by Fernandez and Weissman, and also in those by Clube and Napier).

A. Carusi and G. B. Valsecchi (eds.), Dynamics of Comets: Their Origin and Evolution, 261–278.
© 1985 by D. Reidel Publishing Company.

Planetesimals scattered to orbits half-way between the planetary region and the Oort cloud by Uranus and Neptune during their growth have been proposed by Fernandez and Ip (1983) as a likely source of short-period comets; this suggestion seems to support the above mentioned results by Bailey (1983a) on the structure of the cometary cloud necessary to ensure its survival over the age of the solar system (Bailey, 1983b).

In this review we will first outline the dynamical channels connecting the proposed reservoirs to periodic comets; there have been many studies on various aspects of this problem, using different techniques, and a general framework has emerged, although still needing improvement. We will then examine the numerical work on strong perturbations at close encounters with the giant planets and their consequences on the orbits of short-period comets.

2. FROM THE OORT CLOUD TO PERIODIC COMETS

To fix ideas about the evolutionary channels between Oort cloud and periodic comets, we may refer to the schematic representation of the solar system given in Figure 1. In its left half a plot of the quantity $-1/a$, proportional to the orbital energy per unit mass, versus e, the orbital eccentricity, is given; it covers almost the whole solar system, from the asteroid belt out to hyperbolic orbits, but obviously lacks the information relative to the angular elements. Note that the regions ($0 <$ $e < 1$; $0 < -1/a$) and ($1 < e$; $-1/a < 0$) are forbidden by definition. The plot on the right has the same ordinate, whereas the abscissa is the Tisserand invariant J, given by

$$J = A/a + 2(a/A(1-e^2))^{1/2}\cos(i) \tag{1}$$

(where A is the semimajor axis of the planet perturbing the comet, Jupiter in the case of Figure 1b, and the inclination is with respect to the orbital plane of the planet).

The value of J is related to the unperturbed planetocentric velocity at encounter of the comet U (in units of the planet's orbital velocity) by

$$U = (3-J)^{1/2} \tag{2}$$

The above formulae refer to a circular planetary orbit; in that case J, and hence U, is conserved if the comet is perturbed, even strongly, by the planet under consideration.

Opik (1972) discusses, semi-analytically, the effects of elliptic planetary orbits, which introduce a systematic decrease of J as a conse-

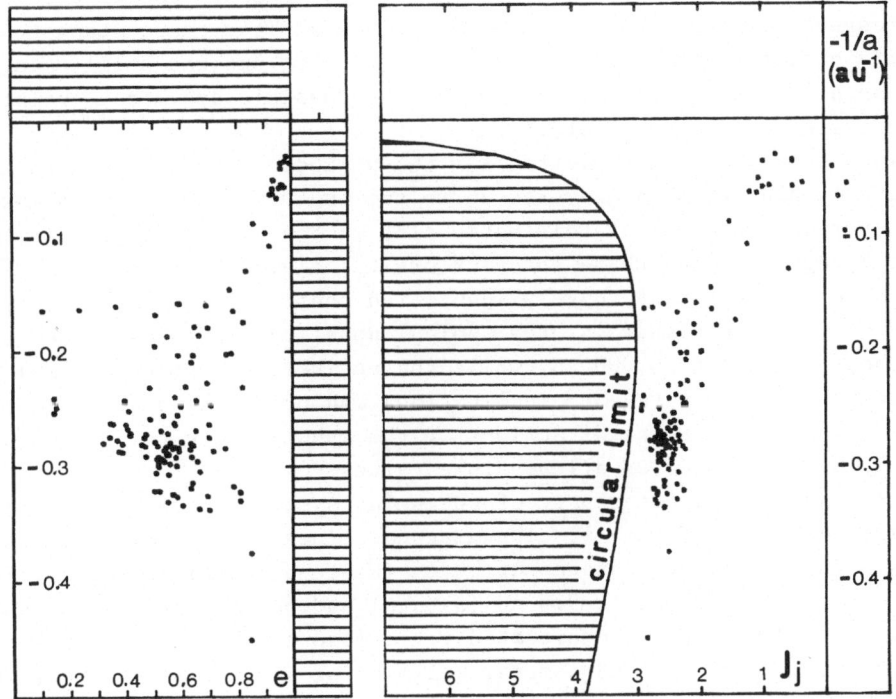

Figure 1. Left: The orbital energy (-1/a) of all periodic comets discovered until the end of 1983 is plotted versus their orbital eccentricity e. Right: The same quantity is plotted versus the Tisserand invariant with respect to Jupiter J_j. Forbidden regions are shaded in both plots.

quence of close encounters; he tabulates the number of revolutions of the small body after which a given decrease of the Tisserand invariant is to be expected. Experimental work, based on long integrations in the restricted elliptical three-body problem, would be very useful both to confirm Opik's results and to extend them to cases in which the orbit of the small body does not cross that of the planet, but still allows close encounters.

From Opik's results one can see that, in the time scales typical for the orbital evolution of periodic comets, the changes in J due to the eccentricity of the controlling planet are rather small; Kresák (1972) has shown that they are anyway much smaller (20 times less, on the average) than those in orbital energy undergone in the same time span by observed short-period comets. Also the error made by using the inclination with respect to the ecliptic is normally negligible; the cause of the greatest changes in J relative to a planet are close encounters with

another one, when they are possible. This prevents the use of J computed
with respect to a specific giant planet, even Jupiter, as a means to
classify orbits permanently (Everhart, 1979), but one can still recognize, on a plot like that of Figure 1b, the possible dynamical evolutionary tracks of comets (see later).

In Figure 1 are also plotted all the periodic comets listed in Marsden (1982), plus those discovered up to the end of 1983. We will make
use of various versions of this basic plot in the following discussions.

The region in the phase space of orbital elements filled by the
Oort cloud comets can be roughly identified considering that the outer
radius of the cloud (that is, the typical aphelion distances of new comets) should be about 50,000 AU, while the perihelion distances should be
outside the planetary region; inclinations and orientations of the orbital planes, as well as the directions of the aphelia, are thought to be
randomly distributed over all possible values.

In Figure 2 the outer part of the solar system is represented, in
the same fashion of Figure 1. The diagonal lines in Figure 2a represent
orbits with perihelion distance of 30 AU (Neptune's orbital radius, leftmost line), 19 AU (Uranus), 9.5 AU (Saturn) and 5.2 AU (Jupiter, rightmost line). New comets are situated very close to $-1/a = 0$ and $e = 1$,
while unobserved Oort cloud comets, still being very close to the line
$-1/a = 0$, are displaced more to the left, not penetrating into the planetary region.

Stars passing through the cloud can impart to the comets small displacements in all directions (but not into the forbidden regions!).
Therefore, if pushed towards the upper right of Figure 2a, through $e = 1$
and $-1/a = 0$, a comet will be lost to interstellar space; if the displacement is horizontal and to the left, it will remain in the cloud,
with a larger perihelion distance; finally, if moved downwards, it may
enter the region below one or more diagonal lines of Figure 2a.

At this stage the further evolution depends essentially on the orbital elements of the comet. In fact the orbit now crosses those of one
or more planets and, for suitable values of the angular elements, planetary encounters become possible. These, unlike the stellar perturbations
mentioned before, can move comets in the phase space only along peculiar
tracks, which depend on the planet involved, and are constrained by the
approximate conservation of the Tisserand invariant relative to the planet with which the comet interacts (Kresak, 1982). These tracks are represented by vertical straight lines in Figure 2b, in the case of Jupiter;
interactions with other planets give other straight lines, some of which
are also shown in Figure 2b. They refer to $J = 2$ and $J = 2.83$ for all
four outer planets.

Using (2) we see that $J = 2$ means that at the encounter the comet's
velocity relative to the planet is as large as the heliocentric velocity

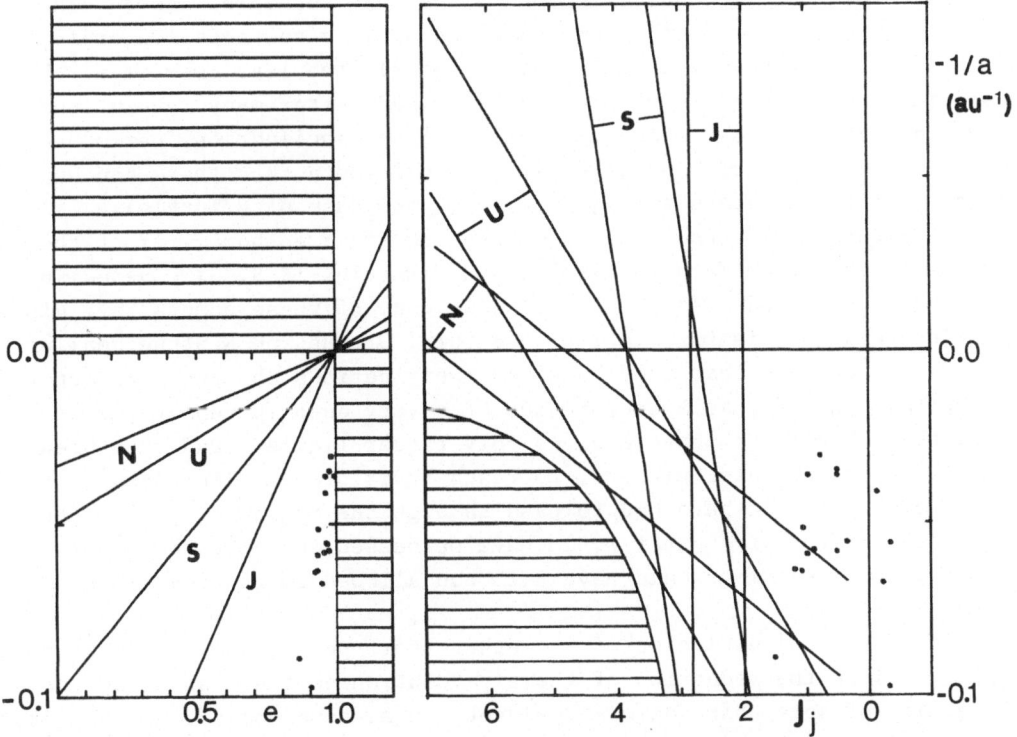

Figure 2. The same as Figure 1, the ordinate being restricted to the outer part of the solar system. The slant lines in the left plot represent orbits tangent in perihelion to those of the planet whose initial labels the line. The pairs of straight lines in the right plot represent orbits whose Tisserand invariant is equal to 2 (line on the right of each pair) and 2.83 (line on the left) with respect to the four outer planets (labels as before).

of the planet itself; it is easily understandable that in this case planetary encounters are not very effective unless they are particularly close. J = 2.83 implies that the encounter velocity of the comet is sufficiently high to allow escape from the solar system if the planetary encounter causes a particular reorientation of the velocity vector. On the other hand, if J is greater than 2.83, the comet cannot be ejected on (or captured from) a parabolic or hyperbolic orbit by encounters with the planet under consideration (Kresák, 1982).

Coming back to the fate of comets penetrating into the planetary region, the orbital elements determine what is (or what are) the planet that can encounter it, if any, and what is the value of J relative to that planet. If close encounters are not possible, and the perihelion distance is great, the comet may remain for a long time stored in its

"parking orbit"; if encounters with a planet are possible, and J relative
to it is greater than 2, the comet can end up in a short-period orbit (we
will return to this case in more detail later); the last case, i.e. the
possibility of encounters with J < 2, can lead, after many revolutions,
to Halley type comets. These are comets of any inclination, period between 20 and 200 years, and J_j smaller than 2 (note that there can be
comets with P < 20 years and J_j < 2, as is the case of P/Tuttle).

In Figure 3 all observed Halley type comets are shown. 17 of them
have been taken from Marsden (1982), and P/Hartley-IRAS, discovered in
1983, has been added. As it can be seen, many of these comets have also
the Tisserand invariants relative to Saturn, Uranus and Neptune smaller
than 2. It appears that they have had their periods shortened directly
by Jupiter, probably as a consequence of a very deep encounter; after
that, the orbit should have evolved only rather little. Had they been
captured by other outer planets, to reach their present locations in
Figure 3 they should have had to pass through the region J_j > 2; in that
region the effect of Jupiter would have surpassed those of the other
planets, and the comet would have not been allowed to reach smaller J_j
values.

The dynamical evolution of the other short-period comets can be more
complex, since the reduction of their revolution periods can be due to
the action of more than one giant planet, in succession; if, at the first
entrance in the planetary region, the perihelion distance allows only
encounters with Neptune, it is this planet that controls the evolution
until the comet reaches a perihelion distance small enough to allow encounters with Uranus, which then takes the control; in the meantime collisions with Neptune, as well as ejections by this planet to interstellar
space will always be a possibility.

Everhart (1977) modelled numerically the process of multistage capture just described. He found that a minority of comets reaches short-
period orbits without being ejected (he did not consider collisions with
planets as possible end states), the fraction being about one in several
thousand when the capture process starts with Neptune, something more in
the case of Uranus, and in the case of Saturn and Jupiter one order of
magnitude more. To quote precise figures is not very wise, since the
results of the numerical experiments are different for different initial
conditions; the time scales found by Everhart for the process are of the
order of hundreds of million years starting at Neptune and Uranus, millions of years starting at Saturn, and hundreds of thousand years if
comets are captured directly by Jupiter.

Examples of possible evolutionary tracks leading to captures, either
multistage or single-stage, are sketched in Figure 4. There it is possible to see that these tracks converge before comets reach perielion
distances small enough to allow their discovery (Kresák, 1982). There-

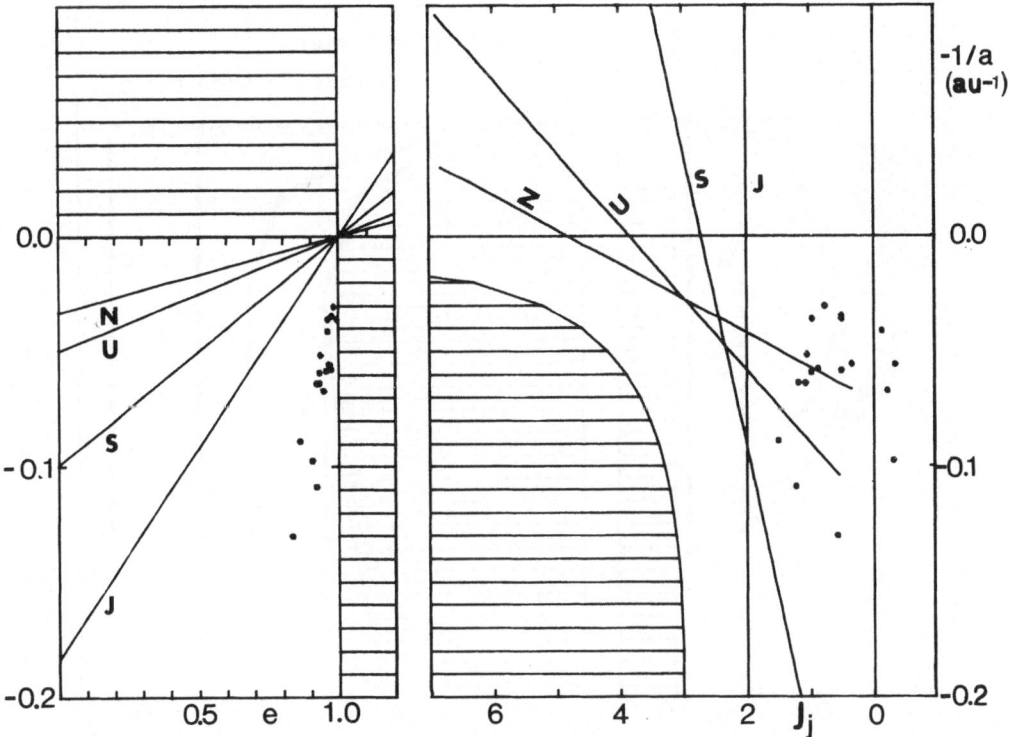

Figure 3. In the same style of Figures 1 and 2, all Halley type comets discovered up to the end of 1983 are indicated. In the right plot only the lines for Tisserand invariant equal to 2 are reported.

fore one should resort to numerical integration of the past motion of comets to establish which of the possible tracks they have actually followed; unfortunately, the motion of a comet cannot in general be reliably computed beyond one or two close planetary encounters (Carusi, Perozzi, Pittich and Valsecchi, this volume), thus rendering this reconstruction practically impossible.

3. FROM URANUS-NEPTUNE PLANETESIMALS TO SHORT-PERIOD COMETS

An alternate source for the short-period comets has been proposed by Fernández and Ip (1983); according to them, the formation of Uranus and Neptune by accretion of planetesimals, in the framework of the accumulation theory of Safronov (1969), would have had as a natural by-product the scattering from the Uranus-Neptune region of an amount of mass larger than the present masses of the two planets, in the form of small bodies mainly composed of volatiles, in other words of possible future

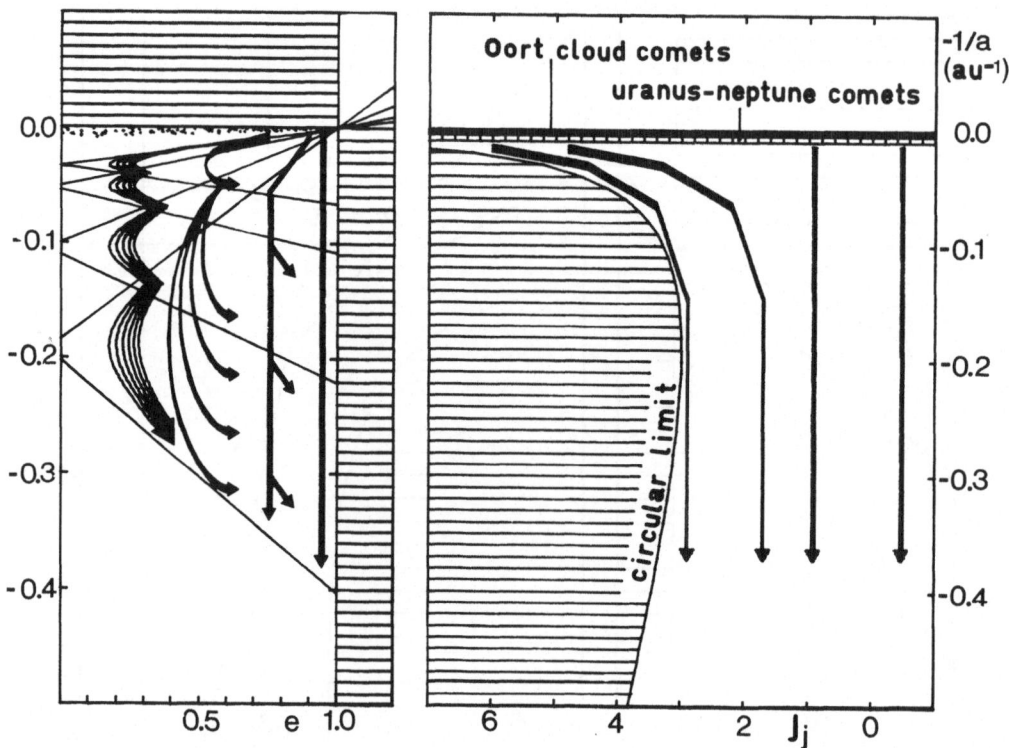

Figure 4. Possible evolutionary paths in the energy-eccentricity and in the energy-Tisserand invariant planes. The leftmost ones in both plots can imply low velocity encounters with all the outer planets in succession; the rightmost ones, leading to Halley type comets, imply mainly encounters with Jupiter; between these two extrema, single-stage captures are possible. The paths are only indicative, and real objects would not follow them strictly. The cometary reservoirs of the outer solar system are indicated by dots in the left plot.

comets.

The majority of the scattered planetesimals would be ejected from the solar system by the giant planets, especially Jupiter and Saturn, (Fernández, this book); another fraction would have their aphelia diffused out to distances allowing passing stars to decouple their perihelia and orbital planes from the planetary region, thus forming the Oort cloud; finally, those bodies not undergoing either of the described fates would remain stored for very long times in orbits with perihelia close to, or slightly outside, Neptune's orbit, and aphelia out to several thousand AU. These comets would be more tightly bound to the solar system than those in the Oort cloud, and their orbital planes would not be completely randomized. As Fernández and Ip point out, quantitative statements are difficult to make because of the uncertainties involved in the model; however, the evolutionary path connecting these comets to the short-period ones is faster than that from the Oort cloud described before.

In terms of the description exemplified in Figure 4, the dynamical evolution in this case is essentially the same, the main difference being that now comets start, as far as the orbital energy is concerned, slightly closer to the arrival than those coming from the Oort cloud. Note also that the orbits in which the Uranus-Neptune planetesimals are stored for very long time resemble those intermediate orbits in which comets from the Oort cloud can be put by stellar encounters, the difference being that the former are more closely coupled to the planetary system and decoupled from the stellar environment.

4. ORBITAL EVOLUTION IN THE PLANETARY REGION

Once comets have come back to orbits in the planetary region, the time scale of their evolution is greatly accelerated. The dynamics can be controlled by one or more planets, depending not only on the perihelion and aphelion distances, but also on the geometry of the orbit. When interactions with, for instance, two giant plenets are possible, in a majority of case it will be the inner one to take sooner or later the control, mainly because the synodic period of the comet with respect to it will in general be shorter than that with respect to the outer one; when encounters with only one planet are possible, this can happen either because the comet's orbit is rather inclined and the nodes are far from from any other planet, or because the perihelion and aphelion distances do not allow crossing of other planetary orbits. In the first case, one has to wait for a sufficient rotation of the argument of perihelion, if this is not prevented by libration mechanisms; in the second, the perturbations of the planet controlling the dynamics of the comet result in a random walk of the perihelion distance that can, after enough time, allow near tangency of the comet orbit to that of the planet immediately inside. In this case, the shorter synodic period already mentioned, together with a great efficiency of the slow encounters on nearly tangent orbits (Carusi and Valsecchi, 1982a) will probably decouple the comet from the outer planet, leaving again the control to the inner one.

It may also happen that a comet orbit crosses (or is tangent to) two planetary orbits, and encounters with one planet are prevented by libration involving the planetary and cometary longitudes and mean motions. This situation cannot last too long, however, since the revolution periods of pairs of neighbouring planets (excluding Neptune-Pluto) are not in exact ratios of small integers, so that a comet cannot be in libration with both the planets whose orbits it crosses.

The multistage capture process ends when the control is passed to Jupiter: at this point, the planets within Jupiter's orbit are by far too little massive to cause significant orbital evolution in the short time span covered by the life of the comet as an active object.

It must be noted that, although the prevailing direction of the dynamical evolution just discussed is inwards, also the evolution outwards, although less frequent, is possible.

For quite a long time in the past people like Laplace, LeVerrier, Tisserand and Callandreau thought that short-period comets come from long-period ones which in a single encounter with Jupiter lose enough orbital energy. Newton (1893) showed that the probability of this process is too low, and Everhart (1969) found that this mechanism would have implied that one quarter of the short-period comets should be in retrograde orbits, a fact in contrast with observations. Everhart (1972), moreover, studied in great detail the evolutionary path starting from parabolic comets; he identified a "capture zone" delimited by low orbital inclination ($i < 9°$) and perihelia close to the orbit of Jupiter ($4 < q < 6$ AU). Decelerating encounters with this planet were shown to be able to produce qualitatively the short-period comets that we observe.

Much of the numerical work done in the last 20 years has been devoted to the last part of the orbital history of short-period comets. Massive computations are in fact necessary to achieve a global picture of the dynamics of observable periodic comets, and the task of doing them has been alleviated by the increase of the performance of electronic computers.

Everhart (1973) has studied the orbital evolution of fictitious comets starting from randomly chosen circular, inclined orbits in the region of Jupiter and Saturn, and following these objects for thousands of revolutions, in a simplified solar system composed only of the Sun and the two planets. He found that many of these objects could enter different regimes of motion, including temporary trojans and horseshoes, generalized trojans and horseshoes, temporary satellite captures (see later), orbits of moderate inclination changing irregularly because of planetary close encounters, nearly circular orbits of longer persistence etc.; he named these orbits "chaotic": an object in a chaotic orbit can pass from one orbital pattern to another, its long-term evolution being generally unpredictable and, if it does not collide with a planet or disintegrate, it will in the end be put by a planet on a hyperbolic orbit.

Pioneering work regarding the dynamics of observed short-period comets was done in the sixties in USSR by Kazimirchak-Polonskaya (1967), Belyaev (1967) and coworkers, and in USA by Marsden (1967). They integrated the orbits of many short-period comets backwards in time for a few centuries and, in the case of Kazimirchak-Polonskaya and Belyaev, also

forwards for one century. Over such time spans they were able to identify many close encounters with Jupiter and a few with Saturn, that changed
in some cases even dramatically the orbits of comets.

Other researchers investigated the problem either using Monte Carlo
techniques or concentrating on the dynamics of close planetary encounters.

The first approach was pursued by Rickman and Vaghi (1976) and by
Froeschlé and Rickman (1980), the latter paper being mainly an improvement of the former. They divided the q-Q plane into a number of cells,
and computed a set of jovian perturbations acting on fictitious comets
in each cell. Then they started populating some cells, which therefore
behaved as source regions, and followed the overall evolution, taking
also into account the limited lifetime of the bodies in the inner part
of the solar system. Rickman and Vaghi (1976) found a too low efficiency
of captures by Jupiter, but it turned out that their numerical procedure
underestimated the largest perturbations, which in this case are the most
important. Froeschlé and Rickman (1980) corrected this shortcoming, and
investigated also the effects of different lifetimes of comets.

The dynamics of objects at close encounters with giant planets has
been extensively studied in Italy (Carusi and Pozzi, 1978; Carusi et al.,
1979; Carusi and Valsecchi, 1979, 1982a,b; Carusi et al., 1983). In
these works various populations of fictitious objects were investigated,
having different distributions of starting orbits; each object was followed through only one encounter with one of the outer planets, disre-
garding completely the previous and subsequent histories of the motion;
the phase space of the orbital elements was filled as uniformly as possible, so that differences in the outcomes of the encounters could be
related to specific regions of that space. The only elements not taken
at random were the true anomalies of the planet and of the fictitious
comet: they were chosen so as to allow the closest possible approach.

The main finding of these studies is that, in order to be most efficient in transforming the minor bodies' orbits, the encounters must be
either very deep or very slow. The last case implies a high value (close
to 3, or even larger) of the Tisserand invariant; as a result, temporary
satellite captures and transformations of the orbits without crossing of
the initial and the final ones are possible (Carusi et al., 1983).

Temporary satellite captures (TSC: i.e. occurrence of elliptical
planetocentric osculating elements of the comet during the encounter)
have been already observed both among real objects (Chebotarev, 1967;
Rickman, 1979) and fictitious ones (Kazimirchak-Polonskaya, 1972; Everhart, 1973), but the initial conditions leading to them were not clearly
recognized.

Carusi and Valsecchi (1981, 1982c) then integrated backwards in time
the orbits of all short-period comets with $J_j > 2.9$ in order to find

other events of this type. In a sample of 22 comets, 7 turned out to
have been temporarily captured as satellites by Jupiter, for time spans
between several months and several years, during the last 120 years. It
must be added that, although a clearcut distinction cannot be made, TSCs
seem to be of two main types (Carusi et al., 1983): one in which the
planetocentric energy has negligible variations compared to those of the
heliocentric energy, and another in which the two energy variations have
similar amplitudes (see Figure 5).

TSCs are possible with all the giant planets; however, the perihelion distances of comets that can be captured as satellites by Saturn,
Uranus and Neptune are far too large to make them observable (Carusi et
al., 1983).

A process often accompanying TSCs is the transition from completely
outside to completely inside the orbit of a planet, or vice-versa, with
the initial and final orbits of the comet far from crossing each other.
The two events are not necessarily related, in the sense that not always
they occur during the same encounter; however, the initial conditions
leading to both of them are essentially the same (Carusi et al., 1983),
characterized also in this case by high values of the Tisserand invariant. This point needs to be stressed, since it can be very important for
the orbital evolution: in fact, the perihelion distances of comets on
these orbits can be substantially reduced even if the pre-encounter minimum distances between the cometary and planetary orbits are large (up
to 0.6 AU in the case of the 1937 encounter with Jupiter of P/Oterma).
This greatly increases the mobility across the planetary system of comets
in these high Tisserand invariant orbits.

5. POSSIBLE EVOLUTIONS FROM PECULIAR ORBITS

A different approach to the study of cometary dynamics involves the integration of the motion of many varied versions of the same orbit, in
order to evaluate the probabilities pertaining to different evolutionary
tracks. Also in this case the progress in the performance of electronic
computers has been crucial in making possible the computations and the
processing of the outputs.

Following the discovery of 2060 Chiron, the first object found between the orbits of Saturn and Uranus, Kowal et al. (1979) integrated
its motion backwards and forwards over nine millennia, showing the possibility of encounters with Saturn and Uranus, and revealing the chaotic
nature of its orbit. This led Oikawa and Everhart (1979) and Scholl
(1979) to investigate the long-term motion of Chiron in more detail.

Oikawa and Everhart made an accurate numerical integration of the
motion of Chiron in a solar system composed by the Sun and the five outer

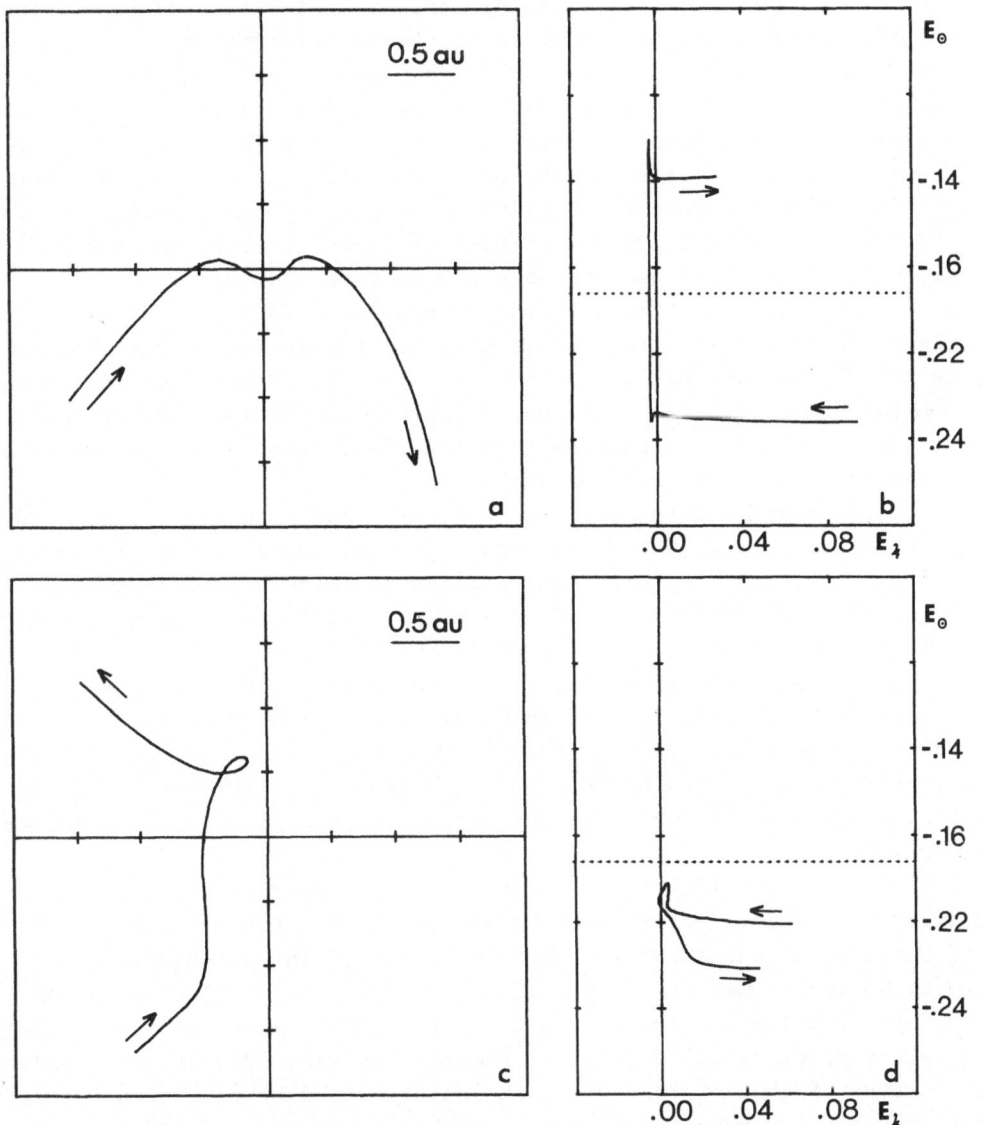

Figure 5. Left: Jovicentric paths in a rotating frame (Sun on the negative x-axis) of P/Oterma (a) during the 1960–1967 encounter and of P/Gunn
(c) during the 1870–1884 encounter; tic marks: 0.5 AU. Right: For the
same encounters (b: P/Oterma; d: P/Gunn) the heliocentric orbital energy
$E_\odot = -1/a$ versus the jovicentric one $E_\downarrow = -m/(Ma')$ are given, where m is
the mass of Jupiter, M that of the Sun, and a' is the jovicentric semi-
major axis.

planets, covering nearly 14 millennia roughly centred on the present epoch. Encounters with Saturn and Uranus were confirmed, showing that it

is the former planet to control mainly the evolution (the current values of the Tisserand invariants relative to the two planets are J_S = 2.9 and J_u = 2.95); successive large perturbations were often found to be correlated in size and sign. Since the accuracy of integration is lost after the first very close planetary encounter, to assess the probability of inward or outward evolution of Chiron these authors integrated 60 fictitious "Chirons", obtained by changing at random the 5th significant digit of the position-velocity vector of the real body, in a simplified 3-body model in which only the Sun and Saturn were retained. In 7 out of 8 cases the evolution was found to be inwards, towards Jupiter's control, showing also the interesting result that, in the region of the phase space of orbital elements in which Chiron is, the largest perturbations of orbital energy, which in the end determine the evolution, are negative; the time scale for passing to Jupiter's control was found to be of the order of a hundred thousand years.

Similar results were obtained by Scholl (1979), who integrated 10 "Chirons" for about 24,000 years, two thirds of the time span being in the future, keeping the four giant planets as perturbing bodies. He found 5 future evolutions inward, 3 outward, and in the remaining 2 cases the objects did not leave the region in which Chiron is; going backwards, the same figures are 0, 5 and 5, confirming that probably Chiron will reach Jupiter's control in tens or hundreds of thousand years.

Other researchers concentrated on some peculiar close encounters of short-period comets, again applying the method of slightly varying the initial conditions, in order to determine the probabilities of specific outcomes.

Carusi et al. (1981) modelled the strong perturbations exerted by Jupiter on a chain of fictitious comets in the orbit that P/Oterma had just before its 1937 encounter with the planet. The 80 objects were equally spaced in mean anomaly, and their motion was followed, in some cases, for 250 years. The main aim was to find if an evolutionary channel exists connecting the orbit of P/Oterma between 1939 and 1961 and one of the type of that of P/Encke, both characterized by very high Tisserand invariant and low aphelion distance, perhaps with the help of nongravitational forces. It turned out that only three objects were placed by Jupiter on rather long lasting orbits of low aphelion distance (4.5 AU) and moderate perihelion distance (2.5 AU); whether or not these values could be further reduced by nongravitational forces, to end in Encke type orbits, is still an open question. This work, however, showed other important features typical of this kind of orbits: first, all the 80 "Otermas" were temporarily captured by Jupiter as satellites – indicating a very high probability per revolution of this event (0.026) – and in one case the duration of the binding was of the order of 100 years; second, in spite of the large unperturbed minimum distances between the pre-encoun-

ter orbits of the "Otermas" and that of Jupiter (> 0.6 AU), the actual
minima of the encounter distances were much smaller, even by one or two
orders of magnitude, thus rendering problematic the definition of a
"sphere of action " of the planet independent of the parameters of the
encountering body; third, the high probability for orbits tangent to that
of Jupiter to change the type of tangency (perihelion to aphelion tangency, or vice-versa) without even approximate crossing of the initial
and final orbits, was further assessed.

In a subsequent work, Carusi, Kresák and Valsecchi (1982) examined
the patterns followed during the encounter with Jupiter by P/Oterma and
its 79 variations. They found that these patterns are arranged in a sequence (see Figure 1 in that paper) in which the first and the last, the
second and the last-but one, etc., are related by symmetries; it appeared
that different patterns were separated, at least in some cases, by values
of the difference in mean anomaly leading to long-lasting satellite captures (up to 100 years). Moreover, the span in M covered by individual
patterns varied widely, and one of them was identical to that followed by
P/Gehrels 3 during its 1963-1976 encounter with Jupiter. The authors
also compared these patterns with those followed by other 100 fictitious
bodies on orbits of different semiaxis and eccentricity but still tangent
either in their perihelion or aphelion to that of Jupiter, and thus having very high values of the Tisserand invariant. Similar patterns were
again found, together with new ones, mainly exhibited by objects with
values of J_j somewhat smaller than that of P/Oterma. They concluded that
the basic types of paths followed by objects having low velocity encounters with Jupiter were included among these 180 cases.

Rickman and Malmort (1981) studied the effects of varying the starting orbital elements of P/Gehrels 3 to strengthen the evidence of its
1967-1974 TSC by Jupiter. For orbits varied up to several degrees in angular elements, up to 0.05 in eccentricity and up to 0.2 AU in semiaxis,
they found an extended interval of gravitational binding to the planet;
moreover, a smaller domain (about 1° in angular elements, 0.01 in eccentricity, and 0.05 AU in semiaxis) in which at least one revolution around
Jupiter was performed, and some cases of long-lasting TSCs (up to 60
years), were found.

Carusi, Kresákova and Valsecchi (1982) followed the motion of particles ejected from the nucleus of P/Lexell at its 1770 perihelion passage through the very close approach to Jupiter in 1779, to examine the
dynamics of the peculiar meteor stream possibly associated with the comet. In some sense this work is analogous to the one on P/Oterma; this
time the modulus of the velocity vector at the perihelion passage of 1770
was slightly varied, and again a sequence of orbital patterns at the
close encounter was recognized. In this case, however, the trajectories
were much simpler than in that of P/Oterma, due to the much higher veloc-

ity of the bodies relative to Jupiter, a velocity sufficiently high (J_j < 2.83) to allow ejection from the solar system on a hyperbolic orbit.

6. CONCLUSIONS

Like other problems regarding comets, e.g. those of their origin, composition, relation to the outer planets, lifetime, etc., also that of their evolution into short-period orbits still demands much work. The last 20 years have seen a great increase of our knowlodge, especially for what concerns the main evolutionary channels and the dynamics of close planetary encounters, a subject that has had to wait the availability of fast computing tools before being tackled. Although the principal evolutionary paths are probably all identified, the efficiencies are not known for all of them, and new numerical experiments, modelling for instance the transfer from low eccentricity orbits just outside Neptune to ones like those of Chiron and P/Schwassmann-Wachmann 1, would be useful.

However, a quantitative assessment of the efficiencies of the various evolutionary channels, as well as a better understanding of the dynamics of close encounters, are still insufficient for a global picture: we do not know how densely the various regions of the phase space of orbital elements are populated, and how the present situation differs from the past ones.

As we have seen, the orbits of periodic comets are chaotic, that is they can pass through different regimes of motion in an unpredictable way. This obviously complicates the numerical work on the reconstruction of past orbital evolutions; moreover, apart from the orbits of Halley type comets, and of those of the so-called Jupiter family, all other possible regimes of motion for chaotic orbits imply large perihelion distances, rendering bodies in these phases of evolution unobservable.

Another factor that needs to be taken into account is the physical nature of the objects; this can have several consequences, since the finite lifetime in the inner regions of the solar system can render practically impossible those transitions between types of orbits whose low probability implies long time scales. Moreover, the number of candidates for potential evolution into short-period orbits can vary also in the other direction, i.e. increase because of splitting events that leave long lasting fragments, like in the case of the recently suggested common origin of P/Van Biesbroeck and P/Neujmin 3 (Carusi et al., 1984).

ACKNOWLEDGMENTS. The authors are grateful to the referees E. Everhart and L.Kresák for useful comments and suggestions which helped to improve the final appearance of the paper.

REFERENCES

Bailey, M.E.: 1983a, Mon. Not. R. astr. Soc. 204, 603-633.
Bailey, M.E.: 1983b, in "Asteroids, Comets, Meteors" (C.-I. Lagervist and
    H. Rickman eds.), Uppsala Univ., Uppsala, Sweden, pp. 383-386.
Belyaev, N.A.: 1967, Sov. Astron. - A.J. 11, 366-373.
Carusi, A., and Pozzi, F.: 1978, Moon Planets 19, 71-87.
Carusi, A., and Valsecchi, G.B.: 1979, in "Asteroids" (T. Gehrels ed.),
    Univ. Arizona, Tucson, U.S.A., pp. 391-416.
Carusi, A., and Valsecchi, G.B.: 1981, Astron. Astrophys. 94, 226-228.
Carusi, A., and Valsecchi, G.B.: 1982a, in "Sun and Planetary System" (W.
    Fricke and G. Teleki eds.), Reidel, Dordrecht, Holland, pp. 379-384.
Carusi, A., and Valsecchi, G.B.: 1982b, in "Sun and Planetary System" (W.
    Fricke and G. Teleki eds.), Reidel, Dordrecht, Holland, pp. 385-388.
Carusi, A., and Valsecchi, G.B.: 1982c, in "The Comparative Study of the
    Planets" (A. Coradini and M. Fulchignoni eds.), Reidel, Dordrecht,
    Holland, pp. 131-148.
Carusi, A., Kresák, L., and Valsecchi, G.B.: 1981, Astron. Astrophys. 99,
    262-269.
Carusi, A., Kresák, L., and Valsecchi, G.B.: 1982, Bull. Astron. Inst.
    Czechosl. 33, 141-150.
Carusi, A., Kresákova, M., and Valsecchi, G.B.: 1982, Astron. Astrophys.
    116, 201-209.
Carusi, A., Kresák, L., Perozzi, E., and Valsecchi, G.B.: 1984, IAUC 3940.
Carusi, A., Perozzi, E., and Valsecchi, G.B.: 1983, in "Dynamical Trap-
    ping and Evolution in the Solar System" (V.V. Markellos and Y. Kozai
    eds.), Reidel, Dordrecht, Holland, pp. 377-395.
Carusi, A., Pozzi, F., and Valsecchi, G.B.: 1979, in "Dynamics of the
    Solar System" (R.L. Duncombe ed.), IAU Symp. 81, Reidel, Dordrecht,
    Holland, pp. 185-189.
Chebotarev, G.A.: 1967, "Analytical and Numerical Methods of Celestial
    Mechanics", Elsevier, New York, U.S.A., pp. 239-241.
Everhart, E.: 1969, Astron. J. 74, 735-750.
Everhart, E.: 1972, Astrophys. Letters 10, 131-135.
Everhart, E.: 1973, Astron. J. 78, 316-328.
Everhart, E.: 1977, in "Comets, Asteroids, Meteorites" (A.H. Delsemme
    ed.), Univ. Toledo, Toledo, U.S.A., pp. 99-104.
Everhart, E.: 1979, in "Asteroids" (T. Gehrels ed.), Univ. Arizona, Tuc-
    son, U.S.A., pp. 283-288.
Fernández, J.A., and Ip, W.-H.: 1983, Icarus 54, 377-387.
Froeschlé, C., and Rickman, H.: 1980, Astron. Astrophys. 82, 183-194.
Kazimirchak-Polonskaya, E.I.: 1967, Sov. Astron. - A.J. 11, 349-365.
Kazimirchak-Polonskaya, E.I.: 1972, in "The Motion, Evolution of Orbits
    and Origin of Comets" (G.A. Chebotarev, E.I. Kazimirchak-Polonskaya

and B.G. Marsden eds.), IAU Symp. 45, Reidel, Dordrecht, Holland, pp. 373-397.

Kowal, C.T., Liller, W., and Marsden, B.G.: 1979, in "Dynamics of the Solar System" (R.L. Duncombe ed.), IAU Symp. 81, Reidel, Dordrecht, Holland, pp.245-250.

Kresák, L.: 1972, Bull. Inst. Astron. Czechosl. $\underline{23}$, 1-34.

Kresák, L.: 1982, in "Sun and Planetary System" (W. Fricke and G. Teleki eds.), Reidel, Dordrecht, Holland, pp. 361-370.

Marsden, B.G.: 1967, Science $\underline{155}$, 1207-1213.

Marsden, B.G.: 1982, "Catalogue of Cometary Orbits", Smithson. Astrophys. Obs., Cambridge, U.S.A..

Napier, W.M., and Staniucha, M.: 1983, Mon. Not. R. astr. Soc. $\underline{198}$, 723-735.

Newton, H.A.: 1893, Mem. Nat. Acad. Sci. (Washington) $\underline{6}$, 8-23.

Oikawa, S., and Everhart, E.: 1979, Astron. J. $\underline{84}$, 134-139.

Öpik, E.J.: 1972, "Interplanetary Encounters", Elsevier, Amsterdam, Holland, chapter 4.

Rickman, H.: in "Dynamics of the Solar System" (R.L. Duncombe ed.), IAU Symp. 81, Reidel, Dordrecht, Holland, pp. 293-298.

Rickman, H., and Malmort, A.M.: 1981, Astron. Astrophys. $\underline{102}$, 165-170.

Rickman, H., and Vaghi, S.: 1976, Astron. Astrophys. $\underline{51}$, 327-342.

Safronov, V.S.: 1969, "Evolution of the Protoplanetary Cloud and Formation of the Earth and Planets", Nauka, Moscow; 1972 transl. Israel Program for Scientific Translation, Jerusalem.

Scholl, H.: 1979, Icarus $\underline{40}$, 345-349.

DISCUSSION

Milani: The variation of orbital behaviour you find in TSCs by varying one angular variable reminds me of the theory of "Cantori", that is invariant subsets in hamiltonian systems whose projection on one axis is a Cantor set. Do you find anything suggesting this kind of behaviour?
Valsecchi: The spacing of our objects was possibly large compared to the scale of the phenomena you mention. However, we found the longest TSCs, and the most complex orbital patterns, between neighbouring "families" of similar, simple patterns, and this may well be due to what you suggest.

Mignard: Among the TSCs by Jupiter, are the comet orbits comparable to those of Jupiter outer satellites?
Valsecchi: They are less tightly bound, and their planetocentric elements vary considerably even during a single revolution. Even when you have a TSC of long duration (10-100 years), the eccentricity is quite large, larger than those of the irregular jovian satellites.

Fernández: Are there possible evolutionary paths that bring retrograde-orbit comets captured by the outer jovian planets to small perihelion distances under the dynamical control of Jupiter?
Valsecchi: I cannot support my opinion quantitatively, since numerical experiments on the process you mention have not been carried out, but I think that this channel should be inefficient.

THE AGING AND LIFETIMES OF COMETS

Ľubor Kresák
Astronomical Institute
Slovak Academy of Sciences
84228 Bratislava
Czechoslovakia

ABSTRACT. The aging of comets is evidenced by a number of observable phenomena: production of gas, dust and meteor particles, splitting of cometary nuclei, nongravitational effects in the comet's motion, sudden and progressive absolute brightness variations, and ultimate disappearance. Statistical data on comet losses, absolute magnitudes and orbits also bear signatures of their aging. The knowledge of potential active lifetimes of individual objects is a prerequisite of any realistic model of the long-term evolution of the whole comet complex. This paper reviews different sources of information on the aging process and summarizes implications for the mean lifetimes of comets, their dispersion and dependence on the orbital parameters. Two alternative end fates of comets - their total disintegration or change into an inactive asteroid-like object - are also discussed.

1. INTRODUCTION

The papers presented at the preceding sessions have shown that we still have a number of competing hypotheses on the origin of comets. The only straightforward way to demonstrate the validity of any of them, is to start from a correct model of the initial state and to trace the evolution forwards to a state compatible with observation. The great progress in modelling experiments with the use of modern computing techniques makes this possible in principle. However, there are three serious impediments :

1. Very incomplete information on the present state. We can only observe comets in the innermost region of their huge system. For the new comets in Oort's sense the period of observation is always less than one millionth of the period of revolution, which again is only about one thousandth of the age of the Solar System. Some short-period comets are observable all around their orbits, but their active lifetimes are apparently less than one millionth of the age of the Solar System. The number of known comet orbits (over 700, including over 100 of short period) constitutes a fairly rich statistical sample. However, in order to compare it with the results of

A. Carusi and G. B. Valsecchi (eds.), Dynamics of Comets: Their Origin and Evolution, 279–302.

modelling experiments, a number of strong selection effects must be taken in account, in particular, those of revolution period, perihelion distance and absolute brightness. We are also not sure whether or not this sample is representative for a quasi-steady state in the inner Solar System, persisting over a considerable part of its lifetime as an equilibrium between source and sink.

2. Uncertainties about the past perturbing environment. These refer to individual perturbing events – encounters of the Solar System with stars and clouds of interstellar matter – the recurrence rate of which may be subject to long-period variations associated with the motion of the Sun within the Galaxy. If the age of the comet system is about the same as that of the planetary system (which is currently the prevailing opinion), then the structural changes of the latter may have played a significant role, especially during the earliest phase of evolution.

3. Progressive disintegration of the comets themselves. This is accompanied by nongravitational effects in their motion, which are often erratic and cannot be extrapolated with confidence outside the period covered by observations. And what is still more important, at some evolutionary stages the rate of physical aging may become much higher than that of the dynamical evolution. Progressive aging of comets introduces a definite asymmetry into the occurrence rate of fundamental perturbing events, which is not borne out by computer simulations of their motion. A comet whose disintegration process was triggered by a decelerating encounter with a planet, may survive not long enough to experience an analogous accelerating encounter. Captures into short-period orbits become more frequent than ejections from them; Jupiter's reflecting barrier thus becomes a partially absorbing barrier, and the quasi-steady state becomes different from a pure dynamical equilibrium.

The purpose of the present paper is to discuss the problems of determination of the active lifetimes of comets, their distribution, and correlation with the orbit type. The other question to be addressed is what happens at the end of the comet's active lifetime and what is the character of its remnants.

2. PHENOMENA ACCOMPANYING THE AGING PROCESS

The aging of comets is borne out by a number of observable phenomena:
-- Production of coma and tail consisting of escaping gas and dust.
-- Splitting of the nucleus and sudden brightness bursts, indicating a temporary acceleration of the mass loss.
-- Nongravitational effects in the comet's motion, produced by jet effects of the escaping matter on the nucleus insolated from one direction.
-- Existence and dispersion of meteor streams occupying the orbits of comets.
-- Progressive decrease of the absolute brightness of short-period comets observed at a number of returns to perihelion.
-- Differences between the absolute brightness of dynamically old (short-period) and young (long-period, especially new) comets.

-- Failure of the rediscovery of some short-period comets at their
   expected returns to perihelion.
-- Untimely disappearance of some long-period comets.
-- An appreciable excess of new comets in Oort's sense as compared
   with other long-period comets.
In principle, each of these effects can assist in estimating the
active lifetimes of comets. Unfortunately, each of them has its specific constraints in this respect.

3. THE ABSOLUTE MASS LOSS

The physical lifetimes of comets are evidently determined by the
rate of their progressive mass loss. The observable products of the
aging process include: (1) gas, (2) icy grains, (3) dust grains,
(4) meteor particles, and (5) larger fragments up to long-lived secondary components producing their own comas and tails.

Estimates of the gas production rates are currently available
for about ten brighter comets observed since 1970. They are based
on quantitative measurements of emission line and band strength,
combined with the lifetimes of individual species. In view of the
high prevalence of the H_2O parent molecules (Delsemme, 1982), the
relevant data are provided by UV measurements from satellites and
sounding rockets. The results summarized by Feldman (1982) and Ney
(1982), if converted under reasonable assumptions into the total
loss of H_2O per revolution, yield 10^{10} to 5×10^{10} kg for short-period
comets (1.5×10^{11} kg for P/Halley according to Delsemme, 1976) and
up to 8×10^{11} kg for long-period comets. The addition of other volatiles should not increase these rates appreciably. Just on the contrary, since the measurements were made on relatively bright comets
at smaller solar distances, one has to expect considerably smaller
mass loss rates for comets of low absolute brightness or large perihelion distance. There is no dependable information as to the loss
of volatiles at larger solar distances, including the formation and
removal of icy grain halos. However, since most of the mass loss
takes place along the perihelion arc of the orbit (e.g., definitely
at r < 2 AU for q = 1 AU, which is about the median value for known
comet orbits), any deviations from the r^{-2} dependence of the mass
loss seem to be immaterial for those comets which are aging rapidly.
Irregular fluctuations may affect the result more seriously.

Estimates of the dust production rates are mainly based on IR
measurements of type II tails of bright comets and refer to micron-
sized particles (Ney, 1982). They indicate that the dust production
per revolution is 2×10^{10} kg to 5×10^{11} kg for bright long-period comets, and less than 10^9 kg for P/Encke. Thus the dust-to-gas mass
ratio is normally between 1:10 and 1:1, the only case where the dust
seems to have prevailed, in a ratio of 5:3, being that of 1957 III
Arend-Roland (Finson and Probstein, 1968).

Estimates of the production of larger solid particles are only
possible when the comet approaches the Earth's orbit close enough
to give rise to a meteor shower. While in such cases it is possible
to determine the space density of meteor particles, problems arise

when it is to be converted into the mass loss. First, we only have one-dimensional sounding along the Earth's path through the stream; in order to reconstruct a three-dimensional picture, extrapolation is necessary. That along the comet's path is possible for permanent meteor streams, making use of the annual recurrence of the shower; that perpendicular to it depends on the distance of the passage from the stream's centre, and on the specific pattern of dispersion produced by perturbations. In the only detailed analysis of this kind published so far and relating to P/Halley, McIntosh and Hajduk (1983) find a highly anisotropic distribution with a flattening of 1:10 perpendicular to the comet's orbital plane. They find $5x10^{14}$ kg as the total mass of the stream, and suggest that this has been assembled in the course of about 1500 revolutions of the comet, i.e. at an average rate of $3.3x10^{11}$ kg per revolution. Taking in account the inevitable losses by progressive destruction and hyperbolic escape, the figure appears unexpectedly high. However, in spite of its long revolution period, P/Halley is a much stronger source of meteoroids than other comets (Whipple, 1967; Kresák, 1979a), and was apparently an exceptionally large object in the past.

Splitting of cometary nuclei is another phenomenon which can reduce at once their potential survival time. Present statistics indicate an average rate of one observed split per 25 apparitions for long-period comets, and one per 170 apparitions for short-period comets (Whipple, 1978a). Taking in account the limited periods of observability, values about twice as large are obtained, relating to non-tidal splits during the whole revolution: one per 12 revolutions for long-period comets, and one per 90 revolutions for short-period comets (Kresák, 1981a). The earlier finding that splits are more frequent for new comets in Oort's sense than for other long-period comets (Stefanik, 1966), is apparently due to their longer periods of observability; the difference becomes statistically insignificant when corrections for observational selection are applied (Kresák, 1981a). As we shall see later, the observed rate implies that many comets, and possibly their majority, experience at least one splitting during their active lifetime. The effect of these events on the lifetime would decrease with increasing mass ratio of the fragments. The differential nongravitational effects indicate that the mass ratio is often very high (Sekanina, 1982); smaller fragments tend to fade out very early, having apparently no major effect on the lifetime of the main component.

The distribution of orbits of long-period comets does not bear out the presence of groups or pairs indicative of persistent components of split comets (Whipple, 1977a; Kresák, 1982b; Lindblad, 1985). The only exception is the Kreutz group of sungrazing comets, which probably experience splitting by solar tides at every perihelion passage (Kresák, 1981a) - a branching process continuing over a number of revolutions of the original object which must have been extraordinarily large. For short-period comets the process is much easier to overview, because checking is possible at every return. A few recorded cases cover the whole range of possibilities. In the case of P/Biela, both components persisted over two revolutions and disappeared almost simultaneously. In the case of P/Brooks 2 (tidal disruption by Jupiter), two minor fragments disappeared just during their first perihelion passage, while the main component is still observable after 14 revolutions. P/Taylor got lost just after its first apparition during which it had experienced splitting, but the main component was rediscovered 9 revolutions later. P/DuToit-Hartley was observed as a single comet in 1945 and as a double comet in 1982. There is only one pair of short-period comets which strongly suggests that we have to do with persistent components of a single progenitor: P/Neujmin 3 and P/Van Biesbroeck (Carusi et al., 1984). Backward computations indicate that both large fragments of the original nucleus have already made 12 and 11 revolutions, respectively, as two separate objects. P/Neujmin 3 was last observed in 1972, P/Van Biesbroeck in 1978. An alternative, less probable explanation is that those two comets originally constituted a binary system of marginal stability, which was dissolved by differential perturbations during the close encounter with Jupiter in 1850.

The existence of such systems was suggested by Van Flandern (1981), and associated by Whipple (1983a, 1984) with the violent double outbursts of P/Holmes in 1892 and P/Tuttle-Giacobini-Kresák in 1973. Whipple interprets these events as a grazing encounter of the components after reduction of their minimum distance by differential nongravitational forces, followed by a collision one revolution later, which destroys the secondary. This scenario may certainly be questioned, but there is so far no other explanation of the recurrence of two bursts in an interval of a few weeks. Collisions of double comets may represent another instantaneous process of aging, though much less frequent than spontantaneous splits. Only two known examples would imply a collision rate of one per 500 revolutions, with a considerable margin of uncertainty. On the other hand, the inherent mass loss may be substantial for the survival time. The integrated radiation output during the two outbursts of P/Tuttle-Giacobini-Kresák was equivalent to that over about 80 normal revolutions (Kresák, 1974b). This comet counts among the faintest known objects; for typical comets with a much higher normal activity level a similar effect might be less drastic but not negligible. Unlike splitting, the binary collisions would only occur in a fraction of comets, without a preference for long-period comets. Both the cases on record refer to the Jupiter family of comets, but it would be too daring to imply that there is any systematic difference on the basis of such poor statistics. Non-catastrophic collisions with small asteroids are evidently much too rare to affect the aging of a significant fraction of comets. The question of unobserved interplanetary boulders (Harwit, 1967; Kresák, 1978) still remains open.

4. THE RELATIVE MASS LOSS

The total masses of comets are even more difficult to estimate than their mass losses. As related to observable quantities, the diameters D and masses M of cometary nuclei can be tentatively expressed by the equations

$$D = c \ 10^{-0.2 \ H} \tag{1}$$

$$M = C \ 10^{-0.6 \ H} \tag{2}$$

where H is the absolute magnitude of the comet (the total apparent
magnitude reduced to a distance of 1 AU both from the Sun and Earth,
assuming brightness variations with the inverse fourth power of the
heliocentric distance and with the inverse square of the geocentric
distance). The scaling factor c involves the brightness ratio of the
nucleus to the coma at 1 AU, and the albedo of the nuclear surface.
The factor C involves, in addition, the mean density of the nucleus,
the shape of which is assumed nearly spherical.

     Using Vsekhsvyatskij's (1958) scale of absolute magnitudes H,
Öpik (1963 and 1973) originally assumed c = 150 km, C = 2.2×10^{18} kg,
ϱ = 2 gcm^{-3}, including deviations from a spherical shape. However,
in his later paper he already pointed out that there is good reason
to revise c to 75 km and C to 2.6×10^{17} kg. A more recent revision
by Whipple (1978a) suggests c = 32 km and C = 2.5×10^{16} kg, with ϱ =
1.5 gcm^{-3}. Thus each of the revisions has put the mass estimates one
full order of magnitude lower which, if related to the mass loss
rates, would put the lifetimes one order of magnitude higher.

     The point is that direct evidence is available neither on the
contribution of the light reflected by the nuclear surface to the
total brightness of the comet, nor on the albedo and density of the
nucleus. The product of its diameter and the square root of its albedo can be determined from observations at extreme solar distances
with large long-focus telescopes, provided that the contribution of
the coma can be neglected under such circumstances (Roemer, 1966;
Kresák, 1973). Another function of diameter and albedo can be determined from the vaporization rate of H_2O production at small solar
distances, provided that the whole surface is covered by water ice
(Delsemme and Rudd, 1973). Solving these two equations, Whipple
(1978a) finds rather high values for the albedos (over 0.6) and determines the diameters of some cometary nuclei, which are in fair
agreement with his scaling factor in Eq. (1). Another approach suggested by him is the mass determination based on the radial acceleration by nongravitational forces. This procedure, applicable only
to short-period comets, requires some assumptions about quantities
which are not measurable directly; nevertheless, results roughly
consistent with other independent estimates could be obtained for
some comets. In their analysis of the rotation, nongravitational deceleration and sublimation of P/Encke, Whipple and Sekanina (1979)
estimate its current relative mass loss by sublimation at 0.09 per
cent per revolution, but point out substantial temporal variations
of this value.

     Now, if we assume the validity of (2) with Whipple's scaling
factor and one half of the nucleus (by mass) being composed of H_2O,
potential lifetimes of comets can be obtained by dividing M by the
H_2O production rate per two revolutions, as discussed in Section
3. There are only ten comets for which this is possible. For seven
long-period comets (1970 II, 1973 XII, 1975 IX, 1976 VI, 1978 XV,

1979 I and 1980 XII) the result ranges from 1 to 50 revolutions,
with the median at 12 revolutions. These are mostly comets of small
perihelion distance, and reducing the mass loss rates to q = 1 AU a
median of 18 revolutions is obtained. For the three short-period
comets the result is entirely misleading: 1.5 revolution for P/Encke,
4 revolutions for P/Tuttle and P/Stephan-Oterma. In fact, P/Encke
has already made 60 revolutions since its discovery, P/Tuttle 15
revolutions, and none of them shows signatures of approaching an
early disappearance.

It must be concluded that the assumptions involved in this mass
determination are invalid. Evidently, both c and C in Eqs. (1) and
(2) are appreciably higher for short-period comets than for long-
period comets, which is most probably due to a considerable reduction of their effective surface area by a non-volatile crust (Shul'-
man, 1972). It may also be noted that the diameter of P/Encke, as
determined from (1) with c = 32 km and H = 9.8 (Meisel and Morris,
1982) comes out D = 0.35 km. Radar detection of this comet (Kamoun
et al., 1982a) sets the limits of D at 0.8 to 8 km. Other estimates,
as assembled by Whipple (1978a) and by Kamoun et al. (1982b) yield
different values, but all within the above radar range. The uncertainty of 1:10 in size means one of 1:1000 in lifetimes computed by
comparison of the total mass with the mass loss. Hence, the resulting limitation to between 20 and 20,000 revolutions covers all possibilities which may be reasonably admitted, and no progress is possible without making the size estimates much sharper. It is hoped
that the spacecraft missions to P/Halley will provide a fundamental
improvement of our knowledge about its size, composition and surface
properties, and will make possible some calibration of the data on
other comets, including the factors c and C in equations (1) - (2).

5. THE SECULAR BRIGHTNESS DECREASE

The aging of comets is inevitably accompanied by a progressive decrease of their absolute brightness. The slow rate of change makes
this only detectable on short-period comets observed at a number of
revolutions. The effect was most thoroughly investigated by Vsekhsvyatskij, who has spent much effort in processing photometric data
on comets, and has produced comprehensive annotated lists of their
absolute magnitudes at different apparitions (Vsekhsvyatskij, 1958,
1966, 1967, 1979; Vsekhsvyatskij and Il'chishina, 1974).

Under some oversimplified assumptions (a homogeneous nucleus of
nearly spherical shape, with a constant depth of the surface layer
removed during each revolution), the trend of the brightness changes
can be predicted by a simple function with a single unknown parameter. Accordingly, the absolute brightness H should decrease by ΔH
= +1.5 magn. during the first half of the comet's lifetime, by the
same amount during the first half of the remaining period, etc. If
the acceleration of ΔH can be measured, the death date can be predicted under the above provisions. Then, assuming that short-period
comets are, on the average, observed in the middle of their active
lifetime, its mean duration can be estimated.

This approach, however, leads to gross underestimates of the computed lifetimes, as evidenced by continuing observations of seven short-period comets which were predicted to disappear between 1958 and 1971: P/Pons-Winnecke, P/Tuttle, P/Wolf, P/Kopff, P/Brooks 2, P/Faye and P/Whipple (Whipple, 1964; Whipple and Douglas-Hamilton, 1966). The reasons of the failure are explained in detail elsewhere (Kresák, 1974a) and can be summarized as follows:

First, the absolute brightness of comets is subject to irregular fluctuations the amplitude of which varies substantially from one object to another. If the comet is decelerated by Jupiter into an orbit of appreciably smaller perihelion distance - and such captures are responsible for about 20% of short-period comet discoveries (Kresák, 1982a) - the change in the insolation regime makes them absolutely brighter for one or two apparitions (Kresák, 1973). Also, if there are major brightness variations from one revolution to another, it is more probable that the comet will be discovered at an increased activity level. Thus the first apparitions cannot be relied upon in determining the general rate of fading. As shown by Svoreň (1979), just the removal of the discovery apparitions is sufficient to reduce the average brightness decrease from 0.36 to 0.22 magnitude per revolution.

Second, and in particular, the photometric data on short-period comets, covering nearly two centuries, bear definite signatures of the development of observing techniques. However paradoxical it may appear, the instrumental effects make the comets fainter with time. This is because the detection threshold is improving, and large telescopes tend to record only the central condensation of the coma. A striking example of instrumental effects producing a systematic difference of over 7 magnitudes, or a ratio of almost 1:1000 in the brightness estimates, is shown in Figure 1 of Kresák (1974a; see also Whipple, 1978a).

The secular trend of the brightness estimates is illustrated by Figure 1, with Vsekhsvyatskij's absolute magnitudes H (exponent 4, H_{10} in his notation) plotted against the year of perihelion passage T. All individual apparitions of comets with periods $P < 20$ years and perihelion distances $q < 1.5$ are included. For P/Encke ($q = 0.3$, open circles) the dashed curve is approximately fitted to the catalogized H-values. While the sample of 51 apparitions of the only 11 comets of $q < 1.0$ (large solid circles) may be affected by random fluctuations of small numbers, the addition of 118 apparitions of 32 other comets of $q < 1.5$ (solid dots) makes the data fairly representative. Only very few short-period comets of $q > 1.5$ were observed during the preceding century, so that their inclusion would make the sample rather inhomogeneous.

It is apparent at first glance that the data points exhibit a progressive displacement towards higher values of H. This refers not only to the lower boundary and the means - an effect to be expected from the improvement of the observing techniques - but also to the upper boundary. There are only two alternative explanations of this. Either the short-period comets are dying out rather rapidly as a family of objects, and after one or two centuries there will be no

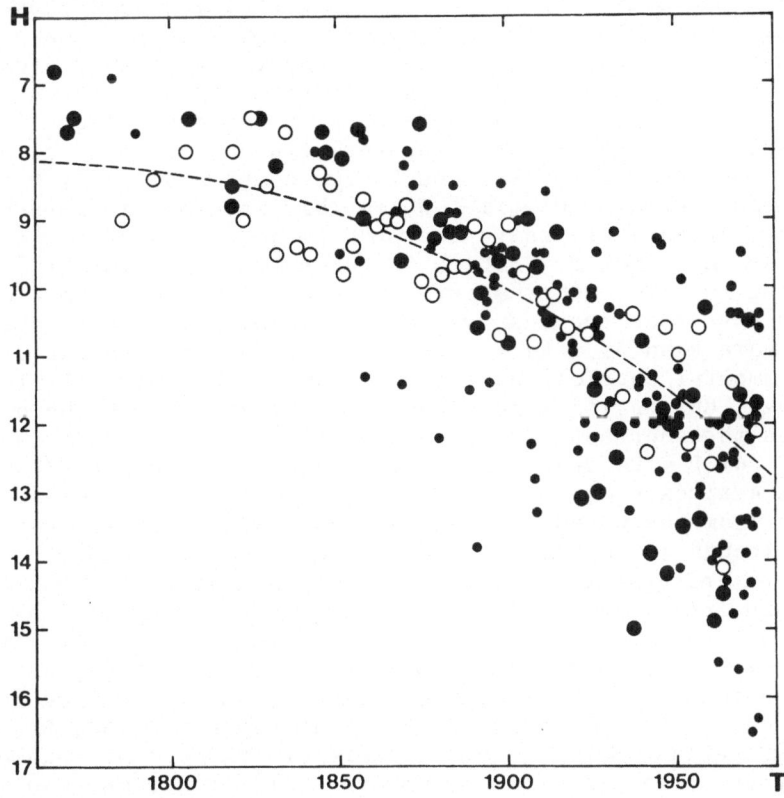

Figure 1. Vsekhsvyatskij's absolute magnitudes H for all individual apparitions of short-period comets of P < 20 years, q < 1.5 AU, as a function of the time of their perihelion passage T. Open circles, P/Encke; solid circles, all other objects of q < 1.0; solid dots, objects of 1.0 < q < 1.5.

detectable objects of this kind anymore; or the H-values are misre-presented by time-dependent instrumental effects. The former expla-nation appears highly unreasonable, the latter being evidently the correct one.

It is instructive that the interpolated curve for P/Encke, with a decline of more than 3 magn. during the last 100 years, deviates from the upper boundary of the populated area by only about 0.6 magn. per century. This is in fair agreement with the decrease determined from the maximum apparent brightness at different returns, which is less affected by instrumental effects because the comet is normally close to the naked-eye limit at maximum (Kresák, 1965). In his cri-ticism of this conclusion, Sekanina (1969) defends a rapid bright-ness decrease accelerating from 2 magn. per century at the time of the comet's discovery to a present value of 4 magn. per century. At the same time, the failure of identifying any observations from the

ancient and medieval Chinese records as pertaining to P/Encke (Ho
Peng Yoke, 1962; Whipple and Hamid, 1972) implies that the secular
fading must have been much less than 1 magn. per century over the
past millennium or two. Whipple and Sekanina (1979) attribute this
discordance to the effects associated with the precession of the
comet's spin axis, leading to long-term changes of the regime of insolation, and thus to differences in the comet's activity over its
heterogeneous surface and time. While this scenario is internally
consistent, it can only solve the problem for this particular object
and not for the comet statistics in general. Vsekhsvyatskij's (1981)
mean fading rate for other short-period comets, 3 magn. per century,
is even higher than the mean value of 2 magn. per century obtained
from the same kind of data for P/Encke (Dobrovol'skij et al., 1983).
The instrumental corrections for this comet are apparently less than
the average, due to its higher apparent brightness. On the other
hand, with the shortest period and smallest perihelion distance of
all comets of $P < 20$ years, the time scale of fading of P/Encke
should be 4 times shorter than the average. Thus there seems to be
no escape from the paradox shown in Figure 1, unless substantial
spurious trends in the absolute magnitudes are admitted.

Current efforts in better calibrated magnitude estimates of
comets, and their unified annotated listing in International Comet
Quarterly, lend promise of improvement in this respect. While this
is obviously a long-term task, just the first processing and summa-
rization demonstrates that many comets previously believed to fade
rapidly, do not indeed exhibit significant changes since their discovery (Meisel and Morris, 1982). There are undoubtedly examples of
a real progressive brightness drop over a limited time span, such
as P/Faye in the second half of the 19th century. On the other hand,
there are comets like P/D'Arrest which was at its last apparition
2 magn. brighter than at the time of discovery 20 revolutions ago,
or P/Perrine-Mrkos whose variations are entirely erratic. For an
overwhelming majority of short-period comets the total systematic
reduction of absolute brightness since their discovery appears to be
definitely within the noise of irregular fluctuations.

The same observational effects apply to long-period comets as
well, being only less pronounced because of their higher apparent
brightness. Many of them would exceed the naked-eye limit, which
would eliminate the instrumental effects entirely. Even so, it is
possible to identify a definite spurious decrease of their mean photometric exponent with time (Table VIII in Kresák, 1974a), which is
due to the extension of the observed orbital arcs by the use of larger telescopes. The widespread opinion that the brightness of short-
period comets exhibits a steeper exponential dependence on helio-
centric distance ($I \propto r^{-6}$) than that of long-period comets ($I \propto r^{-4}$)
seems to result from the same instrumental effect.

As a statistical approach to elucidating the process of comet
aging, Yabushita and Hasegawa (1981) use a correlation between the
dynamical age, as represented by the binding energy $1/a$, and the
physical age, as represented by the absolute magnitude H. Combining
this correlation with the theoretical diffusion rate in $1/a$, they

find that the mean number of 700 revolutions required for a capture
of a new comet into an orbit of P $\sim$ 100 years is accompanied by a
mean brightness decrease of 1.6 magn. This is only equivalent to
0.002 magn. per revolution, or to 0.03 magn. per century for a typical comet of the Jupiter family. By taking in account only comets
of P > 30 years, they essentially eliminate the systematic effect
of the recurrence of discovery opportunities and the underestimates
of the apparent brightness of short-period comets. Unfortunately,
other selection effects remain involved. The demands on the duration
and accuracy of astrometric observations are much more severe if a
comet is to be classified as a new one; and the mean perihelion distance of new comets is larger. Both of these effects tend to increase the mean absolute brightness of new comets, but the former
also makes them more numerous among the comets observed recently.
This selection is reflected by the proportion of the two dynamical
types of comets compared, strongly varying with time: it increases
from 0.9 : 1 before 1900, through 2.9 : 1 between 1900 and 1950, to
7.5 : 1 after 1950! Now, these variations running along with the improvement of the detection techniques, tend to make the new comets
absolutely fainter. As a result, there are two selection effects
operating in the opposite sense, and it is difficult to believe that
they cancel out exactly. Also, the perturbations in the perihelion
distance make individual revolutions unequal in the decay rate, and
a total extinction of a number of comets in the course of their dynamical evolution implies that the present population of comets of
P $\sim$ 100 years is composed of objects which were originally brighter
than those constituting the present population of new comets. The
problem involves so many poorly known parameters that this approach
does not appear promising.

Another indirect source of information on the aging rates was
pointed out by Hughes and Daniels (1982), who have found a significant difference in the absolute magnitude distribution functions of
long- and short-period comets. Just like in the preceding case,
there are two types of selection effects working in the opposite
sense. For fainter objects the statistics of short-period comets are
much more complete, because of the recurrence of discovery opportunities with their revolution periods. This effect tends to increase
the slope of their distribution function, as compared with that of
the long-period comets. At the same time, their much lower apparent
brightness makes the instrumental effects more severe, and the absolute brightness of fainter objects becomes seriously underestimated. This effect tends to decrease the slope of their distribution
function. And finally, the magnitude distribution of long-period
comets is definitely far from exponential, with an abrupt cutoff
near H = 12 magn., i.e., within the range characteristic for short-
period comets (Kresák, 1978; Sekanina and Yeomans, 1984).

From all what was said it can be concluded that the time scale
and irregularity of the systematic decrease of comet brightness, as
well as the instrumental and selection effects involved, are such
that the available photometric data cannot answer quantitatively the
question of comet lifetimes.

6. DIRECT EVIDENCE FROM OBSERVED CASES OF DISAPPEARANCE

The most straightforward way to determine the lifetimes of comets leads through specifying those cases where the death of the comet coincided with its observation. This is substantially easier for short-period comets, which can be repeatedly observed over decades to centuries. Failure of rediscovery at subsequent predicted returns would provide a good check of the comet's ultimate disappearance. As for the long-period comets, we have to restrict ourselves on circumstantial evidence gathered from a few weeks or months of observation around the perihelion.

At present we know 109 comets with revolution periods less than 20 years, and 18 between 20 and 200 years. Only 83 comets (76%) of the former group and 14 comets (78%) of the latter were observed at their last perihelion passage. Taken at the face value, these figures would imply the disappearance of 30 short-period comets, or a mean lifetime of 26 revolutions. However, a more detailed examination of all circumstances shows that in most cases the loss was due to other reasons than the death of the comet: mainly to changing observing geometry and inaccuracy of orbit determination (Kresák, 1981b). In fact, two thirds of the lost comets were discovered under especially favourable observing conditions which normally recur once or twice per century. For almost all of the lost comets there was either no favourable return since the last observation; or between the last observation and transition into an orbit of larger perihelion distance resulting in a considerable reduction of its apparent brightness; or between the last observation and the time when the ephemeris became entirely unreliable. To explain the failure of rediscovery, it even seems unnecessary to assume progressive fading. There are many cases of rediscovery of comets which were long held for hopelessly lost, the most recent examples being P/Denning-Fujikawa, P/Schwassmann-Wachmann 3, P/DuToit-Hartley, and P/Peters-Hartley. Sorting out all cases where the loss is readily explainable by other reasons, there remain only three or four comets for which a virtual disappearance can be inferred: P/Biela (in 1852), P/Brorsen (in 1879), P/Westphal (in 1913), and possibly P/Neujmin 2 (after 1927). The active lifetimes of comets expressed by the number of revolutions should increase approximately with the square root of the semilatus rectum of the orbit - a dependence which can be replaced with good approximation by one on the perihelion distance. Then, observations suggest lifetimes of about $300\ q^{1/2}$ revolutions for comets of P < 20 years, and about $100\ q^{1/2}$ revolutions for comets of 20 < P < 200 years.

An analysis of the observational records, orbital data and observing geometry of all 400 long-period comets discovered since 1840 (Kresák, 1984) helped to identify eight cases where the disappearance was apparently due to the final extinction of the comet (1859, 1897 III, 1903 I, 1926 III, 1954 II, 1954 XII, 1957 IX, and 1974 XV). One cannot be entirely sure with each individual case, but since there are additional cases of weaker evidence, the number of eight seems to be a fair estimate. There is a full agreement with the expectation that the rate of aging is proportional to the rate of change of the true anomaly, and that the lifetime varies with $q^{1/2}$. Comets approaching extinction are absolutely faint, lack in distinct nuclear condensation of the coma, and the ultimate brightness drop is much steeper than expected for a progressive removal of homogeneous spherical layers of volatiles. The mean active lifetime of long-period comets can be estimated at 20 $q^{1/2}$ revolutions. Unfortunately, most of the dying comets were observed too short to allow determination of their binding energies. Nevertheless, there is some indication that the mean lifetime (the number of revolutions) tends to increase with $1/a$, i.e. with the dynamical age of the comet.

7. INDIRECT DYNAMICAL EVIDENCE

While the present review is concentrated on independent information on the physical aging of comets as a tool for improving the interpretation of the dynamical data, some constraints set by numerical modelling of their motions deserve mentioning. In the first place, it is the excess of nearly parabolic orbits ($1/a < 10^{-4}$) which has led to the concept of the Oort cloud (Oort, 1950).

   The relevance of this phenomenon to the problem of comet aging is illustrated by Figure 2. The histogram pointing upwards shows the distribution of binding energies $E = 1/a$ of 111 best determined orbits of long-period comets (Marsden et al., 1978) before entering the planetary zone; that pointing downwards applies to the same comets after leaving it. The sharp peak at $0 < E < 10^{-4}$, formed by the

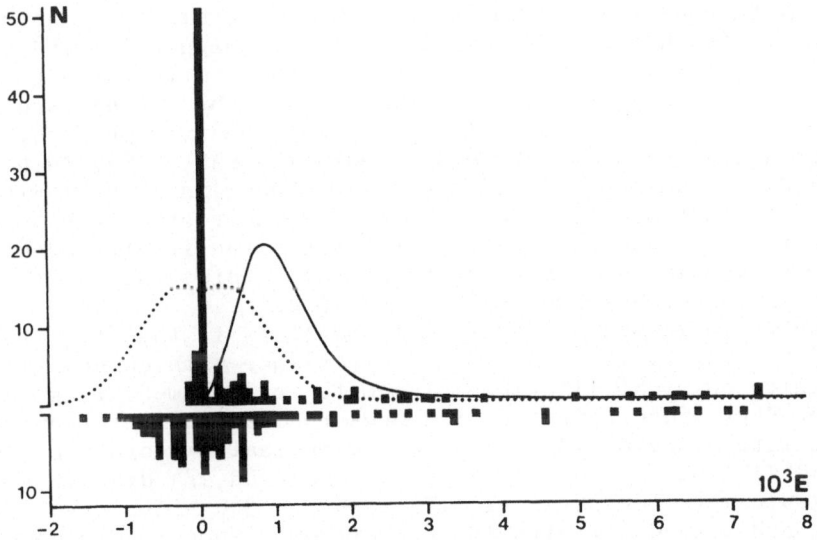

Figure 2. Distribution of the binding energies $E = 1/a$ of long-period comets before entering the planetary region (up), and after leaving it (down). For explanation of the curves see text.

new comets, is smeared out completely just by the first passage between the planets. The dotted curve indicates the expected distribution immediately after the passage (Everhart, 1969), and the full
curve that corrected for the probability of the next apparition
falling into a limited time span (i.e., weighted by $E^{3/2}$). For a
better discrimination, the curves are scaled by a factor of 10 with
respect to the histograms.

While the lower histogram agrees very well with the dotted
curve, there is a striking discrepancy between the upper histogram
and the full curve. This implies that almost all of the new comets
must have been observed at their only passage near the Sun, and will
not return as observable objects anymore. Of course, this conclusion
is tied with the assumption of a statistically steady influx of new
comets, but we have no evidence contrary to it.

The nature of new comets still leaves some open questions.
First, they do not display any apparent destructive changes during
their apparitions, suspected to be the last ones. In view of their
wide range of perihelion distances, one would expect that those with
perihelia closer to the Sun will disappear more rapidly. But just
on the contrary, the proportion of large-q orbits is appreciably
higher among new comets than among the old ones. Second, energy perturbations which would become detectable in the statistics of comet
orbits by smearing out the peak in the 1/a distribution, correspond
to perihelion passages between Saturn and Uranus (Fernández, 1981),
which seems too far for triggering the outgassing activity and aging
of the nucleus. But in the random walk of their perihelia due to
stellar perturbations, many of them should have evolved through this
stage before the discovery apparition.

A simultaneous loss of the dynamical and physical signatures
of new comets is indeed difficult to explain. Whipple (1977b) suggests that their fading away may be due to the removal of a primordial surface frosting of the nucleus, activated by a long exposure
to the cosmic rays. The possibility of a rejuvenation process, effective on a time scale of $10^6 - 10^7$ years (Kresák, 1977) lacks in
the knowledge of an appropriate mechanism. The main problem with the
new comets is that most of them seem to disappear exactly between
their first and second passage near the Sun. Even if all the 8 extinct comets mentioned in Section 6 were new, this would be far from
enough to explain the sharp peak in Figure 2.

Another implication of numerical modelling is that the statistics of dynamical evolutions remain at variance with observation
when infinite physical lifetimes of comets are assumed. This fact
was independently demonstrated by a number of authors. In general
agreement with Dobrovol'skij's (1972) theoretical expectations, Fernández (1981) finds a good fit with the observed 1/a distribution of
long-period comets for physical lifetimes L = 200 - 500 revolutions
at q < 1, and a good fit with the proportion of new to old comets
for L = 210 revolutions. However, his new/old ratio of 1:11 appears
to be strongly underestimated, due to a much higher accuracy of the
orbit determination required for the classification of a comet as
a new one; for the best determined orbits (Marsden et al., 1978) the

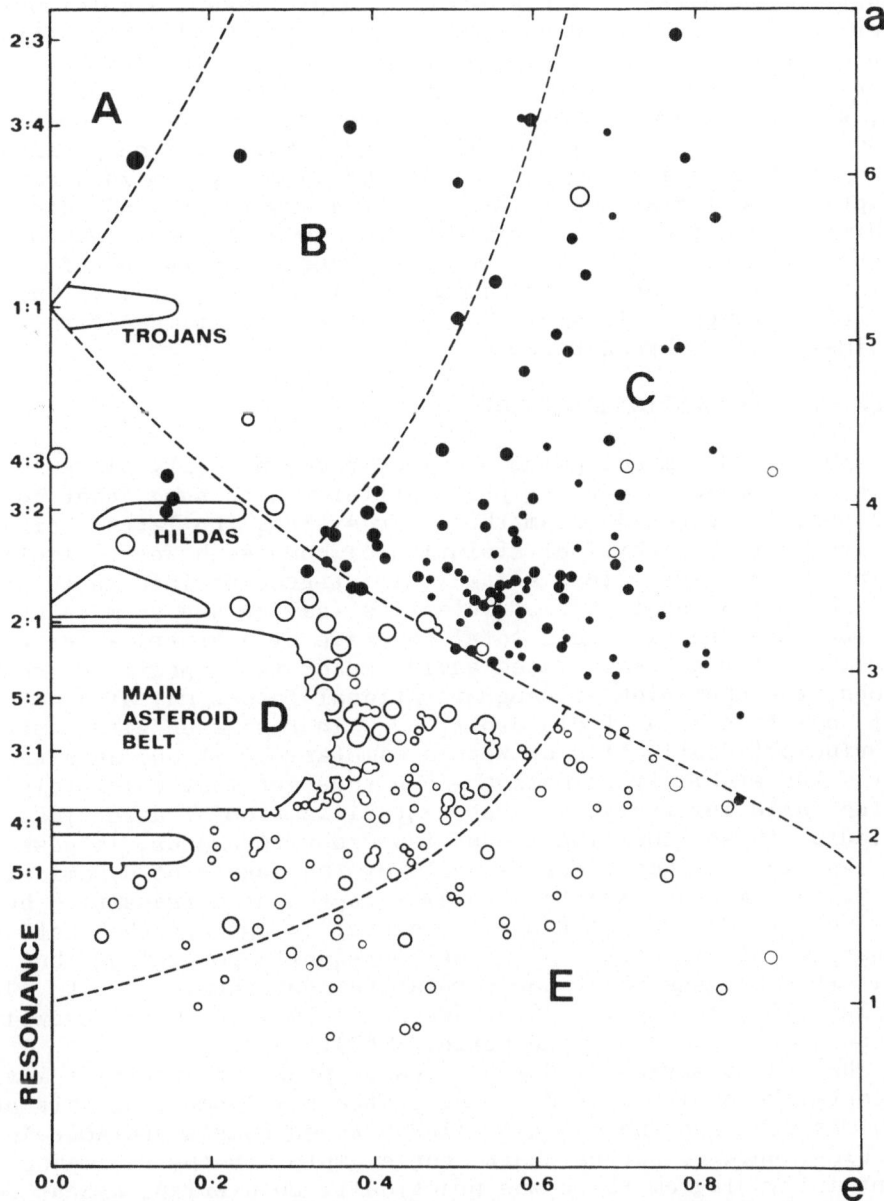

Figure 3. Short-period comets (solid circles) and asteroids (open
circles) plotted in a diagram of semimajor axis vs. eccentricity.
The increasing circle sizes distinguish tentatively the objects by
size : diameter less than 1 km or a lost object, 1 to 3 km, 3 to
10 km, 10 to 30 km, > 30 km. A indicates the transjovian region,
B Jupiter's domain of weak cometary activity, C Jupiter's domain
of strong cometary activity, D the minor planet region, and E the
Apollo region.

proportion is nearly 1:1. Rickman and Vaghi (1976), simulating the
evolution of short-period comets, assume L = 100 $q^{1/2}$ revolutions,
but they point out that longer lifetimes would allow for more realistic replenishment rates from the region between Jupiter and Saturn. Weissman (1980) assumes much higher values of L = 600 to L =
28,000 for q = 1, depending on the albedo. Dividing the physical end
states of long-period comets left in the solar system into random
disruption and formation of an insulating crust, he finds disruption
prevailing by a factor of four. He also finds the best combined fit
of energy and perihelion distance distributions with 85% of comets
subject to disruption and 15% immune to it. Hence, also in this case
different interpretations exist, and the quantitative estimates of
lifetimes cover a broad range.

8. THE FINAL EVOLUTIONARY STAGE

From a purely dynamical point of view, there are only two possible
end fates of comets: catastrophic collision with some other object
and hyperbolic ejection from the solar system, the latter being much
more frequent. The physical evolution of comets, however, implies
alternative possibilities: a total disintegration into meteoroids,
dust and gas; a total loss of volatiles leaving one or more asteroid-
like inactive nuclei; and a total coverage by a crust, after which
the remnant can be reactivated again. Under very specific circumstances, the operation of nongravitational forces may also help the
comet to settle in a stable orbit, e.g. in resonance with Jupiter.
The principal distinction between a cometary orbit and an asteroidal
orbit - the stability of motion - can thus get lost completely. This
is also valid inversely, because destabilization of asteroids, in
particular those librating around low-order resonance, is possible
under special circumstances as well. By the degree of dynamical stability, the Amor and Apollo asteroids represent a transition between
normal asteroids and short-period comets. It appears that both sources participate in maintaining the Amor-Apollo population, but there
is little consensus about their relative contribution (Öpik, 1963;
Whipple, 1967; Wetherill, 1979; Kresák, 1979 and 1981c; Degewij and
Tedesco, 1982; Simonenko and Levin, 1983).
        The recent series of discoveries of peculiar asteroids moving
in comet-like orbits opened an unexpected development of this problem. Not long ago the dividing line between comets and Amor-Apollo
asteroids appeared rather sharp. For example, in the semimajor axis/
eccentricity diagram there was practically no overlap, except for
the asteroid 944 Hidalgo situated deep within the comet region, and
the librating asteroids whose stability is controlled by the resonance rather than by the size and shape of the orbit. The a/e diagram, depicted in its 1978 shape in Kresák (1979b) and Degewij and
Tedesco (1982) is presented in its updated version in Figure 3. One
can clearly recognize a number of new asteroids occupying the comet
region, in particular 5025 P-L (Van Houten et al., 1984), 1983 SA,
1982 YA, 1984 BC and 1983 XF. All of them are faint objects complying in every respect with our expectation of extinct comet nuclei.

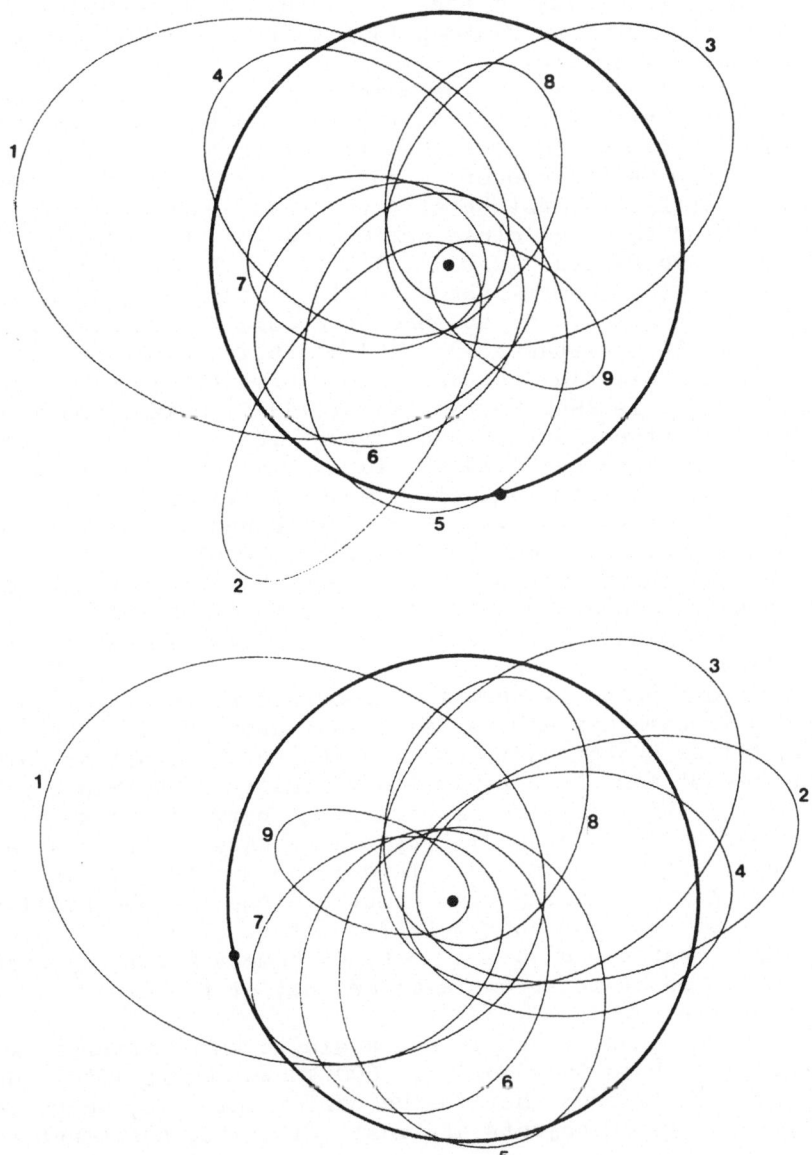

Figure 4. Above, most comet-like asteroid orbits : 1 = 944 Hidalgo,
2 = 5025 P-L, 3 = 1983 SA, 4 = 1982 YA, 5 = 1984 BC, 6 = 1983 XF,
7 = 1983 LC, 8 = 1983 VA, 9 = 2212 Hephaistos. Below, similar comet
orbits : 1 = P/Wild 1, 2 = P/Denning-Fujikawa, 3 = P/Swift-Gehrels,
4 = P/Finlay, 5 = P/Kopff, 6 = P/Tempel 1, 7 = P/Blanpain, 8 = P/
Grigg-Skjellerup, 9 = P/Encke. The two pictures are rotated by 90°
with respect to one another, as indicated by the dots marking the
Sun and the perihelion of Jupiter's orbit (the thick, low-eccentricity ellipse). Vernal equinox is down for the asteroids and to
the left for the comets.

The long-term integrations of these objects, as performed by Benest et al. (1985) and Hahn and Rickman (1985) are entirely consistent with their cometary origin.

The orbits of nine asteroids which are most comet-like are plotted in the upper half of Figure 4; the lower half shows a selection of nine comet orbits which are most similar to them. The two pictures are rotated by 90° with respect to one another to bear out the similarity more clearly; it must be stressed that there are no individually associated pairs of close orbits, as one could infer for Nos 1 (944 Hidalgo and P/Wild 1) or 3 (1983 SA and P/Swift-Gehrels) after the rotation. Another comet-like asteroid is 1939 TN, but this may well be an active comet. The absence of a coma may be simply due to its large perihelion distance (3.4 AU) and brightness near the plate limit of the four existing photographic observations. An entirely exceptional object is 2060 Chiron (Kowal, 1979; Oikawa and Everhart, 1979) revolving between Saturn and Uranus. The asteroid of smallest known perihelion distance, 1983 TB, identified as the parent body of the Geminid meteor stream (Whipple, 1983b; Williams et al., 1985), demonstrates that even inactive objects with aphelia in the inner zone of the asteroid belt may be indistinguishable from active comets as to the production of meteor showers and the observable properties of their members (Jacchia et al., 1967). Transition from the objects of Figure 4 to typical Amor and Apollo asteroids is represented by 6433 P-L, 1979 VA, and others.

An important point is that the number of known asteroidal objects moving in cometary orbits is already about 1/10 of the active comets, as far as objects with aphelia inside the orbit of Saturn are concerned. Except for the big Hidalgo (diameter 28.6 km according to Bowell et al., 1979), their diameters range between 0.5 km and 6.0 km (median 1.8 km) if albedo of S-type asteroids is assumed, and are twice as large for C-type albedo. This is about the same as the size estimates for larger short-period comets. Since the observational coverage must be much more complete for active objects of comparable size, it seems that the number of extinct comets revolving within the orbit of Saturn is at least the same as that of the active ones and possibly even greater.

On the other hand, there is not a single known asteroid moving in an orbit similar to those of over 600 known comets with aphelia beyond Saturn. It is true that in the ecliptical zone, where searches for small solar system objects are most effective, short-period comets would spend more time and move at lower angular velocities, which makes their detection easier. But even so the discrepancy is much too large. It suggests that the end fates of long- and short-period comets and, hence, their structure and physical evolutions are different.

9. CONCLUSIONS

From the various possible approaches to determining the lifetimes of comets, those based on the mass loss rates are limited by our lack of knowledge of the exact sizes of cometary nuclei; and those based on the progressive fading are biased by instrumental and selection effects involved in the determination of the absolute brightness of comets. Indirect methods still give controversial or ambiguous results. Direct observational evidence on the disappearance of some comets, when properly corrected for all interfering effects, seems to be most reliable, yielding estimates of mean lifetimes with an uncertainty presumably within a factor of two.

The main feature of the lifetimes, as expressed by the number of revolutions for which the comet remains active, is their broad dispersion. The physical survival is strongly correlated with the revolution period, and thereby with the dynamical age of the comet, as documented by the following table. For the first entry the period refers to the original orbit, i.e. to the osculating orbit before entering the planetary region.

Revolution period	Mean lifetime
years	revolutions
$> 10^6$	mostly 1
$200 - 10^6$	$20 \ q^{1/2}$
$20 - 200$	$100 \ q^{1/2}$
< 20	$300 \ q^{1/2}$

There is little doubt about an additional considerable dispersion within individual dynamical types; but in general, the number of revolutions for which the comet remains active tends to increase with the number of revolutions required for entering the respective type of orbit. The only feasible explanation of this interrelation is a substantial difference between individual comets just at the time when they enter the inner region of the solar system and become active for the first time. During the number of revolutions required for capturing the comet into a short-period orbit, selection by size and structure becomes effective. If all comets come from the Oort cloud, those which are not abnormally resistive disintegrate long before reaching the short-period stage, unless they are ejected by accelerating perturbations.

After a new comet has passed for the first time near the Sun, its activity may become substantially reduced by the loss of a super-active frosting (Whipple, 1977b). During the subsequent returns the comet would purge its renewing insulating crust (Whipple, 1978b). At this evolutionary stage, the activity is kept at a slowly decreasing level, corresponding approximately to a progressive removal of surface layers to the same depth per revolution. Normally, the comet would become destroyed after a few tens of revolutions. This process may be accompanied by splitting of the nucleus which would occur, on the average, about twice during the comet's active lifetime. However,

only exceptionally the secondary fragments are large enough to reduce substantially the survival time of the primary component. The
progressive disintegration of the parent object of the Kreutz group
of comets by solar tides is possibly one of such exceptions.

Towards the end of the comet's lifetime, the relative mass loss
and the absolute brightness drop would strongly accelerate, giving
rise to an abrupt decrease of the number of active comets at $H > 12$
(Kresák, 1978; Sekanina and Yeomans, 1984). As evidenced by the observed absence of asteroidal objects moving in long-period orbits,
all or almost all long-period comets disintegrate completely at the
end of their active lifetime. They may also evolve into objects of
Halley type ($P \sim 100$ years), provided that they are able to survive
physically $\sim 100 \, q^{1/2}$ revolutions. The latter figure is rather uncertain due to the very limited statistical sample available.

The situation is different for short-period comets of the Jupiter family ($P < 20$ years), typical lifetimes of which amount to
$\sim 300 \, q^{1/2}$ revolutions. A definite disproportion between their mass
loss rates and photometric size estimates indicates the presence of
a shielding crust of non-volatile low-albedo materials. Temporary
activation of isolated surface areas (in rare cases possibly due to
impacts of small satellites - Whipple, 1983a and 1984) makes them
more apt to sudden brightness bursts, which can reduce their lifetimes even more drastically than the splitting (Kresák, 1974b). It
appears that the conversion factor between absolute brightness and
mass (Equation 2) is substantially higher for short-period comets
than for the long-period ones.

The reason for this difference between long- and short-period
comets is puzzling, the more that there are no observable systematic
differences in their radiation mechanism (Donn, 1977). One possible
explanation is that their birthplaces are different, which would
allow for their different internal constitution. The short-period
comets may come, at least predominantly, from the inner condensation
of the Oort cloud, as hypothesized by Whipple (1972), Hills (1981),
and Fernández and Ip (1983). This inner condensation would provide
an adequate source of replenishment not only for the short-lived
family of short-period comets, but also for the outer envelope of
the Oort cloud, stripped away during encounters with the giant molecular clouds (Van den Bergh, 1982; Bailey, 1983). In fact, if the
latter process is also going on, the original source would be essentially the same. But even in this case, or in the case of absence
of the inner condensation, the two types of comets differ substantially by their dynamical history. The comets of the Jupiter family
would have to make a number of revolutions with perihelia not far
from the orbit of Jupiter, within Everhart's (1973) capture zone,
before their perihelion is changed into aphelion. This interlude
could affect their subsequent physical evolution and survival time,
whether by changing their surface structure or by removal of the
less resistive objects, before they become detectable. The IRAS data
files may already contain some information on such objects.

The decelerated physical evolution of short-period comets would
allow at least some of them to leave extinct asteroid-like nuclei.

It remains open whether such objects still include some supply of volatiles under their surface crusts, which would make possible a later re-activation, say, by a non-destructive collision. Otherwise their dynamical lifetimes would be limited mainly by accelerating encounters with Jupiter.

REFERENCES

Bailey, M.E.: 1983, Mon. Not. R. Astron. Soc. 204, pp. 603-633.

Benest, D., Bien, R. and Rickman, H.: 1985, this volume.

Bowell, E., Gehrels, T. and Zellner, B.: 1979, in T. Gehrels (ed.), Asteroids, Univ. Arizona, Tucson, pp. 1108-1129.

Carusi, A., Kresák, L., Perozzi, E. and Valsecchi, G.B.: 1984, IAU Circ. No. 3940, and this volume.

Degewij, J. and Tedesco, E.F.: 1982, in L.L. Wilkening (ed.), Comets, Univ. Arizona, Tucson, pp. 665-695.

Delsemme, A.H.: 1976, Lecture Notes in Phys. 48, pp. 314-318.

Delsemme, A.H.: 1982, in L.L. Wilkening (ed.), Comets, Univ. Arizona, Tucson, pp. 85-130.

Delsemme, A.H. and Rudd, D.A.: 1973, Astron. Astrophys. 28, pp. 1-6.

Dobrovol'skij, O.V.: 1972, IAU Symp. 45, pp. 352-355.

Dobrovol'skij, O.V., Ibadinov, Kh.I., Aliev, S. and Gerasimenko, S. I.: 1983, Dokl. Akad. Nauk Tadzh. SSR 26, pp. 25-29.

Donn, B.A.: 1977, in A.H. Delsemme (ed.), Comets, Asteroids, Meteorites, Univ. Toledo, Ohio, pp. 15-23.

Everhart, E.: 1969, Astron. J. 74, pp. 735-750.

Everhart, E.: 1973, Astron. J. 78, pp. 329-337.

Feldman, P.D.: 1982, in L.L. Wilkening (ed.), Comets, Univ. Arizona, Tucson, pp. 461-479.

Fernández, J.A.: 1980, Mon. Not. R. Astron. Soc. 192, pp. 481-491.

Fernández, J.A.: 1981, Astron. Astrophys. 96, pp. 26-35

Fernández, J.A. and Ip, W.-H.: 1983, Icarus 54, pp. 377-387.

Finson, M.L. and Probstein, R.F.: 1968, Astrophys. J. 154, pp. 327-380.

Hahn, G. and Rickman, H.: 1985, Icarus, in press.

Harwit, M.: 1967, in J.L. Weinberg (ed.), The Zodiacal Light and the Interplanetary Medium, NASA, Washington, pp. 307-313.

Hills, J.G.: 1981, Astron. J. 86, pp. 1730-1740.

Ho Peng Yoke : 1962, Vistas in Astron. 5, pp. 127-225.

Hughes, D.W. and Daniels, P.A.: 1982, Mon. Not. R. Astron. Soc. 198, pp. 573-582.

Jacchia, L.G., Verniani, F. and Briggs, R.E.: 1967, Smithson. Contr. Astrophys. 10, pp. 1-139.

Kamoun, P.G., Campbell, D.B., Ostro, S.J., Pettengill, G.H. and Shapiro, I.I.: 1982a, Science 216, pp. 293-295.

Kamoun, P.G., Pettengill, G.H. and Shapiro, I.I.: 1982b, in L.L. Wilkening (ed.), Comets, Univ. Arizona, Tucson, pp. 288-296.

Kowal, C.T.: 1979, in T. Gehrels (ed.), Asteroids, Univ. Arizona, Tucson, pp. 436-439.

Kresák, L.: 1965, Bull. Astron. Inst. Czechosl. 16, pp. 348-355.

Kresák, L.: 1973, Bull. Astron. Inst. Czechosl. 24, pp. 264-283.
Kresák, L.: 1974a, Bull. Astron. Inst. Czechosl. 25, pp. 87-112.
Kresák, L.: 1974b, Bull. Astron. Inst. Czechosl. 25, pp. 293-304.
Kresák, L.: 1977, in A.H. Delsemme (ed.), Comets, Asteroids, Meteo-
    rites, Univ. Toledo, Ohio, pp. 93-97.
Kresák, L.: 1978, Bull. Astron. Inst. Czechosl. 29, pp. 114-125 and
    135-149.
Kresák, L.: 1979a, IAU Symp. 90, pp. 211-222.
Kresák, L.: 1979b, in T. Gehrels (ed.), Asteroids, Univ. Arizona,
    Tucson, pp. 289-309.
Kresák, L.: 1981a, Bull. Astron. Inst. Czechosl. 32, pp. 19-40.
Kresák, L.: 1981b, Bull. Astron. Inst. Czechosl. 32, pp. 321-339.
Kresák, L.: 1981c, Adv. Space Res. 1, pp. 85-90.
Kresák, L.: 1982a, in L.L. Wilkening (ed.), Comets, Univ. Arizona,
    Tucson, pp. 56-82.
Kresák, L.: 1982b, Bull. Astron. Inst. Czechosl. 33, pp. 150-160.
Kresák, L.: 1984, Bull. Astron. Inst. Czechosl. 35, pp. 129-150.
Lindblad, B.A.: 1985, this volume.
Marsden, B.G., Sekanina, Z. and Everhart, E.: 1978, Astron. J. 83,
    pp. 64-71.
McIntosh, B.A. and Hajduk, A.: 1983, Mon. Not. R. Astron. Soc. 205,
    pp. 931-943.
Meisel, D.D. and Morris, C.S.: 1982, in L.L. Wilkening (ed.),
    Comets, Univ. Arizona, Tucson, pp. 413-432.
Ney, E.P.: 1982, in L.L. Wilkening (ed.), Comets, Univ. Arizona,
    Tucson, pp. 323-340.
Oikawa, S. and Everhart, E.: 1979, Astron. J. 84, pp. 134-139.
Oort, J.H.: 1950, Bull. Astron. Inst. Netherl. 11, pp. 91-110.
Öpik, E.J.: 1963, Adv. Astron. Astrophys. 2, pp. 219-262.
Öpik, E.J.: 1973, Astrophys. Space Sci. 21, pp. 307-398.
Rickman, H. and Vaghi, S.: 1976, Astron. Astrophys. 51, pp. 327-342.
Roemer, E.: 1966, Mém. Soc. R. Sci. Liège 12/1, pp. 23-28.
Sekanina, Z.: 1969, Astrometriya i Astrofizika 4, pp. 54-76.
Sekanina, Z.: 1982, in L.L. Wilkening (ed.), Comets, Univ. Arizona,
    Tucson, pp. 251-287.
Sekanina, Z. and Yeomans, D.K.: 1984, Astron. J. 89, pp. 154-161.
Shul'man, L.M.: 1972, IAU Symp. 45, pp. 271-282.
Simonenko, A.N. and Levin, B.Yu.: 1983, Highlights of Astronomy 6,
    pp. 391-398.
Stefanik, R.P.: 1966, Mém. Soc. R. Sci. Liège 12/1, pp. 29-32.
Svoreň, J.: 1979, Contr. Astron. Obs. Skalnaté Pleso 8, pp. 105-140.
Van den Bergh, S.: 1982, J. Astron. Soc. Canada 76, pp. 303-308.
Van Flandern, T.C.: 1981, Icarus 47, pp. 480-486.
Van Houten, C.J., Herget, P. and Marsden, B.G.: 1984, Icarus 59, pp.
    1-19.
Vsekhsvyatskij, S.K.: 1958, Fizicheskie Kharakteristiki Komet, Gos.
    izd. fiz.-mat. literatury, Moskva, 575 pp.
Vsekhsvyatskij, S.K.: 1966, Fizicheskie Kharakteristiki Komet 1954-
    1960 gg, Nauka, Moskva, 88 pp.
Vsekhsvyatskij, S.K.: 1967, Komety 1961-1965 gg, Nauka, Moskva, 86
    pp.

Vsekhsvyatskij, S.K.: 1979, Fizicheskie Kharakteristiki Komet 1971-
    1975 gg, Naukova dumka, Kiev, 116 pp.
Vsekhsvyatskij, S.K.: 1981, Problemy Kosm. Fiziki 16, pp. 20-25.
Vsekhsvyatskij, S.K. and Il'chishina, N.I.: 1974, Fizicheskie Kharak-
    teristiki komet 1965-1970 gg, Nauka, Moskva, 112 pp.
Weissman, P.R.: 1980, Astron. Astrophys. 85, pp. 191-196.
Wetherill, G.W.: 1979, Icarus 37, pp. 96-112.
Whipple, F.L.: 1964, Astron. J. 69, pp. 152-155.
Whipple, F.L.: 1967, in J.L. Weinberg (ed.), The Zodiacal Light and
    the Interplanetary Medium, NASA, Washington, pp. 409-426.
Whipple, F.L.: 1972, IAU Symp. 45, pp. 401-408.
Whipple, F.L.: 1977a, Icarus 30, pp. 736-746.
Whipple, F.L.: 1977b, in A.H. Delsemme (ed.), Comets, Asteroids,
    Meteorites, Univ. Toledo, Ohio, pp. 25-35.
Whipple, F.L.: 1978a, in J.A.M. McDonnell (ed.), Cosmic Dust, Wiley,
    Chichester, pp. 1-73.
Whipple, F.L.: 1978b, Moon and Planets 18, pp. 343-359.
Whipple, F.L.: 1983a, Highlights of Astronomy 6, pp. 323-331.
Whipple, F.L.: 1983b, IAU Circ. No. 3881.
Whipple, F.L.: 1984, Icarus, in press.
Whipple, F.L. and Douglas-Hamilton, D.H.: 1966, Mém. Soc. Roy. Sci.
    Liège 12/1, pp. 469-480.
Whipple, F.L. and Hamid, S.E.: 1972, IAU Symp. 45, pp. 152-154.
Whipple, F.L. and Sekanina, Z.: 1979, Astron. J. 84, pp. 1894-1909.
Williams, I.P., Fox, K.A. and Hunt, J.: 1985, this volume.
Yabushita, S. and Haségawa, I.: 1981, Mon. Not. R. Astron. Soc. 195,
    pp. 361-370.

DISCUSSION

B. Lokanadham : How the complete disintegration of some comets
could be explained ?

L. Kresák : Unless the nucleus is differentiated with depth,
the only processes which can prevent a complete disintegration are
removal of the perihelion far from the Sun or formation of an insulating crust. Otherwise the nucleus would grow smaller, until no
sizable object remains. The total disappearance of P/Biela and P/
Brorsen, and the absence of asteroidal objects moving in long-period
high-eccentricity orbits, hardly admit an alternative end fate for
a majority of comets.

P. Weissman : I believe that these lifetimes you state are too
short and they are influenced by a variety of physical and observational effects. For new comets from the Oort cloud it is obvious
that they are anomalously bright due to a surface layer of more volatile ices which sublimate away on the first return. This is clear
from the 1/a vs. q scatter diagram, where only new comets are found
beyond about 3 AU, the point at which water ice sublimation becomes
negligible. This step decrease in brightness occurs only on the
first return and only slow fading occurs afterward. Our thermal calculations show that for a 1 km nucleus with perihelion of 1 AU the

lifetime is about 1000 returns against sublimation. In reality, the
governing process on the lifetime of comets is more likely the build-
up of nonvolatile crusts on the nucleus, or random disruption, each
of which gives much shorter lifetimes, but not as short as you state.
Also, the thermal modeling indicates the sublimation or crust build-
up processes have a $q^{0.7}$ dependence (the exponent increasing sharply
beyond 2 - 3 AU), not $q^{0.5}$. Random disruption does not appear to be
strongly correlated with perihelion distance. I do agree that comets
appear to have variable "survivability" as some sort of intrinsic
quality, the more survivable comets being the ones which evolve to
short-period orbits.

       L. Kresák : For the mean lifetime of 1000 returns, the proba-
bility of having just a single observational record of extinction of
a long-period comet would be about 1:8. In fact, there are about five
cases where this seems to be proven beyond any shadow of doubt, and
about ten additional cases of various degree of confidence. From this
point of view, a mean lifetime much longer than 20 returns appears
inconsistent with observational evidence. But I agree that the most
survivable objects, in particular some short-period comets, can remain active for 1000 revolutions or more. Observational evidence also
casts doubts on the assumption that the new comets from the Oort
cloud simply become much fainter for the subsequent returns. In that
case we would have an abundance of absolutely faint, dynamically old
long-period comets. However, what we observe is just the opposite: a
definite lack of long-period comets of $H > 12$ magn. The lifetime dependence on $q^{0.5}$ is the simplest way to take in account the integrated insolation, without considering its efficiency for the destruction processes. More sophisticated thermal models may yield different
and variable values of the exponent, but I think we know too little
about the crusting and purging processes, phase transition effects
etc. to be sure which is correct. The q-distribution of comet disap-
pearances is fully consistent with the exponent of 0.5, but it would
not contradict to that of 0.7 either. The difference is simply too
small, within ± 10% for one half of the known comets, and for statistical mean lifetime estimates it is practically irrelevant. Similarly, the neglect of the reduced irradiation efficiency at larger
solar distances cannot affect the results appreciably, because one
half of these comets have q < 1 AU, and thus receive more than 50% of
the total irradiation at r < 2 AU. The main source of uncertainty is
definitely the limited size of the sample of known comets, and the
limited time span covered by the observations.

PLANETARY PERTURBATIONS, DYNAMICAL ENERGY AND EVOLUTION OF ORBITAL
ELEMENTS OF PERIODIC COMETS.

A. MANARA - L. BUFFONI - M. SCARDIA
Osservatorio Astronomico di Brera
20121 MILANO
Italy

ABSTRACT. Planetary perturbations, dynamical energy and orbital evolution of the elements of three comets are calculated. This paper presents a part of the catalogue that will be issued in the next year.

1. INTRODUCTION

This paper presents a part of the catalogue that will be finished
about the end of 1985.
    The catalogue will contain a chronological list of planetary
perturbations, dynamical energy (Buffoni et al., 1982a), orbital evolutions (Buffoni et al., 1982b) of 83 short-period comets that have at
least a perihelion passage from 1968 through 1996.
    The evolution of orbital elements is calculated only for those
comets that show strong planetary perturbations and the dynamical
energy for those comets that can have a doubtful orbit after a strong
perturbation.
    In addition the catalogue contains the orbital elements approxi-
mately at the epoch of the perihelion passage.
    As an example we study P/Whipple (1978 VIII) and the planetary
perturbations on P/Borelly (1981 IV), a comet with very short period
(about 6.8 years), and P/Crommelin (1956 VI), a comet with only one
perihelion passage in the period considered (period of the comet about
27.9 years).
    The nongravitational forces are not taken into account in our cal-
culations. These forces, probably random and impulsive in nature, are
of small value, and in the graphs of planetary perturbations would not
cause significant changes. They would have more importance in the calculation of the orbital energy of those comets with dubious orbit, but
also in this case, due to their small intensity, there would not be
definitive results.
    For the computation of orbits Encke's method (Buffoni et al.,1971)
has been used.
    The time step of integration can be changed from many days to 1
day. For the three comets under examination it was taken constant,

A. Carusi and G. B. Valsecchi (eds.), Dynamics of Comets: Their Origin and Evolution, 303–310.
© *1985 by D. Reidel Publishing Company.*

namely 7 days.
        Machine time for each planetary perturbation graph is about ten
minutes; for orbital energy and elements variation is about three hours.
        Computations have been made by PDP 11/34 computer of Brera Astro-
nomical Observatory of Milan.

2. SOURCES OF STARTING ORBITS

        Osculating elements, from which we started, have been taken from
Marsden (1982). As a rule, they correspond to the last apparition.
        A few orbits were taken from D.F. Bender (1981) because more accu-
rate orbits were available. We must remember that in Encke's method the
choice of reference orbit is very important.

3. AIM OF THE WORK

        The purpose of this work is to have at hand a quick reference from
which the dynamical characteristic of any short period comet in the
near future be ready available.
        Our work stands out from the other similar works owing to the
short period considered, which permits to study in detail the single
comets passages.
        The graph of planetary perturbations gives an immediate image of
the epoch of possible "close encounters" with the planets. The graph of
orbital energy is very interesting to see if the perturbations by the
planets on the comet can change the sign of the energy; in fact if the
energy remains negative the orbit will be elliptical, if the energy becomes positive the orbit will be hyperbolic. Orbital elements variations
can be used to study the nature and origin of comets.
        The present paper is only a section of our complete research about
a sistematic calculation of comet orbits from observations published in
Smithsonian Astrophysical Observatory and MPC circulars, conducted from
1980 by the Astronomical Observatory of Brera (Buffoni et al.,1981,
1983).
        From its birth our Observatory had a real interest about comets
and astronomers like Boscovich, Oriani, Carlini and Schiaparelli studied
these celestial objects; we have especially to mention G. Celoria (1921)
a pioneer of determination of comet orbits of Brera Observatory
(Marsden, 1982). Both from sistematic study of comet orbits and from an
analysis of the results of this paper, our purpose is to study those
comets particularly useful for an appreciable improvement of the understanding of cometary evolution.

4. P/WHIPPLE

        Fig. 1 shows planetary perturbations on comet Whipple. The per-
turbing forces that the individual planets exert on the comet are shown
on logarithmic scale as the ratio R between the modules of perturbing

force and the solar attractive force.

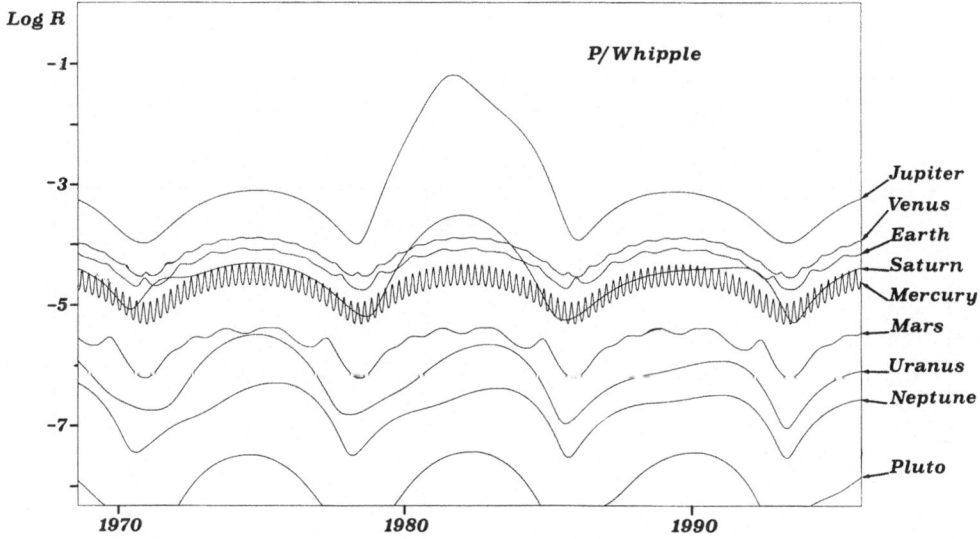

Figure 1. Planetary perturbations on P/Whipple

From an examination of the graph, it appears that in the period considered the planet which exerts the greatest perturbing force on the comet is always Jupiter and the maximum of this force is about 6% of the of the Sun in May 1981. We can see also the perturbing forces of the other eight planets.

We can see in Fig. 2 the orbital energy of Whipple's comet

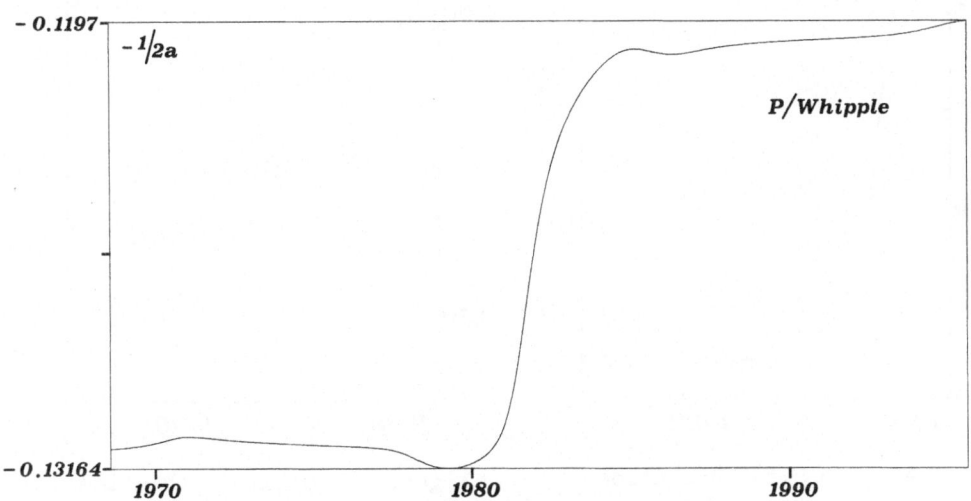

Figure 2. Mechanical energy of P/Whipple

expressed in units of -1/2a . The Jupiter perturbation is also well
shown in this graph. Even though the perturbation is large, the comet
energy always remains negative, consequently the comet orbit is elliptical. The evolution of inclination, in degrees, is shown in Fig. 3 .
All sudden changes of the elements are due to close approaches to Jupiter. As it can be seen, the inclination exibits a sudden decrease in

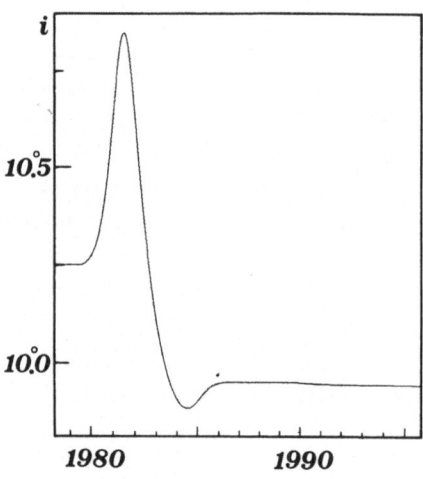

Figure 3. Orbital evolution
of inclination

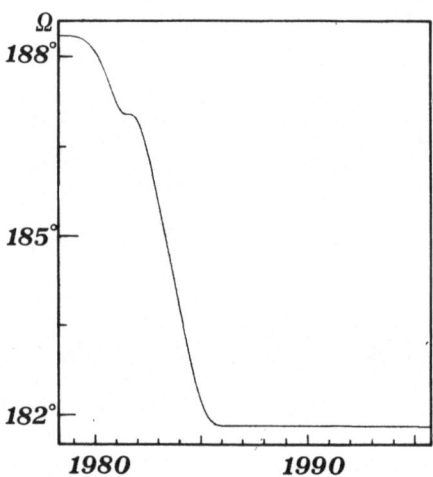

Figure 4. Orbital evolution
of node

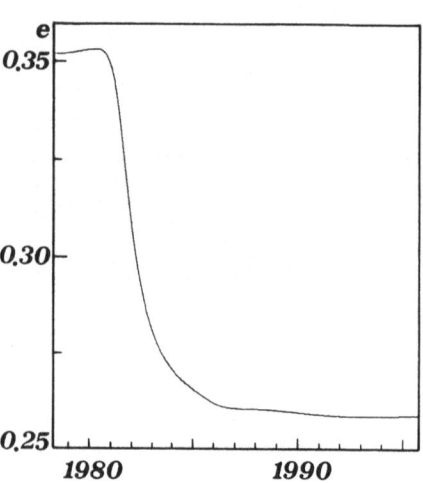

Figure 5. Orbital evolution
of eccentricity

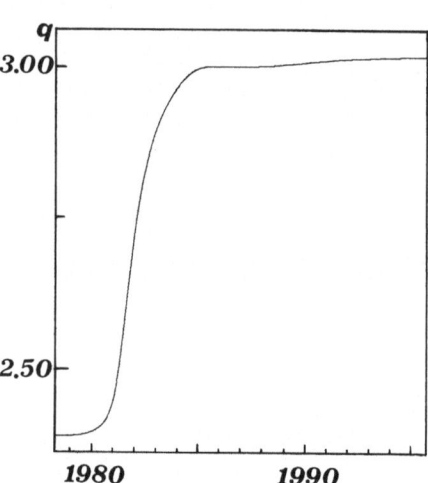

Figure 6. Orbital evolution
of perihelion distance

1981 after which the inclination remains constant.

The node (Fig. 4) and the eccentricity (Fig. 5) both also decrease suddenly and, after 1985, reach a constant value.

The perihelion distance (Fig. 6), also in 1981, grows suddenly when the comet is approaching Jupiter and, after 1985, reaches a constant value.

The argument of perihelion (Fig. 7) shows a sudden increase and finally reaches a nearly constant value.

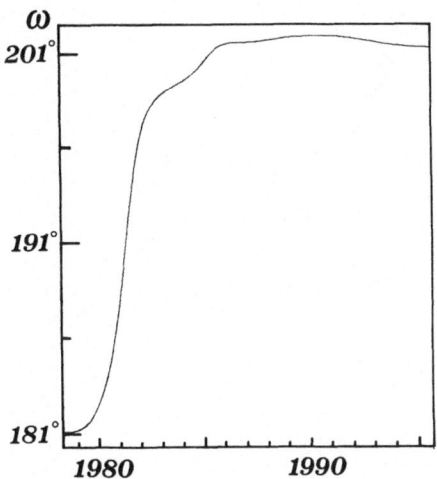

Figure 7. Orbital evolution of argument of perihelion

From the changes of times T of perihelion passage and of true anomaly of comet P/Whipple, which we calculated for the period 1978-1995, we found that the orbital period increases due to Jupiter's perturbation. We do not give the graph of perihelion passage variation because it is difficult for a graphical representation since our computations refer to several passages.

The minimum approach distance to Jupiter is 0.631 AU (1981, July 9). In this case the perturbation time step of integration is one day. Fig. 8 shows the projection of the orbits of Jupiter and P/Whipple into ecliptic plane for an interval of 5500 days around the epoch of approach to Jupiter.

The position of the comet and of Jupiter are given at corresponding epochs every 30 days starting from 1979, May 7 to 1983, June 15.

We can see in Table 1 the orbital elements approximately at the epoch of the perihelion passage. The epoch is given in year, month, day.

For the time of perihelion passage T one decimal of the day is added; the perihelion distance is given in AU.

All the angular elements are in degrees and their fractions, and referred to the mean equinox and ecliptic of 1950.0 .

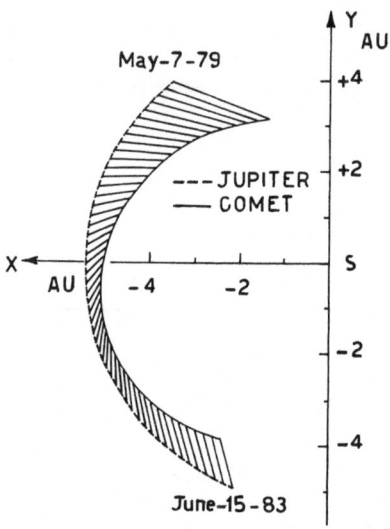

Figure 8.   Projection of Jupiter and
P/Whipple orbits into ecliptic plane

Table   1

Orbital elements of P/Whipple approximately at the epoch of perihelion

Epoch	T	q	e	i	Per.	Node
1978.0402	1978.03275	2.468633	0.352244	10.246	189.976	188.339
1986.0629	1986.06250	3.077550	0.260582	9.943	202.047	181.800
1994.1218	1994.12224	3.093882	0.258705	9.934	201.881	181.792

5. P/BORELLY

   Planetary perturbations on P/Borelly are shown in Fig. 9   . In the
graph we can see the very strong perturbation by Jupiter in 1972; the
perturbation by the Earth in 1988 (the planet which exerts the greatest perturbing force in that short period) and the perturbation by Mars
in 1994.

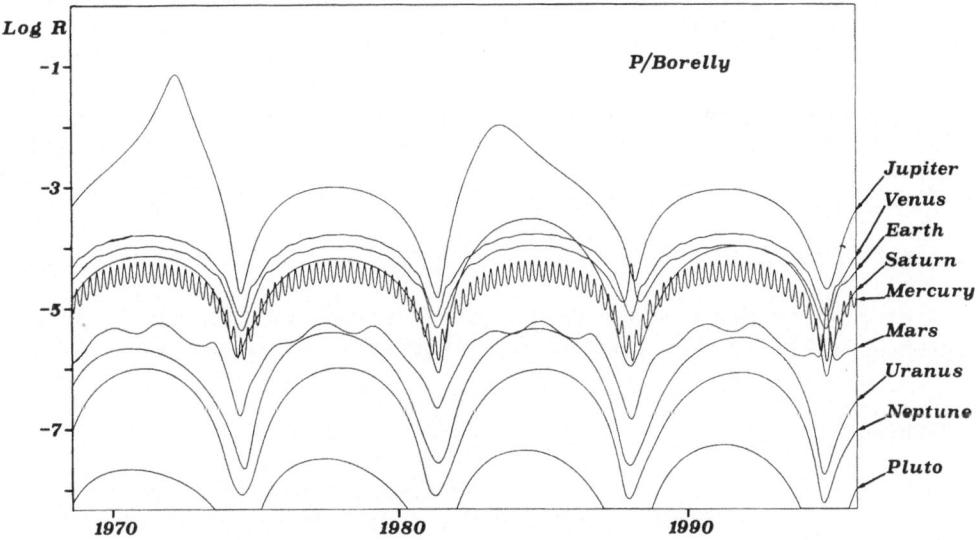

Figure 9.   Planetary perturbations on P/Borelly

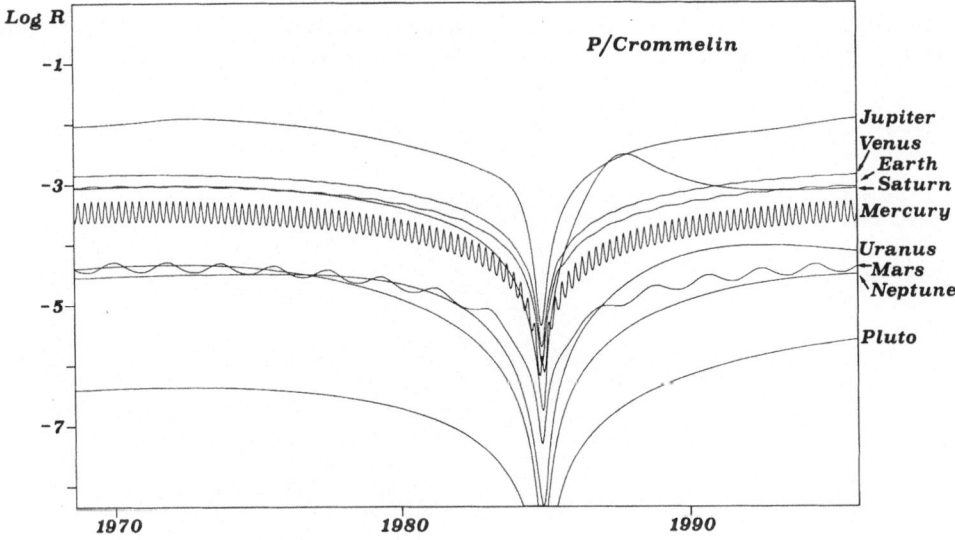

Figure 10.   Planetary perturbation on P/Crommelin

6. P/CROMMELIN

The perihelion passage in 1984 and the largest perturbation due to

Saturn around 1987–1988 are clearly shown in the diagram of planetary perturbations on P/Crommelin (Fig. 10). It is very interesting to observe that for P/Crommelin and generally for most of the comets the Mercury's perturbation is larger than that of Mars, Uranus and Neptune, planets which have a much greater mass; the explanation is that generally the comets pass much closer to Mercury than to Mars, Uranus and Neptune.

7. ACKNOWLEDGMENTS

We wish to thank A. Carusi and G. Valsecchi of Istituto di Astrofisica Spaziale – C.N.R., Roma, for their interest in the work.

REFERENCES

BENDER,D.F.: 1981, J.P.L. Interoffice Memorandum 312/81.3-1686
BUFFONI,L. and MANARA,A.: 1971, Pubbl. Oss. Astr. Milano-Merate n° 24
BUFFONI,L.,MANARA,A. and SCARDIA,M.: 1981, Astron. Nachr. 302, 299
BUFFONI,L.,MANARA,A. and SCARDIA,M.: 1982a, Astron. Astrophys. 108,141
BUFFONI,L.,MANARA,A. and SCARDIA,M.: 1982b, The Moon and the Planet 26,311
BUFFONI,L.,MANARA,A. and SCARDIA,M.: 1983, Astron. Nachr. 304,189
CELORIA,G.: 1921, Pubbl. Oss. Brera 55, 8-38-45 and 56
MARSDEN,B.G.: 1982, Catalogue of Cometary Orbits, 4rd ed., IAU Central Bureau for Astronomical Telegrams, Cambridge Mass.

THE PROBLEM OF THE 1/a-DISTRIBUTION AND COMETARY FADING

M. E. Bailey
Department of Astronomy
University of Manchester
Manchester M13 9PL
England

ABSTRACT. The background to the problem of explaining the frequency distribution of cometary 1/a-values is briefly reviewed and it is emphasised that the explanation in terms of the Oort Cloud model relies on an ad hoc fitting function - the fading/disruption probability per revolution. Assuming an underlying steady-state Oort Cloud and the integral equation formalism developed by Oort and Yabushita to predict the 1/a-distribution for an arbitrary fading probability, we have been able to constrain the unknown fading function by comparison with observations. In agreement with previous work we find that the tendency for fading or disruption should be strong at small 1/a-values and weak at large 1/a-values. The mean fading probability per revolution, $k(x)$, is found to lie within a factor roughly of order 2 about $k(x) \approx 0.3(1+(x/4 \times 10^{-3})^2)^{-3/2}$, where x is $1/a$ in units AU^{-1}. A physical model for fading which might qualitatively account for this behaviour is tentatively proposed. This depends on the thermal shock experienced by a long-period comet nucleus around perihelion passage. It is emphasised that until a viable model for fading has been found, the validity of the steady-state primordial hypothesis remains unresolved.

1. INTRODUCTION

The excess of nearly parabolic orbits amongst the observed comet population has long been a major stumbling block for theory. Thus Laplace (1805, 1816) who could not explain the high eccentricities and near isotropy of aphelion directions within his 'Nebular Hypothesis' for the origin of the Sun and the planets, was forced to adopt an interstellar origin for comets. He showed that this hypothesis naturally explained the parabolic excess, provided that the Solar motion (then in dispute) could be neglected. Despite later arguments (notably by Schiaparelli; see Richter 1963) that this assumption was false and led to erroneous conclusions, the interstellar hypothesis dominated ideas about cometary origins for nearly a century (cf. Newton 1878) and the flaw in Laplace's argument remained generally unrecognised until nearly the end of the period. By the beginning of the present century, however, the interstellar

311

A. Carusi and G. B. Valsecchi (eds.), Dynamics of Comets: Their Origin and Evolution, 311–317.
© 1985 by D. Reidel Publishing Company.

hypothesis had been abandoned, and a new consensus, namely that comets were primordial Solar System material which had somehow been formed originally with random inclinations, began to develop. The observed near-parabolic excess could then be explained as a neat example of Darwinian 'survival of the fittest' - the shorter period comets, presumably formed in equal numbers as the long-period group, having long since decayed (Crommelin 1910).

The agreement lasted for less than twenty years. During the second decade of this century the accepted age of the Solar System increased a hundred-fold (into line with its present value), and the decay argument was then found, around 1920, to apply equally strongly to the parabolic group. This dilemma led Bobrovnikoff (1929) to revive the interstellar hypothesis. Following an earlier suggestion by Nölke he proposed that the observed comets had been recently captured from a dense interstellar cloud, and argued that the lack of observed hyperbolic orbits was due to this occurrence having recently ended. Other authors (cf. Öpik 1932), however, continued to advocate a primordial Solar System model. As noted by Russell (1935) the basic difficulty with both ideas at this time was that they each depended on ad hoc assumptions which could not be tested observationally. Thus the interstellar hypothesis had to postulate an extremely dense nearby cometary cloud, while the primordial Solar System hypothesis required an enormous number of unseen comets with large perihelion distances.

This position of uncertainty lasted for a further ten years, after which it was shown (Van Woerkom 1948) that capture of interstellar comets by planetary perturbations was untenable because it would imply an excess of direct orbits (not observed) amongst the long-period group. Thus by about 1950 it appeared inescapable that comets should somehow be primordial Solar System material (Oort 1950), and the development of the now standard model involving a primordial swarm of comets surrounding the Solar System followed.

However it had also been shown by Van Woerkom that planetary perturbations acting on a 1/a-distribution with an initial parabolic excess would cause this distribution to relax quickly to a form having a flat profile. So, although Oort's (1950) theory had shown how stellar perturbations could produce a quasi-steady influx of 'new' nearly parabolic orbits, it did not naturally explain the subsequent sharp fall-off in the 1/a-distribution that was observed. To surmount this difficulty Oort supposed that 'new' comets contained an excess of more volatile material, which made them brighter and more easily discovered than the dynamically 'old' group making second and subsequent passages through the planetary system. These old comets might largely go undetected, thereby providing a possible explanation for the sharp decrease in the number of comets at larger 1/a-values. Thus the 'fading problem' was born: the theory could explain the steady influx of nearly parabolic orbits, but to explain the detailed shape of the 1/a-distribution, it became necessary to invoke arbitrary strong fading (disruption) of the new-comet population. Later work has confirmed this general result (eg. Whipple 1953, 1962; Dobrovol'ski 1972; Shtejns 1972; Weissman 1979; Everhart 1979), and has shown that approximately half of the new comets must physically be able to survive only one perihelion passage as a

detectable comet. Those that survive once, however, should then go on
to make tens or thousands of revolutions in order to explain the extended
tail of the 1/a-distribution towards relatively large 1/a-values.

2. RECENT WORK

In recent discussions Everhart (1982), Yabushita (1983) and Bailey (1984)
have drawn attention to this problem and have emphasised that the conventional Oort Cloud model does not account naturally for the sharp peak
in the 1/a-distribution. The validity or otherwise of the model therefore depends crucially on the assumption made about fading (cf. Yabushita
1983; Bailey 1985).

In order to quantify the degree of fading necessary on the steady-
state hypothesis, the author (Bailey 1985) has re-worked the problem of
the 1/a-distribution within the integral equation formalism developed
and used by Oort (1950) and Yabushita (1983). This work includes, in an
approximate way, the two principal effects of stellar perturbations:
injection of new comets, and removal of out-going comets with small 1/a-
values into unobservable orbits of large perihelion distances. Inclusion
of the latter effect is important in that it (a) 'mimics' fading, and
(b) allows the integral equation method to be compared directly with
Monte-Carlo studies (Weissman 1979; Everhart 1979). The equation which
is solved is

$$\nu(x,q) = \nu_{inj}(x,q) + (1-P_{rem}(x,q,q_{obs})) \int_{y_L}^{y_U} (1-k(y,q))\nu(y,q)\phi_{pl}(q,x-y)dy$$

(1)

where $\nu(x,q)dx\,dq$ is the number of comets passing perihelion per unit
time with 1/a-values, x, in the range (x, x+dx), and perihelia, q, in
the range (q,q+dq), ν_{inj} is the injection spectrum of new comets from
the Oort Cloud, P_{rem} is the probability that a stellar perturbation will
deflect an out-going comet into an unobservable orbit with $q > q_{obs} \gtrsim$
2 AU, ϕ_{pl} (q, Δ) is the probability that planetary perturbations will
change a comet's x-value by Δ and k(x,q) is the unknown fading/disruption
probability per revolution. Here P_{rem} is calculated in a simple way by
assuming that the random increments in angular momentum per revolution
due to stellar perturbations are governed by a two-dimensional Maxwellian
distribution with dispersion $\sigma_J(a) \simeq 4 \times 10^{15}$ $(a/10^4$ AU$)^2$ m^2 s^{-1}. A
graph of P_{rem} is given by Bailey (1985, Fig. 4); for q in the observable
region it is a function which decreases from order unity at $x \lesssim 5 \times 10^{-5}$
AU^{-1} to zero at $x \gtrsim 10^{-4}$ AU^{-1}. The expression for the planetary perturbation function ϕ_{pl} is assumed to be Gaussian with dispersion $\sigma(a) \simeq$
10^{-3} exp(-q/5.2 AU) AU^{-1}, which gives reasonable agreement with values
for ϕ_{pl} determined numerically by several authors (eg. Fernandez 1981).
The lower and upper limits of the integration are taken to be $y_L \simeq 10^{-5}$
AU^{-1}(corresponding to a cloud with outer radius of order 10^5 AU) and
$y_U \simeq 10^{-1}$ AU^{-1}. The results are insensitive to the precise value of y_U
provided $(y_U-x_{max})/\sigma \gg 1$, where σ is the dispersion of the planetary
perturbation function and x_{max} is the largest 1/a-value for which the

1/a-distribution is required. Comparison of numerical solutions of
eqn.(1) with the observed 1/a-distribution then in principle enables the
fading probability distribution $k(x,q)$ to be approximately constrained.

Unfortunately an accurate observed 1/a-distribution as a function
of q is not available, due to the increase of important selection effects
as q increases and the small numbers of comets actually observed. We
have therefore adopted an average 1/a-distribution $\nu(x)$ using the 225
comets in the catalogues of Marsden, Sekanina and Everhart (1978) and
Everhart and Marsden (1983). Solving eqn.(1) with q = 0 then enables the
mean fading probability distribution $k(x,0)$ to be determined for these
observed comets.

The principle results of this work are two-fold:
(1) The initial very sharp fall-off in the 1/a-distribution (the 'parabolic excess', with $1/a < 10^{-4}$ AU^{-1}) can be attributed to the injection
spectrum of new comets from the Oort Cloud (cf. Bailey 1983). For such
small 1/a-values removal of comets from the observable region by stellar
perturbations occurs with relatively high probability, and from (1) we
see that the appropriate solution is then indeed $\nu(x, q) \simeq \nu_{inj}(x, q)$.
The precise form of the injection spectrum depends on the detailed model
of the Oort Cloud and on the form of the velocity distribution function
within the loss cone.
(2) The subsequent continuing decrease in the 1/a-distribution towards
larger 1/a-values ($\nu(x, q) \overset{\propto}{\sim} x^{-1}$; Yabushita 1983) can only be understood
in this model by invoking strong fading. Assuming a Gaussian distribu-
tion for ϕ_{p1} with dispersion $\sigma \simeq 10^{-3}$ AU^{-1} (appropriate to setting q = 0),
and adopting an Oort Cloud model with energy spectral index $\gamma = 3/2$ (cf.
Bailey 1983), it is found that the fading probability distribution is
constrained to lie within a factor of order 2 about $k(x) \simeq 0.3$ $(1 + (x/4$
$\times 10^{-3})^2)^{-3/2}$. Changing the spectral index of the Oort Cloud model or
allowing the planetary perturbation function to realistically have a
non-Gaussian tail does change this result, but not by a large amount (cf.
Bailey 1985).

Thus we conclude, in agreement with previous Monte-Carlo studies,
that in order to explain the observed 1/a-distribution the required
fading probability per revolution has to be high at small 1/a-values and
low at large 1/a-values. The detailed shape derived for $k(x)$ does however suggest that the initial loss of volatiles may not be the most
important effect (cf. Whipple 1962; Weissman 1979), as strong fading
appears to be necessary both for 'new' comets ($a \overset{>}{\sim} 10^4$ AU) and for
dynamically old comets having a-values $\overset{>}{\sim}$ few hundred AU.

3. A THERMAL SHOCK MODEL FOR FADING

Since it is not our aim to explain the 1/a-distribution purely by assumption, it is important to develop physical models to account for the required fading probability distribution. In the past a number of
qualitative suggestions have been made to account for the required fading
behaviour of comets. These include loss of volatiles during the first
significant warming of the comet, the formation of an inert surface layer
or crust, and physical disruption of comets (see Weissman 1980 and

references therein). Implicit in some of these explanations is that the degree of fading experienced by a comet depends on an ageing process measured, for example, in the number of orbital revolutions. On the other hand, splitting events or major outbursts might allow an 'old' comet on this scheme to rejuvenate (for example the crust might be removed), thereby complicating the relationship between degree of fading and orbit number. A third possibility is that the amount of fading per revolution is unconnected with orbit number, but instead correlates simply with the semi-major axis, as for example could be implied by a straightforward interpretation of the fading probability distribution $k(x)$. Since it is important to investigate all possibilities for fading, we here present a 'thermal shock' model for fading of this third type.

We assume that fading or disruption is related fundamentally to the detailed temperature and physical structure of the nucleus, appealing to a physical process for fading similar to that which causes the cometary outburst phenomenon (eg. thermal stress, release of volatiles, low-temperature phase transitions, release of the energy of frozen-in radicals etc.). It is thus plausible to assume that when the temperature of part of the nucleus reaches some critical value (which may, in fact, be quite low; cf. Greenberg 1982) this part is somehow broken away from the main body, leading either to disruption of the nucleus or strong physical fading. In this way one might expect the amount of fading to correlate with the proportion of the nucleus which is significantly affected by the heat pulse occurring around perihelion passage.

The characteristic skin depth of penetration of a heat pulse of duration τ is typically of order $(k_D \tau)^{\frac{1}{2}}$, where $k_D = \kappa/\rho C$ is the thermal diffusivity of the material, κ is the thermal conductivity, ρ the density and C the specific heat. Assuming the nucleus is made primarily of crystalline water ice, we have at low temperatures the approximate relations $\kappa \simeq 30\ T^{-1.4}$ W cm^{-1} °K^{-1} (Klinger 1975, Fig.1) and $C \simeq 2.43$ x $10^{-5}\ T^{-2.83}$ Jg^{-1} °K^{-1} (Giauque and Stout 1936). With $\rho = 0.94$ g cm^{-3} this gives $k_D \simeq 1.3$ x $10^6\ T^{-4.23}$ cm^2 s$^{-1} \overset{\propto}{\sim} T^{-4}$ ($10 \overset{<}{\sim} T \overset{<}{\sim} 22$ K). If the mean temperature of a comet nucleus in an orbit of semi-major axis a is assumed to be $(L_\odot/16\pi\sigma a^2)^{1/4} \propto a^{-\frac{1}{2}}$, we thus have that the depth of significant penetration of the heat pulse is $\delta \overset{\sim}{\sim} (k_D \tau)^{\frac{1}{2}} \overset{\sim}{\sim} 0.1$ (a/200 AU) km, where we have taken $\tau \overset{\sim}{\sim} 1$ yr. The longest period comets, with nuclear radii of order of a few km, may therefore be substantially affected by the heat pulse associated with perihelion passage, while the comets of a shorter period may only be affected in a thin surface layer. This argument therefore suggests a possible qualitative explanation for the fading probability distribution required by observations. We emphasise, however, that whatever the true explanation for 'fading' (and a combination of factors seems probable) the net effect must be to produce the $k(x)$-distribution found here. The a priori probability of this being the case leads some authors to reject the steady-state assumption!

4. CONCLUSIONS

The fading probability per revolution required in order to explain the

observed 1/a-distribution in the context of a steady-state Oort Cloud
type of model is given approximately by $k(x) \simeq 0.3 \; (1+(x/4 \times 10^{-3})^2)^{-3/2}$.
A qualitative physical explanation for fading has been presented, but
until this or some other model has been shown to work quantitatively the
validity of the primordial Solar System hypothesis remains unresolved.

ACKNOWLEDGMENTS

I should like to thank J. A. Fernandez for useful comments on the
manuscript. This work and my attendance at the meeting were supported
by the SERC, and in part by a grant from the IAU.

REFERENCES

Bailey, M. E., 1983. *Mon. Not. R. astr. Soc.*, <u>204</u>, 603.
Bailey, M. E., 1984. *Observatory*, <u>104</u>, 65.
Bailey, M. E., 1985. *Mon. Not. R. astr. Soc.*, in press.
Bobrovnikoff, N. T., 1929. *Lick Obs. Bull.*, <u>14</u>, 28.
Crommelin, A. C. D., 1910. *Rivista di Scienza*, <u>7</u>, 241.
Dobrovol'ski, O. V., 1972. *IAU Symp.*, <u>45</u>, 352.
Everhart, E., 1979. *IAU Symp.*, <u>81</u>, 273.
Everhart, E., 1982. In *Comets*, L. L. Wilkening (ed), IAU Coll., <u>61</u> 659.
Everhart, E. and Marsden, B. G., 1983. *Astr. J.*, <u>88</u>, 135.
Fernandez, J. A., 1981. *Astr. Astrophys.*, <u>96</u>, 26.
Giauque, W. F. and Stout, J. W., 1936. *J. Amer. Chem. Soc.*, <u>58</u>, 1144.
Greenberg, J. M., 1982. In *Comets*, L. L. Wilkening (ed), IAU Coll.,
      <u>61</u>, 131.
Klinger, J., 1975. *J. Glaciol.*, <u>14</u>, 517.
Laplace, P. S., 1805. *Mech. Celeste*, <u>4</u>, 193.
Laplace, P. S., 1816. *Additions a la Connaissance des Temps*, 213.
Marsden, B. G., Sekanina, Z. and Everhart, E., 1978. *Astr. J.*, <u>83</u>, 64.
Newton, H. A., 1878. *Amer. J. Sci. Arts*, <u>16</u>, 165.
Oort, J. H., 1950. *Bull. Astron. Inst. Neth.*, <u>11</u>, 91.
Öpik, E. J., 1932. *Proc. Amer. Acad. Arts Sci.*, <u>67</u>, 169.
Richter, N. B., 1963. In *The Nature of Comets*, Methuen, London.
Russell, H. N., 1935. In *The Solar System and its Origin*, p.43, New York.
Shtejns, K. A., 1972. *IAU Symp.*, <u>45</u>, 347.
Van Woerkom, A. J. J., 1948. *Bull. Astron. Inst. Neth.*, <u>10</u>, 445.
Weissman, P. R., 1979. *IAU Symp.*, <u>81</u>, 277.
Weissman, P. R., 1980. *Astr. Astrophys.*, <u>85</u>, 191.
Whipple, F. L., 1953. In *La Physique des Cometes*, Proc. 4th Coll. Int.
      d'Astrophys. Liege, p.283.
Whipple, F. L., 1962. *Astr. J.*, <u>67</u>, 1.
Yabushita, S., 1983. *Mon. Not. R. astr. Soc.*, <u>204</u>, 1185.

DISCUSSION

A. H. Delsemme: (1) I want to describe a possible mechanism for the fading problem. If comets have stayed for 5 billion years in the Oort Cloud they will have been irradiated by cosmic rays sufficiently that the first 1 - 3 metres of the nucleus will have been considerably modified. This modification yields highly reactive molecules and radicals that remain frozen in the icy matrix until the comet comes closer to the Sun. Such a layer may be completely vapourised in less than one passage for small perihelion distances ($q < 0.2$ or 0.3 AU), or in a small number of passages if the perihelion distance is larger (the layer removed per passage varies as q^{-2}).

(2) Statistically speaking (and whatever the reason) it is well known that the absolute magnitude of new comets (first passage with $q < 7$ AU) is 3 - 4 magnitudes brighter than that of periodic comets. Is this what your fading model would predict?

(3) I have published (In 'Dynamics of the Solar System' p.265, ed. R. L. Duncombe, Reidel, IAU Symp., 81, 265, 1979) a distribution of the absolute magnitudes of new comets. It is (surprisingly) bimodal; 82% have a peak near $H_O = 5.5$ (true new comets?), whereas 18% show a peak near $H_O = 10.0$ (fragmented comets). Fragmentation is therefore well documented and could certainly be one of the contributing factors to the fading of new comets.

P. R. Weissman: I agree with you that there is a fading factor that varies with cometary age. In my work I modelled it as most comets having a 10% disruption probability, and only 15% having a zero disruption probability. But at the same time I agree very strongly with Delsemme that new comets fade strongly after their first return due to this loss of a surface layer of extremely volatile materials. That is an effect that we should try to include in our models in the future.

M. E. Bailey: (In answer to this and Professor Delsemme's first point.) Yes. I agree that it does seem probable that comets coming in for the first time should be a little brighter, due to loss of volatiles etc, than the others. My point is that it is important that modelling of this process should be put on a quantitative physical basis.

To take Delsemme's other two points in reverse order: Yes; and unfortunately my fading 'model' is too qualitative to make definite predictions. But it does 'predict' that long-period comets should be more prone to disruption than the comets of shorter period if one accepts the thermal shock hypothesis. This should correlate with the absolute magnitude distribution!

FIRST RESULTS OF THE INTEGRATION OF MOTION OF SHORT-PERIOD COMETS OVER 800 YEARS

A. Carusi[1], L. Kresák[2], E. Perozzi[1], G.B. Valsecchi[1]

[1]IAS-Reparto di Planetologia, C.N.R.
Viale Università 11, 00185 Rome, Italy

[2]Astronomical Institute, S.A.V.
84228 Bratislava, Czechoslovakia

ABSTRACT.   All the known short-period comets have been followed by numerical integration over a time span of 821 years, from 1585 to 2406. A preliminary survey of the results of these integrations has shown some interesting features, which become recognizable thanks to the length of the time interval covered, not negligible if compared with the typical evolutionary time scale of comets moving in short-period orbits. Interesting phenomena that have been recognized include: (1) captures from, or ejections into, very elongated ellipses, with perihelia of the parking orbits close to the orbit of Jupiter and aphelia within or beyond the region of outer planets; (2) passages of comets from the control of Saturn to that of Jupiter; (3) orbital evolutions controlled mainly by Saturn; (4) librations of comets around low-order resonances; (5) repeated close approaches of comets to Jupiter, often with the comet being captured as a temporary satellite; (6) an almost perfect coincidence of two comet orbits just before a close approach to Jupiter, suggesting their genetic relationship.

1.   INTRODUCTION

   The orbital evolution of all 126 known short-period comets has been followed, by integrating their motion over a long time span, from 1585 February 1.0 (JD 2300000.5) to 2406 June 17.0 (JD 2600000.5). This work (Long-Term Evolution Project, see Carusi et al., 1985a) has provided an atlas of potential cometary orbital evolutions, that is described in more detail in Carusi et al. (1984).

319

A. Carusi and G. B. Valsecchi (eds.), Dynamics of Comets: Their Origin and Evolution, 319–340.
© *1985 by D. Reidel Publishing Company.*

Here we only want to make some comments about the LTEP, its precision and its reliability as a sample of the dynamical evolutions of real objects. First of all, not all the orbits that have been used to start the integrations are of high quality. Many of them – especially those of one-apparition comets – are known with limited accuracy and the corresponding evolutions simply represent possible behaviours of objects in specific regions of the phase space. Moreover, the objects were supposed to move in a purely gravitational dynamical system, where the action of nongravitational forces was completely ignored. For this reason, even the reconstructed motion of some comets with excellent starting orbits cannot be considered reliable before or after very close approaches to Jupiter that took place far from the starting date (see, for example, Carusi et al., 1985b). It should be clear to the user that, while the accuracy of the integrations is the same throughout, the extent of coincidence with the motion of the real objects is quite different and has to be estimated in each case individually.

Although these inaccuracies – and also those coming directly from the integration method used – render our sample of cometary evolutions not perfectly identical to the real sample, it can be used as a powerful mean for investigating the principal dynamical processes governing the motion of comets under the perturbing action of planets.

In this paper we will review some of the most interesting features singled out during a first analysis of the LTEP data. Some of them have already been pointed out and discussed in more detail by other authors, and are included with the relevant quotations for the sake of completeness. Some others will require further detailed studies, in order to depict a coherent scenario of the motion of comets captured into short-period orbits.

2. GENERAL CHARACTERISTICS

Owing to the different initial dynamical status, individual objects have widely different histories, including comets staying all the time very close to the starting orbit; objects oscillating for the whole integration, or only a part of it, about some resonance with Jupiter; and objects crossing different and widely separated regions. In order to overview the internal mobility of the sample, we have collected all the 126 osculating orbits computed at every 5,000 days, extrema included. This represents a total of 7,686 comet orbits: their distribution is shown in fig. 1, where the distribution parameter is the ratio between the comet's osculating period and the mean period of Jupiter. In fig. 1a all the orbits with periods between 0 and 2 times that of Jupiter are included. The distribution shows remarkable Kirkwood gaps, the

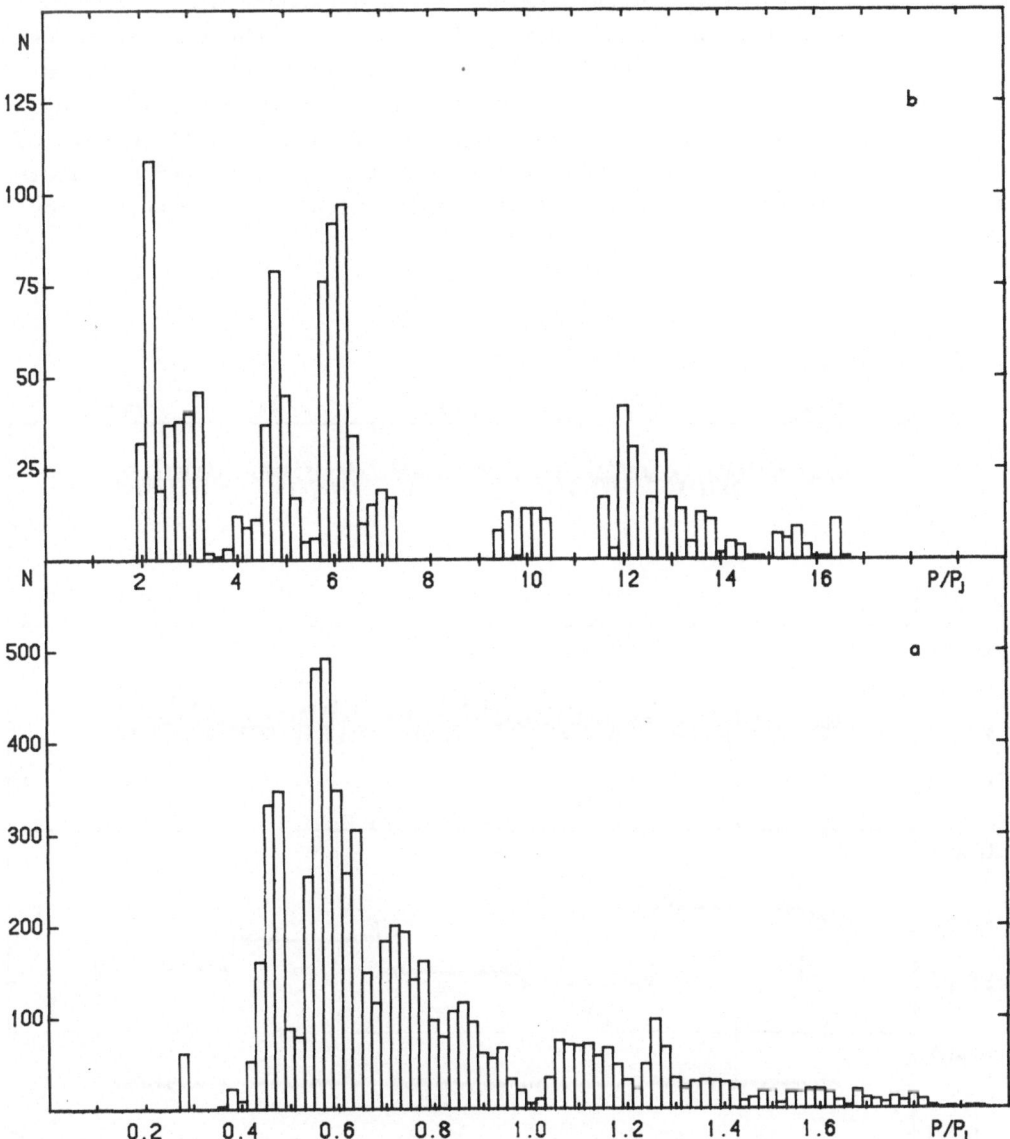

Figure 1. a: histogram of ratios of comet periods to Jupiter's period between 0 and 2. b: same as in plot a, for P/P_j between 2 and 20.

most prominent corresponding to the 1/2, 3/5, 2/3, 3/4, 4/5 and 1/1 resonances with Jupiter's motion. The single 61-points column on the left of the 1/3 resonance is due to P/Encke. The gaps are at least partially produced by comets temporarily librating about the corresponding resonances, especially that of 1/2. On the contrary, a concentration of orbits at 1.25 (5/4 resonance) is evident, representing mainly the contribution of the orbits of P/du Toit and P/Gehrels 1.

Fig. 1b shows the same distribution for ratios of the periods between 2 and 20. Only two comets (P/Lexell after the ejection by Jupiter in 1779, and P/Wild 2 before the capture in 1702) move temporarily along orbits with periods longer than 20 Jupiter's periods. The long tail of the distribution beyond 2 is almost entirely due to 17 comets of inter-mediate periods (Halley type). Their contribution to the distribution,

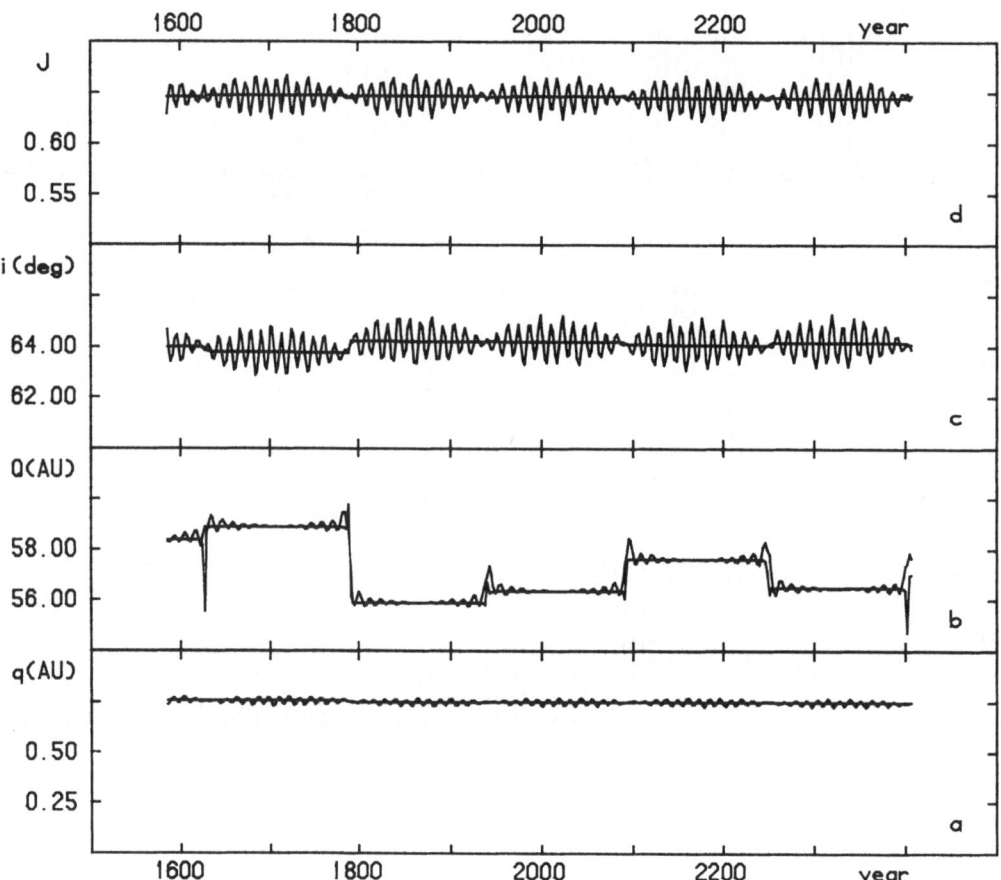

Figure 2. Time evolution of perihelion distance (a), aphelion distance (b), inclination (c), and Tisserand invariant with respect to Jupiter (d) of comet P/Herschel-Rigollet. For explanation see text.

however, is somehow peculiar: just as an example, only four comets of
this type produce the wide-spread portion between 11 and 17, with a
large overlapping of the points corresponding to different comets. The
spread is maximum for P/Wilk, whose points range from 12 to 16.5, implying a variation of semimajor axis exceeding 6 AU.

The mechanism responsible for the spread is connected with the motion
of the Sun about the barycentre of the solar system. A common feature
of comets with aphelia well outside the orbit of Jupiter is a marked modulation of their heliocentric orbital parameters, which disappears
when the orbit is computed with respect to the barycentre instead of the
Sun. In fig. 2, for example, the evolution of four relevant orbital parameters of P/Herschel-Rigollet is shown. The quantities computed in
the heliocentric frame exhibit the mentioned modulation, while their
tracks become almost flat from perihelion to perihelion when computed in
the barycentric frame. The sudden changes in the aphelion distance are
not due to encounters with the major planets, since this comet does not
approach any planet within 1 AU for the whole time span of the integration. They reflect, in fact, the different relative positions of the
Sun and the major planets (mainly Jupiter) each time the comet enters

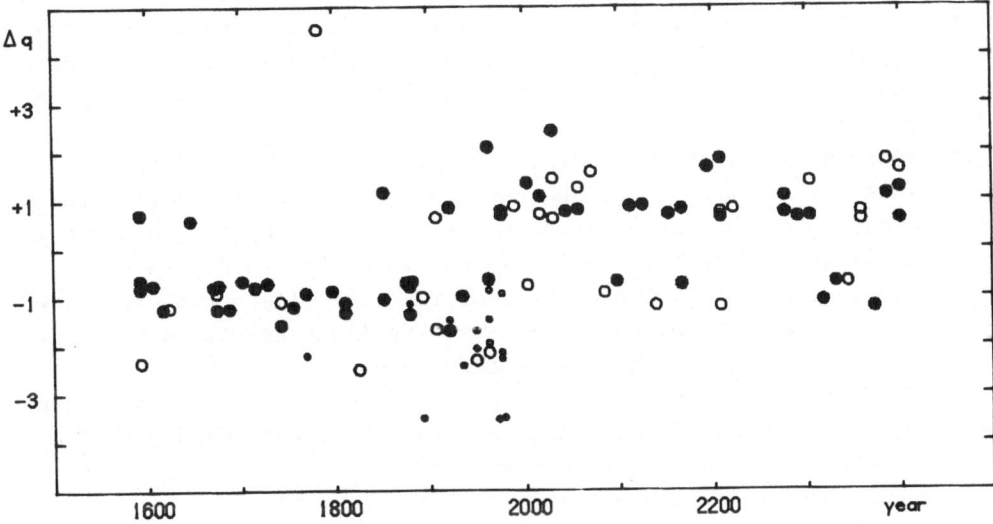

Figure 3. The 100 absolutely largest variations of perihelion distance
Δq. Open circles: one-apparition comets; circles: more-than-one-apparition comets; dots: variations immediately before discovery.

the planetary region, and its motion becomes for a while heliocentric instead of barycentric.

Fig. 3 lists the 100 largest variations (both positive and negative) of the perihelion distance in the 821-year interval. One-apparition and more-than-one-apparition comets are marked by open and solid circles, respectively; small dots indicate that the corresponding variation has taken place immediately before the discovery of the comet, irrespective-ly if now of one or more apparitions. Obviously the latter variations are all negative, and concentrate in the XXth century, when most of the-se comets were discovered. Remarkable is a lack of strong positive var-iations in the past, reflecting a deficiency of accelerating encounters which would have rendered the discovery of such a comet essentially im-possible. On the other hand, decelerating encounters are less frequent than the accelerating ones in the future. This is also understandable, because for a comet that has already suffered a strong deceleration it is more probable to undergo encounters leading to an increase of the perihelion distance instead of a further decrease (that could not be very large, anyway, because most of these comets are already close to Jupiter's barrier).

Only 42 comets contribute to the 100 strongest variations. Among them 16 are present only once; 9 twice; 9 three times; 3 four times (P/Gehrels 3, P/Reinmuth 2, and P/Wolf-Harrington); 4 five times (P/Bus, P/Smirnova-Chernykh, P/Ashbrook-Jackson, and P/Whipple). P/Shajn-Schal-dach contributes as much as 7 times. The four largest variations are due to P/Lexell (4.5 AU, the largest and the only positive among the four) and to P/West-Kohoutek-Ikemura, P/Brooks 2 and P/Wild 2 (-3.5 AU all the three).

A preliminary survey of close encounters with Jupiter, Saturn and Uranus has already been done. No comet encounters Neptune or Pluto, while a statistics of close approaches to the inner planets has not yet been performed. It is certainly possible that some event has remained hidden among the ten million numbers of the core files of LTEP, but we think that the more detailed analyses we are planning will not change the general statistics we are presenting here.

Table 1 lists the number of encounters per century within a sphere of 0.5 AU around each planet. The only encounter with Uranus is due to P/Tempel-Tuttle in 2317 (minimum distance from the planet: 0.29 AU). The eleven encounters with Saturn are listed in table 2.

As regards the encounters with Jupiter, we only list in table 3 the ten closest approaches: note that four of them are due to P/Gehrels 3. In Section 3.5 we will examine the history of these encounters in great-er detail.

In fig. 4 the distribution of encounters with Jupiter to within 0.5 AU between 1601 and 2400 is shown (dates correspond to minima of dis-

TABLE 1. Number of encounters within a sphere of radius 0.5 AU

Period	Jupiter	Saturn	Uranus
1585–1600	9	0	0
1601–1700	67	2	0
1701–1800	63	2	0
1801–1900	82	1	0
1901–2000	91	1	0
2001–2100	78	0	0
2101–2200	80	1	0
2201–2300	76	1	0
2301–2400	73	2	1
2401–2406	3	1	0
1585–2406	622	11	1

TABLE 2. Encounters with Saturn

Comet	Date	Minimum distance (AU)
P/IRAS	(1) 1950	0.06
		Orbit of Phoebe
P/IRAS	(2) 1627	0.17
P/Wild 1	(1) 2260	0.26
P/Tempel-Tuttle	(1) 1630	0.34
P/Chernykh	(2) 1749	0.36
P/Bowell-Skiff	(2) 1816	0.37
P/IRAS	(2) 2333	0.40
P/Wild 1	(1) 2113	0.41
P/Oterma	(2) 1770	0.41
P/Kowal-Vávrová	(2) 2324	0.43
P/Bowell-Skiff	(3) 2403	0.48

(1) Quite reliable   (2) Less reliable   (3) Unreliable

TABLE 3.   The ten deepest encounters with Jupiter

Comet	Date		Minimum distance (AU)
P/Brooks 2	(1)	1886	0.0010
P/Gehrels 3	(1)	1970	0.0014
P/Lexell	(1)	1779	0.0015
			Orbit of Io
P/Gehrels 3	(3)	2305	0.0055
P/Wild 2	(1)	1974	0.0061
			Orbit of Callisto
P/West-Kohoutek-Ikemura	(1)	1972	0.0138
P/Gehrels 3	(2)	2062	0.0159
P/Gehrels 2	(2)	2029	0.0182
P/Brooks 1	(3)	1739	0.0184
P/Gehrels 3	(3)	2400	0.0190

(1) Quite reliable   (2) Less reliable   (3) Unreliable

tance). The highest concentration corresponds to the period 1951–1975, and is apparently due to the high percentage of comets discovered in the

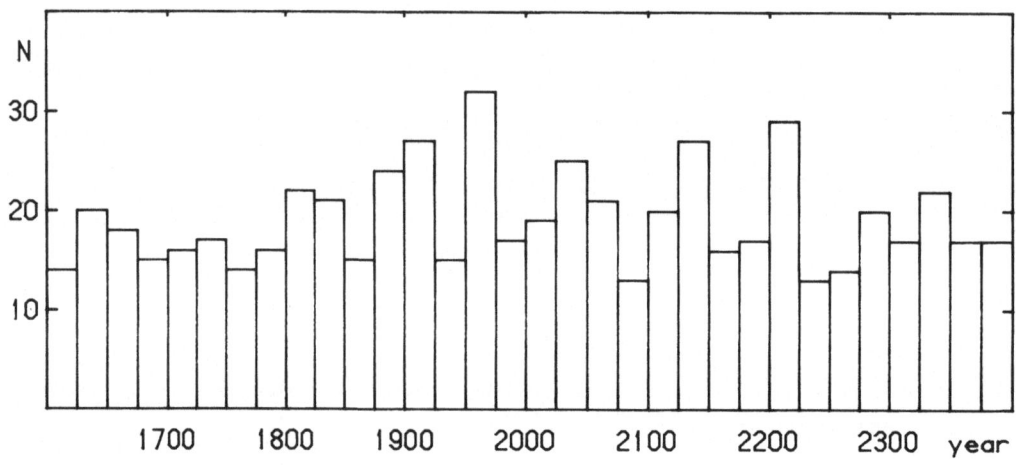

Figure 4.   Time distribution of the encounters with Jupiter within a sphere of 0.5 AU around the planet.

sixties and seventies after a considerable reduction of their perihelion distance (see also fig. 3).

On the average there are 0.76 encounters with Jupiter per year: since in our sample the mean duration of an encounter within 1 AU from Jupiter is about 1.6 years, it appears that, between 1601 and 2400, there were, on the average, more than one of the known short-period comets within that distance from the planet at any time. For comparison, in the period 1951-1975 there were, on the average, two or three of these comets within 1 AU from Jupiter at a time.

3. SPECIAL FEATURES

In this Section we will examine some special features recognizable in the LTEP data that are relevant to the study of the dynamical evolution of short-period comets. They are:

1) captures of comets from, or ejection into, very elongated ellipses caused by the gravitational influence of Jupiter;

2) comets passing from the complete control of Saturn to that of Jupiter;

3) comets mainly governed by Saturn;

4) comets with persisting librations around a resonance with Jupiter;

5) comets undergoing repeated, long and deep encounters with Jupiter, leading often to temporary satellite captures;

6) comets with orbits almost coincident before a close approach to Jupiter.

3.1. Captures and ejections

Only three from among the 126 integrated comets have been captured by Jupiter from, or ejected into, orbits of very high eccentricity, with aphelia well outside the planetary region. The first of them was the well known case of P/Lexell, ejected in 1779 from an orbit of period 5.6 yr (close to the 1/2 resonance with Jupiter) into an orbit of period around 280 yr. P/Lexell is the only case in our sample of a comet that, still remaining bound to the Sun, reaches an orbit of period exceeding the conventional limit of 200 years for short-period comets: in this sense it should not be regarded as of short-period anymore. Before 1767 P/Lexell moved on a somewhat larger orbit, with a period of 9.2 years; it is unique by three close planetary encounters in rapid succession: 1767 Jupiter - 1770 Earth - 1779 Jupiter. The unique dynamical history of P/Lexell was already recognized and studied in detail by Lexell (1778), Leverrier (1857) and Kazimirchak-Polonskaya (1961). An investigation of the fate of the possible meteor stream associated with it (Ca-

rusi et al., 1982) has shown an extreme sensitivity of the orbital parameters after 1779 to the initial conditions: a variation of -1 m/sec
in the velocity of the comet at the perihelion passage before the encounter caused a final orbit with a period of only 23.6 years, whereas a
variation of +1 m/sec led to a hyperbolic orbit.

P/Kearns-Kwee (fig. 5a) starts its evolution in an elongated orbit
with a period of 170 yr and aphelion 57 AU from the Sun. The comet is
then captured by Jupiter into a parking orbit of about 50 yr period
around 1700, and finally transferred inside the orbit of Jupiter in
1962. The final orbit, the one in which P/Kearns-Kwee is moving at present, has a period of only 9 years. The previous 50-yr period was quite

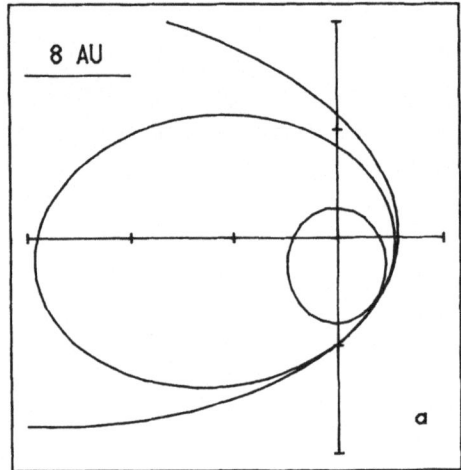 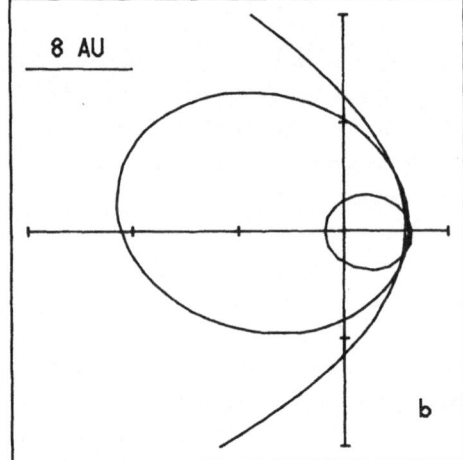

Figure 5. a: successive orbits of P/Kearns-Kwee. Inner orbit: after
1962; intermediate orbit: between 1700 and 1962; outer orbit: before
1700. b: successive orbits of P/Wild 2. Inner orbit: after 1974; intermediate orbit: between 1702 and 1974; hyperbolic orbit: before 1702.

evident from the one-apparition orbit; Kazimirchak-Polonskaya (1967) gave a different scenario that produced it (an earlier capture from a definitely hyperbolic orbit); but Marsden and Aksnes (1967) wisely concluded that it was then unknowable. Even after the later orbit improvement, the circumstances before the 1700 encounter remain rather indeterminate.

The third case of this type is that of P/Wild 2 (fig. 5b). At the beginning of our integration period, in 1585, the comet moves along a hyperbolic orbit of eccentricity 1.03. Given the uncertainty of the orbital elements, which are based on the discovery apparition alone, it can only be said that this comet may have come from an orbit of very long period. The value of the eccentricity of P/Wild 2, 1.03, is comparable with the one of the fictitious Lexell already mentioned: 1.05. A velocity increment of +1 m/sec at perihelion reflects, in the case of Lexell, into a variation of semimajor axis on the fourth decimal digit, which - in turn - causes a relative variation of the orbital period of the order of 0.0001. This is also the uncertainty in the present period of P/Wild 2. In this respect, however, it should be noted that the capture of P/Wild 2 into the present orbit proceeds, like that of P/Kearns-Kwee, through an intermediate parking orbit, with a period of 38 years, where the comet is placed by Jupiter in 1701. The second deep encounter, leading to the present orbit, takes place immediately before the discovery in 1974. During this last encounter, as we have already shown in table 3, the comet passes Jupiter at only 0.006 AU - one of the deepest encounters on record. This fact, together with the orbital uncertainty, makes the evidence of a capture from a hyperbolic orbit entirely unreliable.

It may appear strange that integrations over a grand total of over 100,000 years have not revealed any other case of comparably strong perturbation, all the three extreme cases falling within two revolutions of the comet's discovery, and of the osculating date of its starting elements. In interpreting this finding from the statistical point of view, two effects have to be taken into account. First, it is just a close decelerating encounter which can render a comet observable for the first time, by reducing appreciably its perihelion distance. And second, the limited accuracy of the starting orbits would make the encounter conditions far from the osculating date rather uncertain. The computation may reveal a deep encounter which in fact did not occur, but it can also fail to identify a real one. While the probability of positive and negative errors in the comet's position on the orbit is the same in principle, the window of strong interactions may become much narrower than the error dispersion (see, e.g., Carusi et al., 1981a). This is why more real encounters may remain hidden than fictitious encounters found. Therefore, the rate of very close encounters resulting from our long-term integrations must be taken with due precautions.

3.2. The case of P/Oterma

Comet P/Oterma has been studied by many authors (Oterma, 1958; Fokin, 1958; Marsden, 1961; Kazimirchak-Polonskaya, 1967; Carusi et al., 1981a) due to its high dynamical interest. It has been the first recognized case of a real comet undergoing temporary satellite captures by Jupiter, and the study of its two encounters with the planet in 1937 and 1963, separated by a residence in the 2/3 resonance with Jupiter, has opened a number of new investigations about the dynamical behaviour of comets of moderate eccentricity and very high value of the Tisserand invariant (i.e., a low encounter velocity) with respect to Jupiter.

The LTEP has further emphasized the relevance of this object for the study of comet dynamics. At the beginning of the integration P/Oterma moves along an orbit of very high perihelion distance (6.33 AU), that does not allow the comet to approach Jupiter to less than 0.87 AU. The

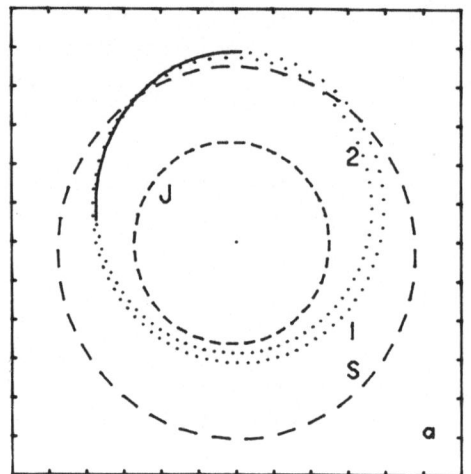 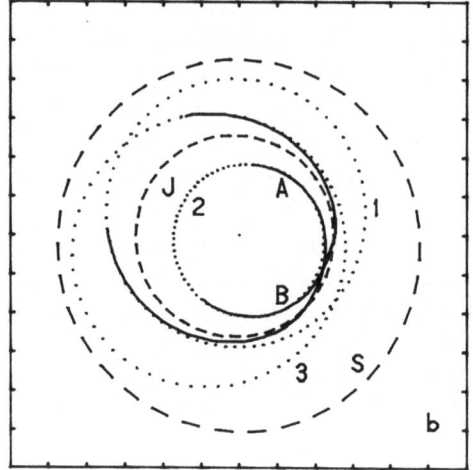

Figure 6.   Transfer of P/Oterma from Saturn to inside Jupiter. Dashed orbits in both plots refer to Jupiter and Saturn. a: orbit 1: before 1770; solid line: path of the comet inside a 2 AU sphere around Saturn in 1770; orbit 2: after 1770. b: orbit 1: before 1933; solid line A: 1933-1941 encounter; orbit 2: between 1941 and 1960; solid line B: 1960-1967 encounter; orbit 3: after 1967.

motion of the comet is, at that stage, completely dominated by Saturn; then, an approach to this planet in 1770, with a minimum distance of 0.4 AU, reduces the perihelion distance to 5.77 AU (see fig. 6a). During the XIXth century two rather shallows encounters with Jupiter reduce the aphelion distance by almost 1.5 AU, detaching the comet from the control of Saturn. Finally, the mentioned encounter with Jupiter around 1937 puts P/Oterma in the inner orbit, exactly in the 2/3 resonance with the planet, so that a symmetric encounter takes place three revolutions later, sending the comet back to the original place (see fig. 6b). In the present orbit, after the return into the region between Jupiter and Saturn, P/Oterma can encounter both planets, but no approaches within 0.5 AU are found until the end of the integration.

3.3. Comets mainly governed by Saturn

The orbit of P/Oterma before 1770 represents in the LTEP the only case of avoidance of encounters with Jupiter because of a large perihelion distance. Among the other comets, however, there are many cases in which Saturn shares with Jupiter the control of the orbital evolution. Only in two cases, namely P/Wild 1 and P/IRAS (discovered after the last edition of Marsden's Catalogue, 1982), Saturn totally controls their perturbed motion. Jupiter cannot be encountered because of the unfavourable geometrical configuration of the orbits. Table 4 lists the

TABLE 4. Geometrical data on comets controlled by Saturn

	P/Wild 1		P/IRAS	
	min	max	min	max
Distance from the Sun at				
Ascending node	8.4	9.3	1.5	2.2
Descending node	1.9	2.1	9.3	9.6
Distance from the ecliptic at				
r = 5.2 AU	1.0	1.3	2.6	2.9
r = 5.2 AU	1.4	1.7	2.8	3.5
r = 9.5 AU	–	–	0.02	0.92
r = 9.5 AU	–	–	0.6	1.7

distances from the Sun when the two comets cross the ecliptic plane, and
the vertical distances from the ecliptic when the radius vectors are
equal to 5.2 AU (orbit of Jupiter) and 9.5 AU (orbit of Saturn). Note
that the latter data are not reported for P/Wild 1, since its aphelion
is always less than 9.539, the mean semimajor axis of Saturn.

3.4. Librators between Jupiter and Saturn

The LTEP has shown that many short period comets spend at least part
of the 821 years librating about a low-order resonance with Jupiter.
This behaviour was discovered by Marsden (1970), and discussed also by
Kresák (1974), Franklin et al. (1975), and Vaghi and Rickman (1982).
As an example, comets spending a part or the whole time span librating about the 1/2 resonance with Jupiter are listed in table 5, together
with the number of cycles performed. As shown in the table, 12% of the
comets of our sample are, at least temporarily, librating about the 1/2

TABLE 5. Comets temporarily librating about the 1/2
          resonance with Jupiter

Comet	Number of cycles
P/Forbes	1
P/de Vico-Swift	1
P/Tsuchinshan 1	1
P/Tsuchinshan 2	1
P/Kopff	2
P/Tempel-Swift	2
P/Clark	2
P/du Toit-Neujmin-Delporte	2
P/Harrington-Wilson	2
P/Pons-Winnecke	3*
P/Tempel 1	3
P/Pigott	3*
P/Howell	3
P/Haneda-Campos	4*
P/Tritton	4*

* Comets librating from 1585 to 2406

resonance, and 3% during the whole 821-years interval; these numbers increase, of course, if we take into account librations about other low-
order resonances.

Of particular interest are the objects with semimajor axes between
those of Jupiter and Saturn which, as it is well known, have a ratio of
orbital periods close to 2/5.

The most impressive case is that of P/du Toit, which librates all the
time about the 5/4 resonance with Jupiter. This, in turn, is also very
close to the 1/2 resonance with Saturn. Figs. 7a,b show the jovicentric
and saturnocentric patterns of the comet for the whole time interval:
the frames are rotating with the corresponding planet, which is in the
centre, so that the Sun is always located on the negative x-axis. While
P/du Toit is almost perfectly locked in the mentioned resonance with Jupiter, the 8-shaped pattern with respect to Saturn rotates slowly, but
continuously, in the clockwise direction. The integration covers only
about one fourth of this last cycle, and the possible influence by Saturn on the jovicentric pattern remains open.

P/Gehrels 1 is another example of this type of libration: in this

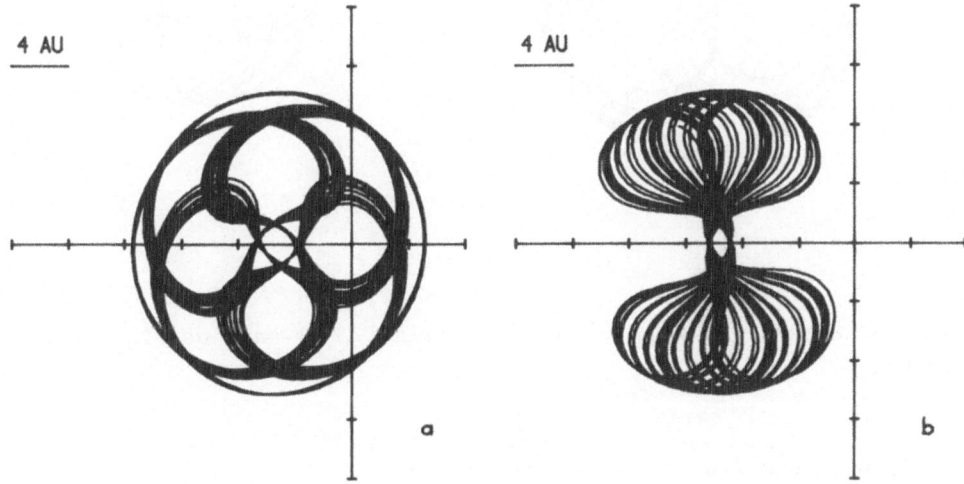

Figure 7. a: path of P/du Toit in a frame centred on Jupiter and rotating with the angular velocity of Jupiter's revolution. b: path of P/du
Toit in a similar frame centred on Saturn. Both paths are between 1585
and 2406. For further explanations see text.

case, however, the jovicentric pattern (see fig. 8a) is much less regular and there are indications that the dynamical situation is changing in the last century of the integration.

A third example is the well known case of P/Neujmin 1, librating for the whole period about the 3/2 resonance with Jupiter, which is also quite close to the 3/5 resonance with Saturn. The jovicentric pattern of P/Neujmin 1 is also much less regular than that of P/du Toit (see fig. 8b), while the comet is definitely circulating across the 3/5 resonance with Saturn. The libration of P/Neujmin 1 was discovered by Marsden (1970) who found it to persist over 4,000 years.

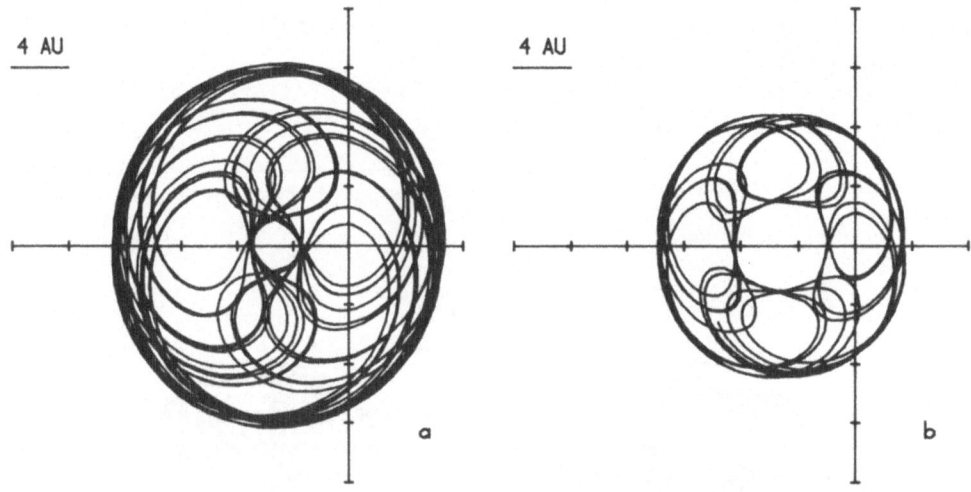

Figure 8.   a: same as fig. 7a for P/Neujmin 1.   b: same as fig. 7a for P/Gehrels 1.

3.5. Close encounters and satellite captures

As we have already shown, during the 800-years period from 1601 to 2400, the 126 short-period comets undergo a total of 610 encounters with Jupiter at a distance of less than 0.5 AU. Only 75% of the comets are responsible for all the encounters: 25% of them do not approach Jupiter within that distance, or do not have even moderate approaches, as is the case of P/IRAS and P/Wild 1; 40% of the comets have more than 5 encounters, and 13% more than ten. The highest number of encounters is due to P/Pons-Winnecke (21) which, as already mentioned, librates about the 1/2 resonance with Jupiter. P/Gehrels 3 alone accounts for 40% of the ten closest approaches (see table 3). This comet is exceptional in many respects: it has one of the highest Tisserand invariants (3.02) and, hence, a very low encounter velocity; it experiences five approaches with minimum distances from Jupiter smaller than those of the direct irregular satellites (one of them inside the orbit of Io, and another one well inside the orbit of Ganymede); it is captured as a temporary satellite at each encounter, spending a total of 31.4 years bound to the planet (3.8% of the total time of the integration, 7% of the time from the first to the last encounter). It is also noteworthy that the only close encounter with Jupiter that occurred in the backward part of the

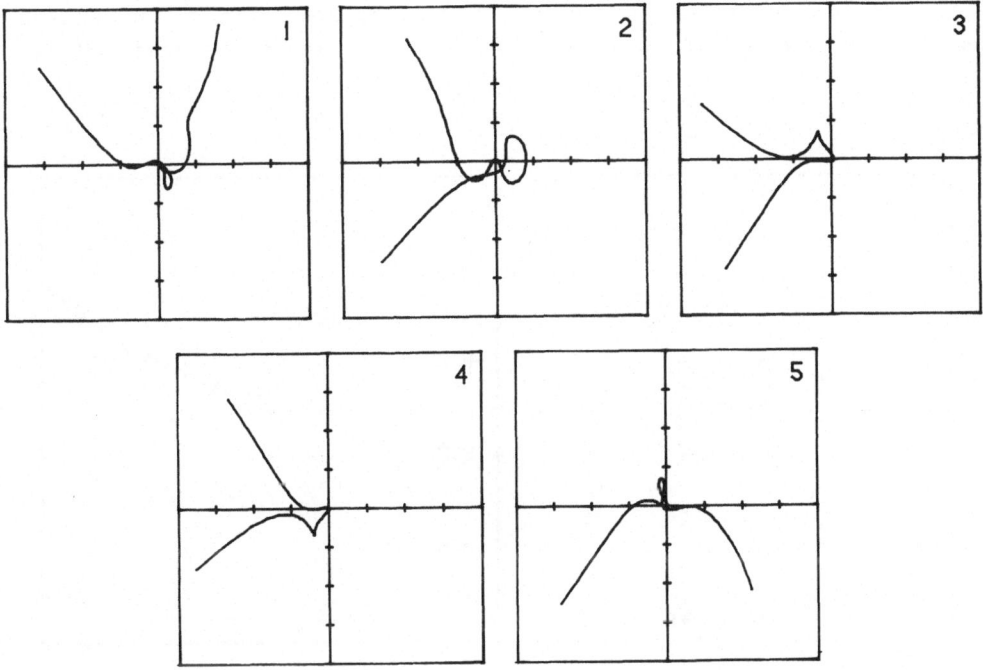

Figure 9. Jovicentric trajectories for the five very close encounters of P/Gehrels 3 with Jupiter. The dates (corresponding to minimum distances) are: 1-1970, 2-2062, 3-2203, 4-2305, 5-2400. Tick marks: 0.5 AU.

integration is the one that transferred P/Gehrels 3 from outside to inside the orbit of the planet. In the forward part the other four encounters occur, the last one placing the comet again beyond Jupiter.

The five jovicentric patterns corresponding to the five encounters
are shown in fig. 9; all of them are typical of low-velocity encounters
with Jupiter (see, for comparison, Carusi et al., 1981b, objects CAP 21,
OTE 14, 61 and 37). Quite impressive is the similarity between the encounters of 2200 and 2300, which can be almost exactly obtained from
each other by a rotation of π about the x-axis. Symmetries of this type
are quite common among low-velocity encounters, as can easily be noted
in the last cited reference.

As already noted (see also Carusi et al., 1985b), very close encounters invalidate to a large extent the integrations beyond them: in the
present case only the two events nearest to the starting date (one backwards and one forwards) can be accepted with a sufficient confidence.
Nevertheless, this evolution is interesting in itself, as a potential
behaviour of a comet with a high value of the Tisserand invariant with
respect to Jupiter.

P/Gehrels 3 is fairly unique in our sample; other comets undergoing
less frequent or more distant encounters, sometimes being captured as

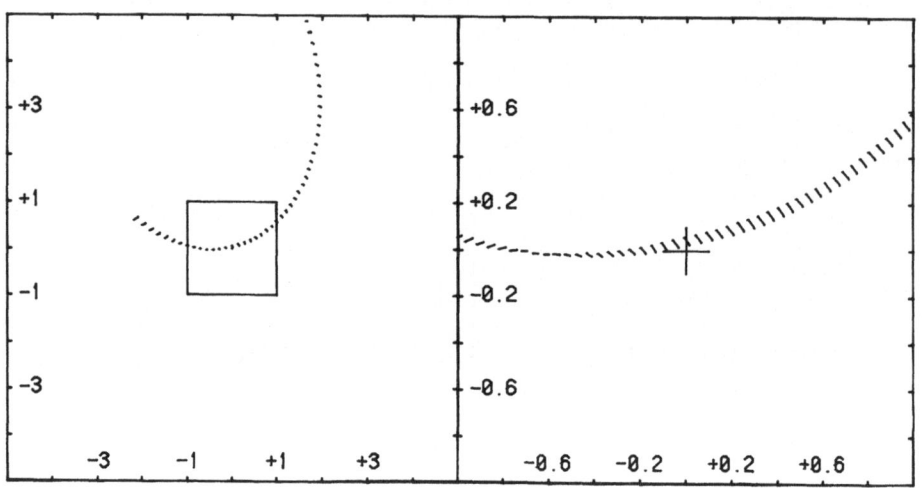

Figure 10. Jovicentric trajectories of P/Neujmin 3 and P/Van Biesbroeck
in a sphere of 5 AU radius (left plot) and 1 AU radius (right plot) centred on Jupiter during the encounter of 1850. The frame is rotating with
the planet. The positions of the two comets are connected by a segment,
whose upper end in the right-hand plot is P/Neujmin 3.

temporary satellites, are P/Smirnova-Chernykh and P/Oterma.

3.6. P/Neujmin 3 and P/Van Biesbroeck

P/Neujmin 3 and P/Van Biesbroeck, two comets with very well determined orbits - currently not very similar to one other - have rather
complex and interesting histories when traced backward in time. Both of
them, in fact, approach Jupiter in 1850, passing it quite closely, and
almost perfectly join prior to this encounter. The striking similarity
between the orbits prior to 1850, and the physical proximity of the two
comets indicate that, very probably, they represent fragments of a parent body that split before the mentioned encounter, as already announced
by the authors in the IAUC No. 3940 (1984). The event is shown graphically in figs. 10 and 11. The two plots of fig. 10 are the jovicentric
trajectories of the two comets during the approach to Jupiter. The po-

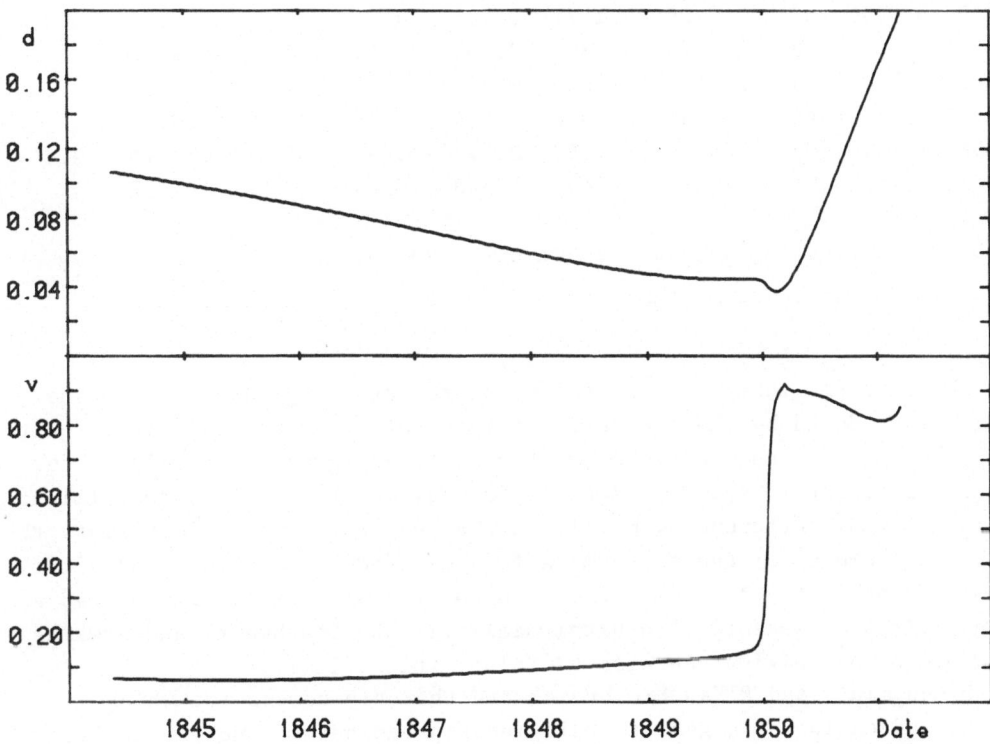

Figure 11. Time evolution of the distance (upper plot) and relative
speed (lower plot) of P/Neujmin 3 and P/Van Biesbroeck between 1844 and
1851. Units are AU and km/sec.

sitions of the objects, projected onto the ecliptic plane, are connected by a segment, whose upper end before the encounter represents P/Neujmin 3. As the segment rotates counter-clockwise, just after the closest approach the two trajectories do cross. Fig. 11 gives the temporal variation of the mutual distance and of the relative velocity of the two objects: they remain for a long time much closer than 0.1 AU, with a relative speed less than 0.1 km/sec.

TABLE 6. Orbital data of P/Neujmin 3 and P/Van Biesbroeck

Date		v	a	e	i	ω	Ω	D
1845 03 10.0	N3	185.71	6.1046	0.5494	4.71	290.90	348.54	0.013
	VB	186.03	6.1352	0.5504	4.11	289.96	349.14	
1849 10 05.0	N3	240.09	6.1477	0.5615	4.74	291.14	348.45	0.017
	VB	240.20	6.1860	0.5621	4.13	290.05	349.08	
1850 05 03.0	N3	239.05	5.1978	0.5814	3.14	131.85	162.66	0.285
	VB	246.46	5.6038	0.5642	7.04	123.09	163.98	
1858 11 17.0	N3	164.16	5.1259	0.5756	3.35	136.49	159.13	0.344
	VB	175.24	5.5443	0.5483	7.05	125.57	162.52	

Orbits referred to equinox and ecliptic 1950.0

Owing to the inherent uncertainty of the starting data, and to the noise introduced by the 100 years of integration after 1850, it is quite surprising that the orbits before that date do agree so closely; a further refinement of starting data is needed, in order to try to eliminate some residual differences, mainly in the inclination. Table 6 lists the orbital elements of the two comets for selected dates before and after the encounter, together with the value of the D-parameter introduced as a quantitative measure of orbital similarity by Southworth and Hawkins (1963; for definition see also Lindblad, this volume).

P/Neujmin 3 and P/Van Biesbroeck are the only comets for which a clear indication of a genetic relationship has been found in our first analysis of LTEP data. Nevertheless, another pair of comets would deserve, in this respect, a deeper study. It is the pair constituted by P/Kearns-Kwee and P/Wild 2, whose evolution is shown graphically in fig. 5. The two objects experienced the most effective captures by Jupiter of all comets examined here (see section 3.1). The timing of the two events agrees to within 14 months (1700 December - 1702 February), a coincidence which is striking in itself when compared with the interval of 821 years covered by our integrations.

At first glance, the pre-capture elements do not appear similar at all because the nodes are reversed. However, the perihelion longitudes differ only by 1.8°, the perihelion latitudes by 1.4°, and the small inclinations make the orbital planes deviate by only 7.2°. The largest difference is that in the eccentricity, the orbit of P/Wild 2 being formally hyperbolic and that of P/Kearns-Kwee elliptic with a period of 170 years. One must bear in mind, however, that the one-apparition orbit of P/Kearns-Kwee (Kazimirchak-Polonskaya, 1967) was also indicative of an originally hyperbolic orbit.

Further difficulties are given by the 1974 encounter of P/Wild 2 with Jupiter, that renders the computed orbit prior to that date very uncertain; by the probable presence of nongravitational forces (Forti, 1983); by the intrinsic uncertainty of the orbit of P/Wild 2, still based on one apparition only. The comet, however, has already been recovered as 1983s and will pass the perihelion this August. A further improvement of its orbital elements and integration of a manifold of neighbouring orbits, both for P/Wild 2 and P/Kearns-Kwee, will be needed to settle the question of their possible genetic relationship with some confidence. We intend to do this when more observations from the current return of P/Wild 2 become available. First tentative computations work against this hypotesis but, owing to the numerous uncertainties, there is still room for a remote possibility that needs a more careful investigation. An identification would probably imply that the XIIth century progenitor comet did not come directly from the Oort cloud, and that there is not a single case of that kind in our sample.

ACKNOWLEDGMENTS. The authors are grateful to B.G. Marsden for useful suggestions on the manuscript.

REFERENCES

Carusi, A., Kresák, L., Valsecchi, G.B.: 1981a, Astron. Astrophys., 99, 262-269.

Carusi, A., Kresák, L., Valsecchi, G.B.: 1981b, IAS-Internal Report 12.

Carusi, A., Kresáková, M., Valsecchi, G.B.: 1982, Astron. Astrophys., 116, 201-209.

Carusi, A., Kresák, L., Perozzi, E., Valsecchi, G.B.: 1984, IAS-Internal Report 12.

Carusi, A., Kresák, L., Perozzi, E., Valsecchi, G.B.: 1985a, this volume

Carusi, A., Perozzi, E., Pittich, E., Valsecchi, G.B.: 1985b, this volume.

Fokin, A.V.: 1958, Byull. ITA Leningrad, 7, 89-119.

Forti, G.: 1983, Astron. Astrophys., 126, 307-310.

Franklin, F.A., Marsden, B.G., Williams, J.G., Bardwell, C.M.: 1975, Astron. J., 80, 729-746.

Kazimirchak-Polonskaya, E.I.: 1961, Trudy ITA Leningrad, 7, 19-190.

Kazimirchak-Polonskaya, E.I.: 1967, Astron. Zh., 44, 439-460.

Kresák, L.: 1974, Asteroids, Comets, Meteoric Matter (C. Cristescu, W.J. Klepczynski eds.), Edit. Acad. Bucharest, 193-203.

Leverrier, U.J.J.: 1857, Ann. Obs. Paris, 3, 203-270.

Lexell, A.I.: 1778, Acta Acad. Sci. Petropol., 1, 317-352.

Marsden, B.G.: 1961, Astron. J., 66, 246-248.

Marsden, B.G.: 1970, Astron. J., 75, 206-217.

Marsden, B.G.: 1982, Catalogue of Cometary Orbits, IAU Central Bureau for Astron. Telegrams, Cambridge, Mass.

Marsden, B.G., Aksnes, K.: 1967, Astron. J., 72, 952-954.

Oterma, L.: 1958, Turku Informo, 17, 1-3.

Southworth, R.B., Hawkins, G.S.: Smithsonian Contr. Astrophys., 7, 261-285.

Vaghi, S., Rickman, H.: 1982, Sun and Planetary System (W. Fricke, G. Teleki eds.), Reidel, Dordrecht, 391-394.

DISCUSSION

F. Mignard: What about using analytical ephemerides elaborated at Bureau des Longitudes, to make an atlas in terms of a fixed number of revolutions for each comet, instead of a preassigned time span?

A. Carusi: I think it is a good idea; the philosophy of the work, however, and its statistical meaning would change, since in that case one would have integrations of very different length.

SECTION VI

NONGRAVITATIONAL FORCES

NONGRAVITATIONAL FORCES ON COMETS: THE FIRST FIFTEEN YEARS

B. G. Marsden
Harvard-Smithsonian Center for Astrophysics
60 Garden Street
Cambridge, MA 02138
U.S.A.

ABSTRACT. The recent and current situation with regard to our analysis and understanding of the nongravitational effects in cometary motions is reviewed. Comets can be categorized according to the different physical situations that may exist in their nuclei. Further experimentation with theoretical models and empirical fits to observations is encouraged.

1. INTRODUCTION

Just 15 years ago this month, I published the first detailed paper that attempted to make a systematic study of the way in which nongravitational forces influence the motions of comets (Marsden 1969). The existence of such forces had been discussed on numerous occasions and in many contexts during the preceding 15 decades, but this was the first time that computations were presented in a uniform and mathematically rigorous manner for a number of different comets, using relatively general equations of motion that are capable of at least some physical interpretation. My motivation was twofold: (a) to improve the accuracy with which positions of comets could be predicted, and (b) to obtain insight into the physical nature of the forces.

This is not to say that earlier researchers did not have the same motivation. Encke (1820) was clearly concerned with the need for making accurate predictions when he introduced a nongravitational secular-acceleration term into his computations on the comet that bears his name, and he subsequently developed in considerable detail a theory (Encke 1831) relating this secular acceleration to the resistive coefficient of the medium in which the comet was supposed to move. Encke's theory was applied to other comets, and the resisting-medium hypothesis was finally abandoned only when it began to be suspected, a century later, that some comets experienced secular decelerations rather than accelerations. Although these secular variations in mean motion were available for only a handful of comets, Whipple (1950) put them to good use in his brilliant paper that introduced the concept of the icy-conglomerate model for a cometary nucleus. Since most of the predictions for the returns of comets were clearly affected by computational approx-

343

A. Carusi and G. B. Valsecchi (eds.), Dynamics of Comets: Their Origin and Evolution, 343–352.
© *1985 by D. Reidel Publishing Company.*

imations and sometimes by downright errors, several astronomers (e.g., Roemer 1961) questioned whether physical nongravitational influences on cometary motions really existed. Clearly there was a need for a con- certed attack on the problem, and the rapid advances in automatic compu- tational capabilities during the 1960s provided the opportunity.

2. COMPUTATIONAL PROCEDURE

The n-body computer program by Schubart and Stumpff (1966) pro- vided an excellent starting point, and with programing help from K. Aksnes it was combined with a differential-correction procedure in which the partial derivatives required are replaced by the differences in the comet's calculated coordinates that arise when small changes are made successively in each of the assumed orbital parameters. In a purely gravitational situation n would therefore be 17, the integration being done for the sun, the planets Mercury to Pluto, the nominal orbit of the comet and the six variations. This process is more accurate than the traditional one of calculating the partial derivatives analytically from the basic equations of the two-body problem, since the two-body equa- tions are frequently a poor approximation to the true motion of a comet, particularly one that makes a close approach to Jupiter during the span of time covered by the observations under consideration. By iterating the solution, terms of higher order than the first are automatically eliminated, and the differences converge to true partial derivatives. The beauty of this process is that it can be readily adapted to include a nongravitational force. The nongravitational force can be any con- tinuous function of the comet's position and velocity vectors and the time. Additional integrations must be done for each new nongravita- tional parameter introduced, and in a few of our computations n has been as high as 21. There is of course a danger that some of the additional parameters will be strongly correlated, and our experience has been that it is undesirable to introduce more than two additional parameters.

Accordingly, we considered the nongravitational force in the form of additional acceleration components F_1 directed along the comet's instantaneous radius vector outward from the sun, F_2 parallel to the line from the sun to the point in the instantaneous orbit 90° ahead of the comet, and F_3 perpendicular to the plane of the instantaneous orbit (to the "northern" side in the case of direct motion). Although the transverse component F_2, which can most closely be related to the secular variation of the earlier computations, obviously turned out to be the best determined of the components, it soon became clear that the radial component F_1 tended to be positive and an order of magnitude larger. This is of course precisely what is to be expected from the Whipple model, F_2 arising solely because there is a lag between the direction of maximum ejection of material from the comet nucleus and the subsolar point. The computations suggested that this lag is generally on the order of 5°-10°. When solutions were made for the normal compo- nent F_3, the results were comparable in magnitude to the transverse component, but since the determinacy of F_3 is poor it has generally been

found advisable to ignore it. Since there seemed to be evidence that
the secular acceleration of P/Encke was decreasing, perhaps exponentially to zero with time, exponential decay terms were introduced
into our early nongravitational-force computations. The existence of
P/Arend-Rigaux and P/Neujmin 1, each of which has both a generally inert
appearance and no detectable nongravitational effect in its motion,
added support to the hypothesis. However, with the recognition (Marsden
1970a) that the transverse component of P/Pons-Winnecke had changed sign
the practice was abandoned.

In computations of this type there is in any case usually a large
correlation between a parameter and its rate of change, and more
insight, as well as greater determinacy, can be obtained by comparing
values of the parameter from discrete solutions over spans of time long
enough that the parameter can be reasonably determined, but short enough
that there is no injustice to the residuals. Although the gradual deac-
tivation of a comet with time presumably plays a role, at least for some
comets, it is apparent that the nongravitational parameters as we have
defined them can also be influenced to some extent by large changes in a
comet's orbit following a close approach to Jupiter, and presumably even
more so by variations in the orientation of the comet's axis of rotation.

Although the idea that some of the Apollo objects and other unusual minor planets are defunct cometary nuclei has been somewhat out of
vogue, it has recently received new fuel with the discovery of the Geminid parent 1983 TB, as well as the discoveries of 1982 YA, 1983 SA, 1983
XF and 1984 BC and their anticipated close approaches to Jupiter. The
possibility that some objects that appear to be asteroidal should show
nongravitational effects (Ziołkowski 1983) is of course interesting, but
the necessary computations are extremely intricate, and there is always
the danger that the small residuals from the gravitational solutions
have some other cause. A "nongravitational effect" suspected in the
motion of (944) Hidalgo could be much more logically explained by an
adjustment to the adopted mass of Saturn (Marsden 1970b). In any case,
not all cometary nuclei may deactivate. If a comet nucleus should instead completely disperse into meteoroids, the nongravitational force,
representing relative mass loss, should show a large increase. Erratic
behavior of the rotation axis in a comet's dying stages would be expected to produce even wilder changes in the nongravitational parameters. Such changes were apparent in the motion of P/Brorsen, and they
appear to have become extreme between the comet's 1873 return and its
last observed appearance in 1879.

From a physical point of view it is highly desirable to consider
the variation of the nongravitational force with the comet's heliocentric distance r. After utilizing for some time an entirely arbitrary
dependence on a combination of an inverse power and a diminishing exponential, we adopted for the F_i what has been variously termed "Style
II" or the "standard model", namely (Marsden et al. 1973):

$$F_i = A_i \, g(r) \qquad (i = 1, 2, 3),$$

$$\tag{1}$$

where the A_i are constants and

$$g(r) = C \; (r/r_o)^{-m} \; [1 + (r/r_o)^n]^{-k}, \tag{2}$$

with $m = 2.15$, $n = 5.093$, $k = 4.6142$, and the normalizing constant $C = 0.1113$. The dependence is essentially that of an inverse-square law out to the vicinity of $r_o = 2.808$ AU, and beyond that distance there is a much more rapid decrease. The form of $g(r)$, suggested by Z. Sekanina, was fitted by A. Delsemme to a curve (Delsemme and Miller 1971) showing the vaporization flux of water snow with heliocentric distance. Since other likely constituents of cometary ices are much more volatile than water, their corresponding transition distances r_o are much larger--almost 8 AU even in the case of ammonia. Application to actual nongravitational orbit solutions for several comets suggested that r_o was not larger than 4 AU, thereby apparently confirming the general belief that water is the principal icy constituent of a comet nucleus.

3. RESULTS

Although the majority of the available solutions for nongravitational parameters are still those published by myself, frequently in collaboration with Sekanina, generally in a series of papers in the <u>Astronomical Journal</u> up to 1974, D. K. Yeomans has also been a very important contributor to this work, beginning with his study of P/Giacobini-Zinner (Yeomans 1971). The day has not yet arrived when predictions for the returns of periodic comets routinely allow for nongravitational effects, but since about 1978 a number of nongravitational orbit solutions, all using the "standard model", have also been made by S. Nakano, by W. Landgraf, and most recently by G. Forti. Because F_2 (or A_2) is a measure of the rate of change of revolution period, nongravitational orbit solutions generally become appropriate when a comet has been observed at a third perihelion passage.

The first five sections of Table I list the 63 comets that fall into this category and summarize the situation with regard to the computation of nongravitational forces. In each section the comets are arranged in order of increasing perihelion distance. Section 2, which is the largest, contains comets that have non-zero transverse parameters A_2 that are either steady or decreasing in magnitude with time. For the most part these are well-behaved, predictable comets, generally consistent with the deactivation/rotation-variation scenario. Some of the large values of A_2 seem to be associated with objects that had recently been thrown in by Jupiter from orbits of larger perihelion distance, but the extreme value for P/Gunn might be influenced by the fact that in this case the solution actually included a pre-approach, prediscovery observation. Section 1 shows comets for which no definite nongravitational parameters have been detected. In addition to the afore-mentioned "asteroidal" comets, this section contains the three three-apparition comets with perihelion distances above the transition distance for water--another indication that water is the principal cometary constituent; and in spite of the outbursts in brightness of P/Schwassmann-Wachmann 1 that might at first sight be expected to influence the comet's motion. The apparent absence (or near absence) of non-gravitational forces acting on P/Crommelin over an interval of more than a century is surprising in view of this comet's rather small perihelion distance and known large brightness surge near perihelion. Computations by Kamieński (1959) indicated that P/Wolf did experience a secular de-

TABLE I. SUMMARY OF NONGRAVITATIONAL INVESTIGATIONS

Comet		q	P	N	Interval	A_1	A_2	Comp.
1. No-nongravitational effects								
Crommelin		0.7	28	5	1873–1984	0.0	0.000	L,M,Y
Arend–Rigaux	a	1.4	7	5	1951–1978	0.0	0.000	M
Tempel 1		1.5	6	7	1967–1983	0.0	0.000	M
Tsuchinshan 1		1.5	7	3	1965–1978	0.0	0.000	L,S
Neujmin 1	a	1.5	18	5	1913–1966	0.0	0.000	M
Reinmuth 2		1.9	7	6	1947–1967	0.0	0.000	M
Neujmin 3		2.0	9	3	1929–1972	0.0	0.000	M
Holmes	o	2.2	7	6	1964–1980	0.0	0.000	M
Wolf		2.5	8	13	1925–1967	0.0	0.000	P,Y
Oterma		3.4	8	3	1942–1962	0.0	0.000	M
Smirnova–Cher.		3.6	9	3	1967–1983	0.0	0.000	N
Schwass.–W. 1	o	5.4	15	5	1902–1983	0.0	0.000	M
2. Stable or decreasing nongravitational effects								
Encke		0.3	3	53	1967–1980	−0.1	−0.004	M
Halley		0.6	76	30	1835–1982	+0.1	+0.015	L,Y
Tempel–Tuttle		1.0	33	4	1865–1965	0.0	+0.009	Y
Grigg–Skjellerup		1.0	5	14	1952–1972	0.0	−0.001	M,S
Tuttle		1.0	14	10	1939–1980	+0.1	+0.013	Y
Finlay		1.1	7	10	1960–1974	+0.3	+0.020	Y
Olbers		1.2	70	3	1815–1956	+0.2	+0.065	Y
Wirtanen		1.3	6	5	1948–1975	+0.5	−0.087	M
d'Arrest		1.3	6	14	1963–1977	+0.6	+0.120	M,Y
Churyumov–G.		1.3	7	3	1969–1982	0.0	+0.012	M
Borrelly		1.3	7	10	1932–1975	+0.1	−0.038	Y
Wolf–Harrington		1.6	7	6	1951–1978	+0.2	−0.049	N,S
Stephan–Oterma		1.6	38	3	1867–1981	+0.2	−0.003	Y
Daniel		1.7	7	6	1937–1964	+1.1	+0.078	M
Tsuchinshan 2		1.8	7	3	1965–1978	−1.2	−0.004	L
Arend		1.8	8	5	1951–1976	+0.1	−0.029	M
Brooks 2	s	1.8	7	12	1946–1961	+1.1	−0.191	M
Reinmuth 1		2.0	8	7	1928–1973	+0.2	−0.028	M
Schwass.–W. 2		2.1	6	9	1956–1981	+2.0	−0.174	F,M,N
Johnson		2.2	7	6	1956–1977	+0.8	−0.027	N
Kearns–Kwee		2.2	9	3	1963–1982	0.0	−0.404	F
Ashbrook–Jackson		2.3	7	5	1948–1979	0.0	−0.012	F
Gunn		2.5	7	4	1954–1982	+2.4	+0.613	M
Whipple		2.5	7	7	1947–1978	+0.6	−0.044	M

3. Slightly increasing nongravitational effects

Honda–Mrkos–P.	0.6	5	6	1969–1980	+0.1	−0.046	M
Giacobini–Zinner	1.0	7	11	1965–1978	−0.2	−0.046	Y
Schaumasse	1.2	8	6	1944–1960	+0.4	−0.041	M
Pons–Winnecke	1.3	6	19	1951–1976	0.0	+0.002	M
Tempel 2	1.4	5	17	1956–1978	+0.1	+0.002	M
Kopff	1.6	6	12	1958–1977	+0.3	−0.084	Y
Faye	1.6	7	17	1954–1977	+0.1	−0.003	L,M,N
Comas Solá	1.9	9	7	1960–1979	+0.8	−0.093	F,M

4. Significantly increasing or wild nongravitational effects

Brorsen	d	0.6	5	5	1868–1879	+1.3	+0.134	L,M
Pons–Brooks	op	0.8	71	3	1812–1954	−0.1	−0.027	Y
Biela	ds	0.9	7	6	1832–1852	+1.2	−0.094	L,M
Tuttle–G.–K.	o	1.1	6	6	1951–1973	+0.7	+0.022	M
Tempel–Swift	d	1.2	6	4	1869–1908	+0.1	−0.113	M
Perrine–Mrkos	d	1.3	7	5	1896–1955	−0.1	−0.060	M
Forbes		1.5	6	6	1961–1980	+0.5	−0.078	M

5. Generally unstudied comets of more than two apparitions

Swift–Gehrels		1.4	9	3	(1889–1982)	+
Jackson–Neujmin		1.4	8	3	(1936–1978)	−
Clark		1.6	6	3	(1973–1984)	0
Harrington		1.6	7	3	(1953–1980)	+
de Vico–Swift		1.6	6	3	(1844–1965)	+
du Toit–N.–D.		1.7	6	3	(1941–1983)	−
Harrington–Abell		1.8	8	5	(1955–1984)	0
Väisälä 1		1.8	11	5	(1939–1982)	0
Taylor	s	2.0	7	3	(1915–1984)	−
Shajn–Schaldach		2.2	7	3	(1949–1978)	0
Van Biesbroeck		2.4	12	3	(1954–1979)	0
Slaughter–B.		2.5	12	3	(1958–1981)	+

6. Two-apparition comets, possible nongravitational effects

Brorsen–M.	p	0.5	72	2	(1847–1919)
Denning–F.		0.8	9	2	(1889–1978)
Schwass.–W. 3		0.9	5	2	(1930–1979)
Gale	dp	1.2	11	2	(1927–1938)
du Toit–H.	s	1.2	5	2	(1945–1982)
Westphal	dp	1.3	62	2	(1852–1913)
Neujmin 2	d	1.3	5	2	(1916–1927)
Peters–Hartley		1.6	8	2	(1846–1982)

The columns q, P and N give the perihelion distance (in AU), period (in yr) and number of apparitions. A_1 and A_2 are representative values, valid for the Interval shown. The column Comp. shows the orbit computers: F = G. Forti, L = W. Landgraf, M = B. G. Marsden, N = S. Nakano, P = E. I. Kazimirchak–Polonskaya, S = G. Sitarski, Y = D. K. Yeomans. The notes preceding the column q are: a = asteroidal, d = disappeared, o = outbursts, p = poor fit, s = split.

celeration, particularly before the comet's close approach to Jupiter in
1922. Nongravitational effects whould also presumably be detectable for
objects like P/Tempel 1 and P/Holmes if the older observations of these
comets (each of which was lost for a long time) were tied in with the
recent ones.

Section 3 of Table I shows comets for which A_2 is currently
slightly increasing, and Section 4 shows those where there has perhaps
been a larger rate of increase, or possibly some particularly erratic
behavior. Although ascription of a comet to one or the other of these
sections could be debated, it is thought that the comets in Section 3
are well-behaved objects whose A_2 values have increased due to
straightforward variations in the axis of rotation, whereas Section 4
contains comets that have split, experienced irregular outbursts,
disappeared—or are perhaps about to do so.

Section 5 lists comets for which satisfactory nongravitational-
force studies have not yet been made, although in some cases the
probable sign of A_2 can be indicated. Section 6 lists eight comets
that have been observed at only two perihelion passages but that might
merit nongravitational investigation. Among these are potential Section
4 comets gravitational orbit solutions for which are known to be unsat-
isfactory (P/Gale, P/Westphal, P/Neujmin 2), as well as cases where the
apparitions are widely separated in time.

Nongravitational effects are sometimes evident in the motions of
comets that have been observed at only a single perihelion passage.
There are in fact at least eight known long-period comets for which one
cannot make satisfactory orbit determinations if nongravitational forces
are ignored, and when nongravitational solutions have been attempted the
results are not out of line with those for the short-period comets.
Since a positive radial nongravitational component has the effect of
making a comet's inverse semimajor axis larger than it would otherwise
be, allowance for nongravitational effects could clearly eliminate those
few cases of comets that otherwise seem to have "original" orbits that
are hyperbolic. This has implications for the size of the Oort cloud
(Marsden et al. 1978).

Two other expressions have been discussed for handling the cometary
nongravitational effects in terms of equations of motion in rectangular
coordinates. The first of these (Brady and Carpenter 1971) consisted
simply of a radial component proportional to the inverse-square of
heliocentric distance and changing linearly with time. Although this
expression amply described the motion of P/Halley since 1682 and produced a perfectly satisfactory prediction for 1986, it is difficult to
ascribe any physical meaning to it, and the explanation (Brady 1972) in
terms of perturbations by a Jupiter-sized planet in a highly-inclined
orbit 65 AU from the sun is obviously unacceptable. Although he continues to postulate that the nongravitational effect consists solely of the
traditional secular variation in the mean motion, Sitarski (1981) has
expressed this directly in terms of the equations of motion in rectangular coordinates. He has also developed an accurate and highly ingeneous
procedure for solving simultaneously for the Keplerian elements and
secular variation and has successfully applied it to a few comets. A
rigorous comparison between Sitarski's model and ours is not possible,

but to assume a constant secular variation in mean motion roughly
supposes that F_2 is proportional to $1/r$.

4. NEW DEVELOPMENTS

In recent years several groups (e.g., Weissman and Kieffer 1981)
have attempted to model the distribution of temperature over a cometary
nucleus, and Rickman and Froeschlé (1982) have used such a thermal model
to examine the variation of the nongravitational force with heliocentric
distance in the case of P/Halley. They have found the "standard model"
to be lacking in the sense that the reaction of the comet to variations
in surface temperature can cause the parameters A_1 and A_2, which we
have taken to be constant (or at least not to have any short-term
variations), to vary by perhaps as much as a <u>factor</u> of 100 as r
ranges from 0.6 to 4.0 AU! The actual variation--in fact, whether A_1
and A_2 are increasing or decreasing over this range--is very dependent
on the comet's thermal inertia, but it would seem that, near the sun,
A_1 is essentially constant and A_2 varies as r^2, with the result that
F_2 is then constant. Beyond some transition distance the nongravitational force would diminish much more rapidly with distance than given
in the Delsemme formula. There is also some asymmetry with respect to
perihelion. Landgraf (1984) has very recently applied the Rickman-
Froeschlé model in an exhaustive examination of the orbit of P/Halley
over 1607-1984, solutions being made over a range of a factor of eight
in thermal inertia and a factor of five in the comet's rotation period.
In addition to considering the thermally induced effect, he assumed the
presence of a factor $1 - Bt$ in both F_1 and F_2. The mean residuals
of the observations from his various computations are identical, and he
in fact obtained a result with precisely the same mean residual when he
applied the "standard model" (modified again with the factor $1 - Bt$) to
the same data. The spread of the values he obtained for A_1 and B
according to the Rickman-Froeschlé model is not large and essentially
encompasses the values given by the standard model. On the other hand,
his various Rickman-Froeschlé values for A_2 differ by up to a <u>factor</u>
of six, and the smallest value is 70 times greater than the A_2 given
by the standard model!
       Since A_2 (or the corresponding secular variation of the mean
motion) is the basic quantity appearing in all studies of cometary nongravitational motion, it might appear that some drastic rethinking is
necessary. More experimentation with the nongravitational forces associated with various thermal models is clearly very desirable, but in
spite of the current interest in P/Halley and in the need for accurate
predictions to ensure the success of the space missions in 1986, we
should not delude ourselves into thinking that we shall be able to come
up with <u>the</u> model that will give an extremely accurate prediction.
The success of the probe will still be governed by the accuracy of the
astrometric observations made just before encounter. If the function
$g(r)$ is systematically deficient in the way a comet nucleus actually
reacts to the ejection of material, it would be appropriate to abandon
it in favor of something else. That something else would presumably be

different for F_1 and F_2, and there might be dependence on $\dot{r}$ as well as on r. If a satisfactory, not overly complicated, continuous function, applicable to all comets, cannot be produced, however, the "standard model" will continue to be useful in predicting the orbits of comets and in allowing some kind of comparison to be made between the characteristics of one comet and another.

REFERENCES

Brady, J. L.: 1972, Publ. Astron. Soc. Pacific **84**, 314.
Brady, J. L. and Carpenter, E.: 1971, Astron. J. **76**, 733.
Delsemme, A. and Miller, D. C.: 1971, Planet Space Sci. **19**, 1229.
Encke, J. F.: 1820, Berliner Astron. Jahrbuch für 1823, p. 222.
Encke, J. F.: 1831, Astron. Nachr. **9**, 311.
Kamienski, M.: 1959, Acta Astron. **9**, 53.
Landgraf, W.: 1984, private communication.
Marsden, B. G.: 1969, Astron. J. **74**, 720.
Marsden, B. G.: 1970a, Astron. J. **75**, 75.
Marsden, B. G.: 1970b, Astron. J. **75**, 206.
Marsden, B. G., Sekanina, Z., and Yeomans, D. K.: 1973, Astron. J. **78**, 211.
Marsden, B. G., Sekanina, Z., and Everhart, E.: 1978, Astron. J. **83**, 64.
Rickman, H. and Froeschlé, C.: 1982, in T. I. Gombosi (ed.), Cometary Exploration, Hungarian Acad. Sci., Budapest, vol. 3, p. 109.
Roemer, E.: 1961, Astron. J. **66**, 368.
Schubart. J. and Stumpff, P.: 1966, Veröff. Astron. Rechen-Inst. No. 18.
Sitarski, G.: 1981, Acta Astron. **31**, 471.
Weissman, P. R. and Kieffer, H. H.: 1981, Icarus **47**, 302.
Whipple, F. L.: 1950, Astrophys. J. **111**, 375.
Yeomans, D. K.: 1971, Astron. J. **76**, 83.
Ziołkowski, K.: 1983, in C.-I. Lagerkvist and H. Rickman (eds.), Asteroids, Comets, Meteors, Uppsala University, Uppsala, p. 171.

DISCUSSION

Weissman: We have just modeled P/Halley using Sekanina's new rotation rotation pole and 41.5-hr period, and we find that the temperature distribution is so symmetrical that it is difficult to believe there could be any transverse nongravitational force. Thus, if we assume any physically reasonable surface material, we have trouble believing the long rotation period. We would prefer the 10.3-hr or 14-hr periods that have been suggested.

Marsden: In view of the discouragingly large range of parameters suggested by Rickman and Froeschlé and considered by Landgraf, it is nice to know that there are perhaps a few constraints on the situation. Unfortunately, the recent photometric data do not seem to give an unambiguous value for the rotation period, and I am inclined to agree with R. M. West that the light curve is being significantly affected by the comet's intrinsic activity, even at its present great distance.

DO COMET GROUPS EXIST?

B.A. Lindblad
Lund Observatory
Box 43
S-221 00 Lund
Sweden

ABSTRACT. The phenomena of comet groups, i.e. sets of comets that exhibit similarity in their orbital elements, is investigated. A computer program based on the D-criterion of orbital similarity is used to search for comet pairs and groups. The reality of the groups is tested by making computer searches in random samples of comet orbits.

The data base for the study is 599 long-period comet orbits. The degree of orbital similarity within a comet group was first assumed to be identical to that encountered in meteor streams. The computer search at this level produced five comet pairs plus two groups with four and seven members, respectively. The latter two represented the eleven known members of the Kreutz group of sun-grazing comets. A comparison with searches in random samples showed that the two Kreutz groups were significant. There is a probability of 0.2 that the five comet pairs found in the real sample could be accidental formations.

In a second study the orbital similarity parameter D_s was varied and the number of comet groups found in the real and synthetic comet populations was compared at each level of D_s. Apart from the Kreutz group of comets, the number of groups detected in the real comet sample was for all levels of orbital similarity only slightly higher than the average found in the random samples. At the 2σ confidence level we conclude that comet groups exhibit similarity in their orbital elements, that is no greater than might be expected by chance.

INTRODUCTION

Hock (1865, 1866, 1867) found that there exist different comets with nearly identical orbital elements (time of perihelion-passage excepted). Such comets form a comet group. A remarkable example of such a group is the sun-grazing group (Kreutz, 1888, 1891, 1901). The group consisted of comets 1843 I, 1880 I, 1882 II and 1887 I. All of these comets were bright with highly eccentric orbits and very small perihelion distances. It was obvious that no two of them could be appearances of the same comet. It was therefore assumed that they were individual parts of a primitive comet which has disrupted in the past. The existence of such

353

A. Carusi and G. B. Valsecchi (eds.), Dynamics of Comets: Their Origin and Evolution, 353–363.
© 1985 by D. Reidel Publishing Company.

comet groups is therefore a question of considerable interest to
cometary physics.

Pickering (1911) in a study of long-period comet orbits listed 66
comet groups with from two to five members in each. Porter (1952) analysed about 500 long-period comet orbits and selected on the basis of
similarity in the orbital elements and proximity of aphelia 19 "clearly-
-defined" comet pairs and groups with from two to six members in each.
A revised list with 15 comet groups was presented in Porter (1963).
Although Porter's groups showed similarity in the orientation of the
orbital planes and major axes there was a considerable spread in the
perihelion distances.

A fundamental but somewhat controversial question is the statistical
significance of the pairs and groups. Öpik (1971) analysed 472 comets
with aphelion distances beyond Saturn. He selected 97 groups that showed
similarities in the angular orbital elements. Öpik calculated an overall
probability of 10^{-39} that these similarities could have occurred by
chance. Öpik's conclusions have been criticized by Whipple (1977).
Whipple repeated the statistical analysis of Öpik and tested it on a
random sample of cometary orbits. Approximately the same number of pairs
and groups were found in the random sample as in the original sample.
Whipple concluded that, except for a few pairs, the groups listed by
Öpik exhibit similarity in their orbital elements that is no greater
than might be expected by chance.

An obvious problem in the previous studies is the somewhat qualitative nature of the selection criteria for orbital similarity. The number
of comet groups can be much enlarged, if the investigator imposes less
severe restrictions as to the allowable spread in each orbital element.
An objective measure of orbital similarity is needed. Such a criterion
applied uniformly to a real and a random sample of comet orbits would
provide a definitive test of the reality of the proposed comet groups.

The problem of classification based on orbital similarity is well
known in meteor astronomy, where the study of meteor streams has
necessitated the use of sophisticated computer techniques for the detection and classification of streams. A mathematical definition of orbital
similarity, the so-called D-criterion, has been introduced by Southworth
and Hawkins (1963). It has been used by Lindblad (1971a, 1971b, 1974),
Sekanina (1973), Porubcan (1968, 1977) to search for meteor streams,
and by Zausajev and Galimova (1982) to describe the evolution of a
meteor stream. It has also been used by Lindblad and Southworth (1971)
to determine the membership of the Hirayama asteroid families, by Kramer
et al. (1979) to study clustering effects in Jupiter's family of short-
-period comets and by Kresák (1982a) to study the Tisserand invariant
for various types of planetary orbits. In an investigation of comet
groups Kresák (1982b) applied the D-criterion to all long period cometary
orbits listed in Marsden's 1972 catalogue.

D-criterion and rejection level D_s. Southworth and Hawkin's criteria
compares two sets of orbital elements. Let A and B represent two individual comets to be tested for orbital similarity. Let the orbital elements
be represented by the five quantities q, e, i, Ω and π, where $\pi = \omega + \Omega$
is the longitude of perihelion. A quantitative measure of orbital simi-

larity (or difference) is then given by the expression

$$D_{AB}^{\;2} = (e_A - e_B)^2 + (q_A - q_B)^2 + (2 \sin \tfrac{1}{2} I_{AB})^2 +$$

$$+ (e_A + e_B)^2 (\sin \tfrac{1}{2} \Pi_{AB})^2 \qquad\qquad (1)$$

where $(e_A + e_B)$ is a weight function, I_{AB} is the angle between the orbital planes and Π_{AB} is the difference between the longitudes of peri-helion measured from the intersection of the orbital planes. The reason for using the perihelion distance q in the comparison instead of the semimajor axis a is that the perihelion distance q for meteor (and comet) orbits is better defined than a.

The D-criterion is an objective method of classification on the bases of the orbital elements, i.e. it can be used to search for concen-trations in five-dimensional (q, e, i, ω, Ω) space. However, it is left to the investigator to choose the appropriate rejection level D_s. The rejection level D_s will vary with sample size and to some extent also with the accuracy of the orbits. For precise photographic meteor orbits Lindblad (1971a, 1971b) found

$$D_s = 0.80 \cdot N^{-1/4} \qquad\qquad (2)$$

where N is sample size. A preliminary study (Lindblad, 1970 unpublished) showed that it is possible to use eq. (2) for cometary orbits as well provided 1) these orbits have approximately the same distribution in five-dimensional (q, e, i, ω, Ω) space as meteors, 2) the errors in the cometary orbits are not larger than those encountered in meteor studies.

Since meteor streams originate from comets, the orbital distribu-tions of the two populations should be rather similar. A comparison of the 1/a distribution of the meteor and comet populations, however, indicates a higher percentage of long-period orbits in the comet popula-tions (Lindblad, 1974). A possible objection to the use of the D-criterion in a study of long-period comet orbits is thus the extremely high eccen-tricity of some of the comet orbits. This objection has been discussed in detail by Kresák (1982b), and is found to be irrelevant. However, it is not à priori known what value of the numerical constant in eq. (2) we should use if the sample is restricted only to long-period orbits.

PART I. SEARCH FOR COMET GROUPS AT FIXED D_s-VALUE

Data base. The data base for the present study was an updated tape version of the comet orbits listed in Marsden (1982). The tape included 1139 cometary apparitions of 724 individual comets, the remainder repre-senting earlier appearances of periodic comets. Of the individual comets 125 were of short period (P < 200 years) and 599 were parabolic or of long period (P > 200 years).

Search at D_s = 0.155 in real comet sample. We now assume that equation (2) defines the appropriate rejection level to use in a computer search

for comet groups. In order to retain a strict similarity with previous
meteor stream searches, we included in the search both short-period and
long-period orbits. Inserting N = 724 in equation (2) we obtain
D_s = 0.155. The computer search at D_s = 0.155 produced a total of twelve
comet groups. We found ten groups with 2 members each, one group with
4 members and one group with 7 members. Five of the ten comet groups
with only 2 members consisted of short-period comets. For the purpose of
the present study these groups were rejected. The remaining five two-
-member groups consisted of long-period comets as did also the two groups
with 4 and 7 members, respectively. The detected groups are listed in
Tables I and II. In passing we note that the comet pairs listed in Table I
are the same as the first five pairs listed in Table IV of Kresák (1982b).

All eleven members of the two largest groups (Table II) were identified as belonging to the sun-grazing group of comets. It is interesting
to note that the computer search at D_s = 0.155 clearly separates the
Kreutz group into two subgroups. This division has been discussed in
some detail by Marsden (1967).

TABLE I. Comet pairs at D = 0.155

1532	1911 VI	1973 VII	1881 IV	1979 X
1661	1790 III	1846 V	1898 X	1770 II

Search at D_s = 0.155 in random samples. To test the reality of the
proposed comet groups the D-criterion was applied to searches in a random
sample of cometary orbits. The main constraints we impose on a random
sample are: 1) the sample should have the same size as the real sample,
2) the sample should have the same overall distribution of the orbital
elements as the real sample, i.e. the frequency functions f(a), f(e),
f(i), f(ω) and f(Ω) should be very nearly the same in the synthetic and
in the real comet population, and 3) existing correlations between the
orbital elements a, e and i should not be destroyed by the randomization
procedure. The latter constraint is, of course, mainly of importance
when both short-period and long-period orbits are included in a search.
A fourth requirement is that the randomization process should be carried
out in such a way that it is feasible to develop a large number of
synthetic samples.

To develop a random sample we first excluded from the data base the
eleven members of the Kreutz group. The remaining 713 orbits were then
randomized by jumbling the nodes, i.e. the longitudes of nodes Ω were
randomly distributed amongst the orbits. In order to preserve the frequency functions f(Ω + ω) and f(Ω - ω) the longitude of perihelion ω was
simultaneously distributed. The randomization procedure was repeated on
a number of data decks, which were shuffled and cut in various ways. In
all twenty different random samples were produced.

The twenty random samples were searched for comet groups at the
same rejection level D_s as in the real comet population. No comet groups
were found which included both short-period and long-period orbits. The
detected short-period comet groups were removed from our study. Table III
lists the number of long-period comet groups found in the random comet

TABLE II. Kreutz comet group

Comet	Y M D of perihelion passage	q	e	i	ω	Ω	Ω-ω	Sub-group	L	B	D
1843 I	43.02.28	0.0055	0.999914	144°35	82°64	2°83	280°19	1	281°86	35°31	0.025
1880 I	80.01.28	0.0055	1.0	144.66	86.25	7.08	280.83	1	281.67	35.25	0.019
1887 I	87.01.12	0.0048	1.0	144.38	83.51	3.89	280.38	1	281.86	35.36	0.014
1963 V	63.08.24	0.0051	0.999946	144.58	86.16	7.24	281.08	1	281.95	35.36	0.020
1882 II	82.09.18	0.0078	0.999907	142.00	69.59	346.96	277.37	2	282.24	35.24	0.006
1945 VII	45.12.28	0.0075	1.0	141.87	72.06	350.50	278.44	2	282.87	35.97	0.033
1965 VIII	65.10.21	0.0078	0.999915	141.86	69.05	346.30	277.25	2	282.26	35.22	0.013
1970 VI	70.05.14	0.0089	1.0	139.07	61.29	336.32	275.03	2	282.26	35.07	0.133
1979 XI	79.08.31	0.0016	1.0	142.68	72.07	350.10	278.03	2	282.24	35.23	0.031
1981 I	81.01.27	0.0049	1.0	142.74	71.98	349.97	277.99	2	282.20	35.15	0.030
1981 XIII	81.07.20	0.0043	1.0	143.18	73.73	352.23	278.50	2	282.26	35.12	0.055

Note

Comets 1979 XI, 1981 I and 1881 XIII have been observed by the SOLWIND coronagraph onboard an Earth-orbiting satellite. Parabolic orbits have been computed on the assumption that the comets perihelion direction agrees with that of the Kreutz group. (IAU Circ. no:s 3647, 3716 and 3719). The orbit of comet 1887 I has been computed under a similar assumption.

populations. There is a surprisingly large dispersion in the number of
cometary groups detected in the various random samples. The number of
comet pairs found varied from 0 to 7, with a mean number of 3.4. It
follows that a study of comet groups which is based on only one or two
random samples may produce very misleading results.

TABLE III. Number of comet groups at D_s = 0.155
in real and random samples.

No. of pairs	No. of triplets	Total groups
2		2
5		5
4		4
3	1	4
2		2
4	1	5
3		3
5		5
3		3
4		4
4		4
2		2
6		6
2		2
4		4
3		3
2		2
0		0
4		4
7		7
Mean random sample	3.4	0.1	3.5
Real comet sample	5	0	7*

* Including two groups with 4 and 7 members

Comparison. It is interesting to note that none of the twenty random
searches produced a group with four or more long-period comets. We
therefore conclude that the two Kreutz groups listed in Table II are
significant, i.e. they cannot be accidental formations. In view of the
extremely small perihelion distances of these comets, this result is
not very surprising.

    The significance of the five long-period comet pairs detected in
the real comet population was investigated as follows. From Table III we
derived a histogram which depicts the number of cases vs the number of

comet pairs. Cumulative relative numbers as derived from the random
searches are plotted in Fig. 1 vs the number of comet pairs detected.
From Fig. 1 we conclude that there is a twenty per cent probability that
a collection of five or more comet pairs could be accidental formations.
It is thus very doubtful if any of the pairs listed in Table I should be
considered as real.

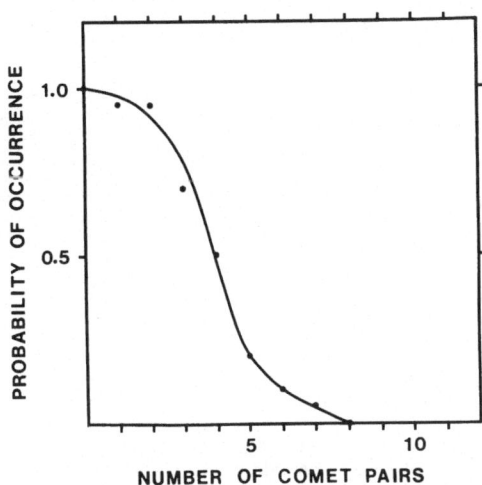

Figure 1. Probability of occurrence of N or more comet pairs as determined from searches in 20 random samples of long-period comet orbits.

PART II. SEARCH FOR COMET GROUPS AT VARIOUS D_s-VALUES

A possible objection to the previous analysis is that equation (2) is
not appropriate for cometary studies, and that one should accept less
severe restrictions as to the allowable spread in the orbital elements
within a comet group. It is obvious that most of the previous investigators have been very tolerant in this respect. In view of these considerations the reality of comet groups was next investigated without making
any à priori assumptions as to the appropriate rejection level D_s. In
this study we only considered long-period comets, i.e. comets with
periods > 200 years. For simplicity in the data handling the eleven
members of the Kreutz group were removed. The data base for part II thus
consisted of 588 parabolic or long-period orbits. This sample was
searched for groups using D_s-values in the interval 0.12 -0.30. The value
0.12 corresponds approximately to the lowest D_s-value used in meteor
stream searches - a larger value of D_s indicating a more liberal definition of orbital similarity. For comparison studies twenty random samples
were developed and searched at the same sequence of D_s-values.
    Fig. 2 compares the number of comet groups found in the real and
synthetic samples. Searches were made at nine different values of the

rejection level D_s as indicated in the diagram. The solid curve depicts the number of comet groups detected in the real comet population. At the lowest D_s-value (0.12) only three comet pairs were detected. Thus only 1 per cent of the comets were in groups at the lowest rejection level. At the highest D_s-value (0.30) 93 comets formed 41 groups of various sizes. Thus at this level 16 per cent of the comets were in groups.

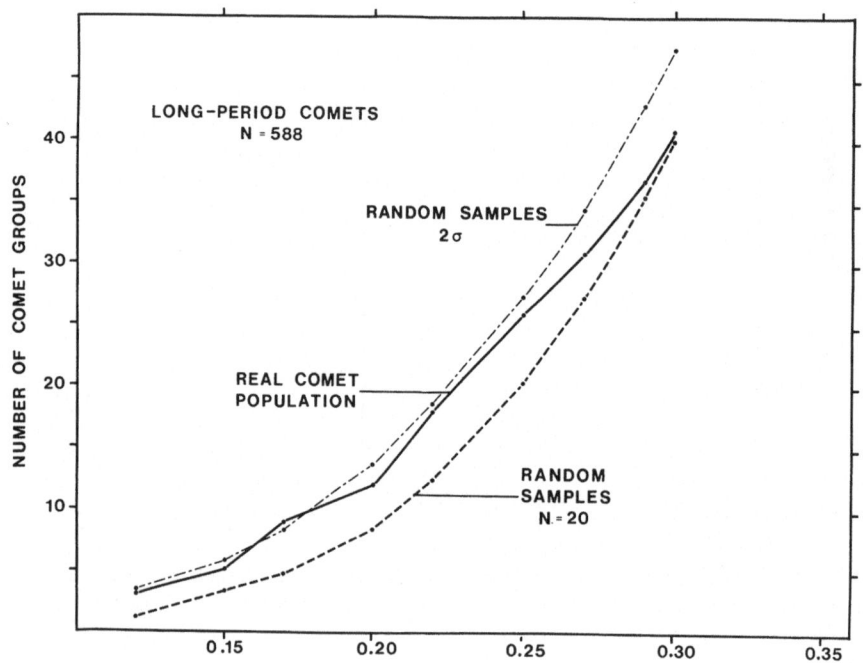

Figure 2.   Solid curve: Number of comet groups found in real comet population (Kreutz group removed).   Dashed curve: Mean number of comet groups found in 20 random samples of comet orbits.   Dot-dashed curve: Upper 2σ confidence limit of number of comet groups as determined from searches in 20 random samples.

The lower, dashed curve in Fig. 2 depicts the mean number of comet groups detected in searches in the twenty synthetic samples. We note that the number of groups is slightly higher (2-5 groups) in the real comet population than in the average of the random samples. We further note that this difference does not vary appreciably with the degree of orbital similarity that is imposed. Although the number of comet groups increases drastically when less severe restrictions are imposed as to the spread in the orbital elements (increased value of D_s), it is evident that the number of possibly significant groups does not increase.

The upper, dot-dashed curve in Fig. 2 shows the 2σ variation in the random samples. We note that the number of comet groups in the real sample generally does not reach the upper 2σ confidence level. Hence, at

any level of orbital similarity there are no significant comet groups at
the 0.05 probability level. By ordinary statistical procedures we must
therefore reject the existence of comet groups. In agreement with previous
work of Whipple and Kresák we conclude that all comet groups (except the
Kreutz groups) are accidental.

Discussion. Our search at D_S = 0.30 was compared with Kresák (1982b). In
Kresák's investigation the discriminant D was calculated for all possible
pairs of orbital elements by computer techniques, while the various
possible combinations of orbits with D_S < 0.30 were selected manually.
This was very time consuming, and there was an obvious risk that some
pairs and groups could be missed. Kresák's Table IV lists 25 pairs, three
groups with 3 members and two groups with 4 members. Our search confirmed
24 of the 25 pairs and also the five larger groups. In addition our
search identified 11 groups that were not included in Kresák's table,
owing either to new or improved orbits in our data base or to omissions
in Kresák's table.
     Kresák tested his analysis on three independent synthetic samples,
which were based on a random distribution of the angular elements and
the observed distribution of perihelion distances. Statistics of D_S-
-values less than 0.30 was computed for each random sample and the results
compared with those obtained in the real comet population. After excluding four pairs in the real population Kresák found that the number of
cases with D_S < 0.30 was exactly the same in the real population as in
the mean of the random samples. Kresák hence concluded that the proposed
comet groups were accidental formations. Although Kresák's results are
based on a precariously small number of random samples his conclusions
are entirely confirmed by our study.

KREUTZ GROUP OF COMETS

The two largest groups detected by the search at D_S = 0.155 consisted of
the 11 known members of the Kreutz family. Table II compiles values of
the orbital elements (1950.0 equinox), perihelion latitudes and longitudes and the D-values (computed from the mean orbit of each group). With
one exception the two sun-grazing groups persisted unchanged in all
searches at all rejection levels, i.e. no members were added or subtrac-
ted to the two Kreutz groups, when the orbital similarity parameter D_S
was varied. Comet 1668, which sometimes has been suggested as a member
of the Kreutz group was not classified as a member in our searches.
     The D-values listed in Table II represent the individual deviations
from the mean orbits of the two subgroups. The mean D-values for the subgroups are 0.020 and 0.043, respectively. If one outlying member (1970
VI) is excluded, the mean D-value of the second Kreutz group is lowered
to 0.028. These mean D-values are markedly lower than those of other
comet groups, i.e. the degree of orbital similarity in the Kreutz group
is much higher than in any of the other comet groups detected in our
searches. The positions of the perihelion points of the Kreutz comet
orbits are located in a limited sky area of about 1° x 1°. However, it
is important to note that four of the orbit computations have been based

on the à priori assumption that the perihelion direction agrees with
that of the mean of the Kreutz group.

Detailed studies of the Kreutz group have been made by Marsden
(1967) and Sekanina (1967). The discovery in a period of less than two
years of three new members which apparently have collided with the sun
(Michels et al., 1982, Sheeley et al., 1982 and IAU Circ. no:s 3647,
3716 and 3719), is an unusual event and suggests that the group may
consist of far more comets than was previously assumed. A difficult
problem in celestial mechanics is to explain how the perihelion distances
could be perturbed to values less than the solar radius. Possible mechanisms have been discussed by Weissman (1983).

ACKNOWLEDGEMENTS

I am indebted to Dr. B. Marsden for kindly placing at my disposal an
updated tape version of the 1982 catalogue of cometary orbits. This
study was supported by the Swedish Board for Space Activities through
contracts DFR 34/82, DFR 28/83 and DFR 15/84.

REFERENCES

Hock, M.: 1865, Mon. Not. R. Astr. Soc. 24, 243.
Hock, M.: 1866, Mon. Not. R. Astr. Soc. 26, 204.
Hock, M.: 1867, Astr. Nachr., 70, 203.
Kramer, E.N., Musij, V.L. and Shestaka, I.S.: 1979, Astron. Vestn.,
      13, 42.
Kresák, L.: 1982a, Bull. Astron. Inst. Czechosl., 33, 104.
Kresák, L.: 1982b, Bull. Astron. Inst. Czechosl., 33, 150.
Kreutz, H.: 1888, Publ. Sternwarte Kiel, no. 3.
Kreutz, H.: 1891, Publ. Sternwarte Kiel, no. 6.
Kreutz, H.: 1901, Astron. Abhandl., 1, 1.
Lindblad, B.A.: 1971a, Smithson. Contr. Astrophys. 12, 1.
Lindblad, B.A.: 1971b, 'Meteor Streams', in Space Res. XI, Akad. Verlag,
      Berlin.
Lindblad, B.A.: 1974, in Asteroids, Comets, Meteoric Matter, IAU Coll.
      22 (Nice), Cristescu and Klepczynski (Eds.), Bukarest, pp. 269-281.
Lindblad, B.A. and Southworth, R.B.: 1971, in Gehrels (Ed.), Physical
      Studies of Minor Planets, Washington, pp. 337-352.
Marsden, B.G.: 1967, Astron. J., 72, 1170.
Marsden, B.G.: 1982, Catalogue of Cometary Orbits, Fourth Ed., Cambridge.
Michels, D.J., Sheeley, N.R., Howard, R.A. and Koomen, M.J.: 1982,
      Science, 215, 1097.
Öpik, E.S.: 1971, Irish Astron. J., 10, 35.
Pickering, W.H.: 1911, Harvard Ann., 61, 163.
Porter, J.G.: 1952, 'Comets and Meteor Streams', Chapman and Hall,
      London.
Porter, J.G.: 1961, Mem. British. Astron. Assoc., 39, No. 3.
Porter, J.G.: 1963, in Middlehurst and Kuiper (Eds.), The Solar System
      IV, Chicago Univ. Press. pp. 550-572.

Porter, J.G.: 1966, Mem. British. Astron. Assoc., 40, No. 2.
Porubcan, V.: 1968, Bull. Astron. Inst. Czechosl., 19, 327.
Porubcan, V.: 1977, Bull. Astron. Inst. Czechosl., 28, 257.
Sekanina, Z.: 1967, Acta Universitatis Carolinae - Mathematica et
      Physica, No. 2, 33-84, (Publ. Astron. Inst. Charles Univ., no. 51).
Sekanina, Z.: 1973, Icarus 18, 253.
Sheeley, N.R., Howard, R.A., Koomen, M.J. and Michels, D.J.: 1982,
      Nature 300, 239.
Southworth, R.B. and Hawkins, G.S.: 1963, Smithson. Contr. to Astrophys.
      7, 261.
Weissman, P.R.: 1983, Icarus 55, 448.
Whipple, F.L.: 1977, Icarus 30, 736.
Zausajev, A.F. and Galimova, A.G.: 1982, Bull. Astrophys. Inst. Acad.
      Sci., Tadjik SSR, 72, 20.

A STUDY OF THE PRE-DISCOVERY MOTION OF
THE TWO ASTEROIDS 1983 SA AND 1983 XF

Daniel Benest
Observatoire de Nice, F-06003 Nice Cedex, France

Reinhold Bien
Astronomisches Rechen-Institut, Mönchhofstrasse 12-14
D-6900 Heidelberg 1, Federal Republic of Germany

Hans Rickman
Astronomiska Observatoriet, Box 515
S-75120 Uppsala, Sweden

ABSTRACT. Our investigation demonstrates that the asteroidal objects
1983 SA and 1983 XF can be considered as temporary visitors of the 4/3
and 2/1 resonance with Jupiter, respectively. Evidence is given that
both objects are first rank candidates for a cometary origin. The case
of 1983 SA is remarkable in so far as (279) Thule is up to now the only
known asteroidal 4/3 librator.

1. INTRODUCTION

Object 1983 SA has a perihelion distance q = 1.21 AU and an aphelion
distance Q = 7.25 AU. Equivalently, the semi-major axis is a = 4.23 AU,
and the eccentricity is e = 0.71. The orbital inclination i amounts to
31°. In case of object 1983 XF, one gets q = 1.45 AU, Q = 4.78 AU
(or a = 3.12 AU, e = 0.54), and i = 4°.
    The aim of the present paper is to investigate the question of a
dynamical relationship to comets. In fact, the orbital characteristics
of the two objects, as given above, are not usual for asteroids. The
values for a and e, and for the Tisserand invariant T (1983 SA: T = 2.49;
1983 XF: T = 2.98), are rather typical for a considerable number of short-
period comets, cf. Kresák (1979, Fig.1). Additionally, the orbital
features are dynamically interesting, since 1983 SA is obviously close
to the 4/3 resonance with Jupiter, whereas 1983 XF is not far from the
2/1 resonance.

2. BASIC MATERIAL

According to the Minor Planet Circular 8678, the orbital elements of
1983 SA have been determined from 54 observations during the period

A. Carusi and G. B. Valsecchi (eds.), Dynamics of Comets: Their Origin and Evolution, 365–370.
© 1985 by D. Reidel Publishing Company.

1983 Sep. 10 to 1984 Mar. 3;   the mean residual is 1!'2.   In case of
1983 XF the published orbit is based on 35 observations during 1983 Nov.
28 - 1984 Mar. 8; mean residual: 1!'5 (see M.P.C. 8679).   In the following, we call these elements "new initial values" for an integration.
     "Old initial values" are taken from M.P.C. 8394 (1983 SA, 41 ob-
servations 1983 Sept.10 - Nov. 9, mean residual 1!'1) and M.P.C. 8467
(1983 XF, 24 observations 1983 Nov.28 - 1984 Jan.4, mean residual 1!'5).

3. MODELS

We used three different dynamical models. The first one is the three-
dimensional elliptic three-body problem Sun-Jupiter-massless object.
The eccentricity of Jupiter's orbit is 0.048061 (cf.Astron.Ephem.1984)
and Jupiter's mass is equal to 1/1046.390 Solar masses. Details of the
integration technique (fourth-order Runge-Kutta method with variable
step length) are given by Benest (1974). In the second model, Saturn's
perturbations are taken into account. Finally, the third model includes
all major planets from Venus to Neptune; Mercury is added to the Sun
and Pluto's mass is neglected. The integration technique in the last
two cases (four-body model and nine-body model) is Bulirsch and Stoer's
method (1966), where the step length is also variable; masses and initial
values for the last two models are taken from Schubart and Stumpff (1966).
     We consider the old initial elements as a reasonable variation of
the new ones. In any case, one might expect that an integration of both
sets in three different models, using two different integration techniques, reveals a reliable picture of the dynamics of 1983 SA and 1983 XF.

4. RESULTS FOR 1983 SA

The new initial values of 1983 SA have been integrated backward over
1 000 yr in the four-body model. During this period the object is in
4/3 resonance with Jupiter. This result is shown in Fig.1, where the
"critical argument" $\sigma_{4/3}$ (according to Kresák, 1974) is plotted versus
time. For convenience the definition of $\sigma_{p/q}$ is as follows:

$$\sigma_{p/q} = (p - q)\, \tilde{\omega} - p\ell_J + q\ell \quad .$$

Here, p and q are relative prime integers with $p/q \approx n/n_J$, and n, ℓ, $\tilde{\omega}$
denote mean motion, mean longitude, longitude of perihelion of the mass-
less body, and the subscript J refers to the corresponding Jovian
quantities.
     The evolution of a is given in Fig.2. The jump of the semi-major
axis at - 64 yr (counted from the present) is probably caused by a close
encounter with Jupiter at a minimum distance of 0.64 AU. During the
integration interval from - 1 000 yr to the present the argument of perihelion, ω, increases from 275° to 317°, see Fig.3, whereas e decreases
from 0.79 to 0.71, and i increases from 18° to 31°.
     A backward integration of the new values over 400 yr in the nine-
body model shows no significant discrepancies; in particular, the

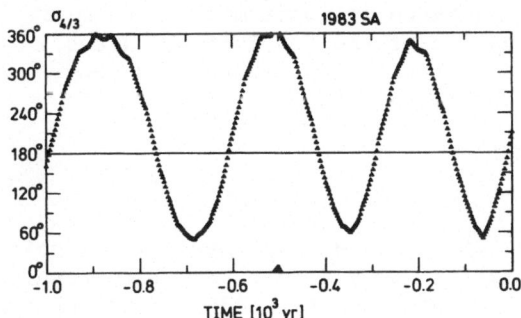

Fig.1.   1983 SA: evolution of the critical argument $\sigma_{4/3}$. Time is given in years counted from the present.   Every 2.7 yr a small triangle is plotted.

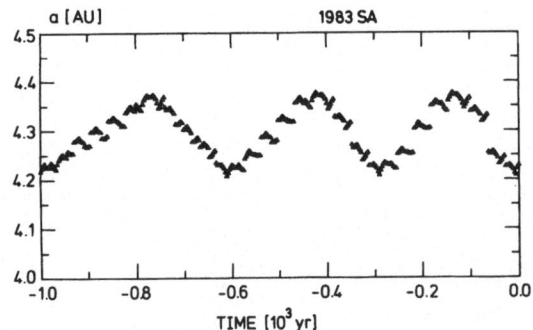

Fig.2.   1983 SA: semi-major axis a versus time.

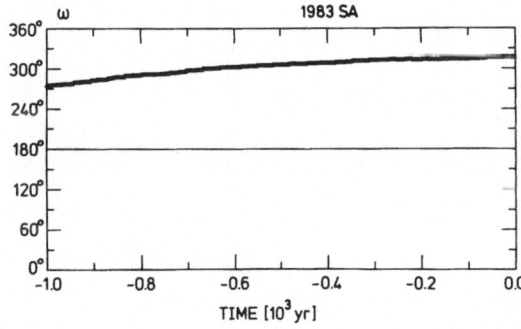

Fig.3.   1983 SA: argument of perihelion ω versus time.

libration around the 4/3 resonance is confirmed.

Moreover, an investigation of the new initial values in the three-body model, and of the old initial values in the four-body model lead to practically the same librational feature.

5. RESULTS FOR 1983 XF

Analogously, we have integrated backward the new initial values of 1983 XF over 1 000 yr in the four-body model. The main result is that since about 900 yr the object has been trapped into a libration around the 2/1 resonance. The critical argument $\sigma_{2/1}$ is plotted in Fig.4, and Fig.5 shows the semi-major axis. Before and after the beginning of libration we found close encounters with Jupiter (minimum distance at -986 yr: 0.37 AU, at -819 yr: 0.36 AU). The smallest minimum distance was reached at -558 yr, namely 0.29 AU.

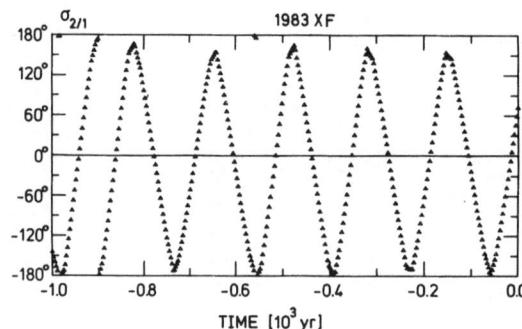

Fig.4.   1983 XF: critical argument $\sigma_{2/1}$ versus time.

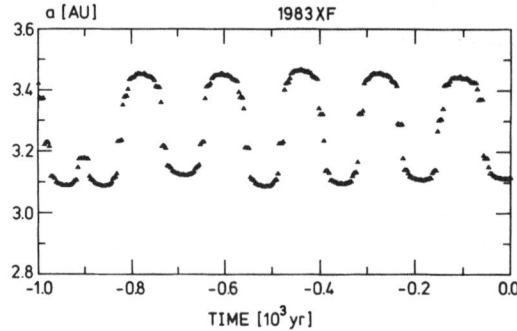

Fig.5.   1983 XF: semi-major axis a versus time.

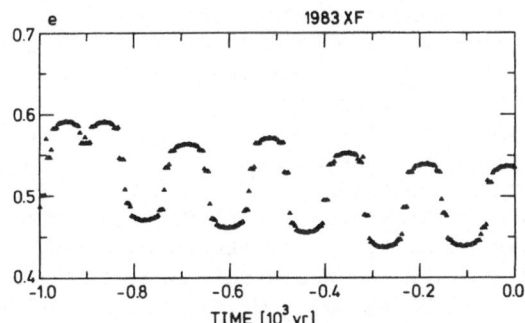

Fig.6. 1983 XF: eccentricity e versus time.

     The eccentricity shows an oscillation of large amplitude following
the period of libration, while the mean value of e decreases (roughly
spoken) from 0.54 to 0.48; see Fig.6. During the whole time interval
between -1 000 yr and the present, the inclination takes values between
10° and 3°.
     As in the case of 1983 SA, the nine-body model (new values, back-
ward integration over 400 yr) confirms the results of the four-body model.
The three-body model confirms the librations.
     Larger deviations have been found in the evolution of the old in-
itial values (four-body model). After a more or less chaotic behaviour,
where a takes values up to 5.5 AU, the libration around the 2/1 resonance
started only about 300 yr ago. It is interesting that at the same time
the object approached Jupiter up to 0.14 AU.

6. REMARKS ON THE FUTURE MOTION OF 1983 SA AND 1983 XF

A forward integration over 2 000 yr of both new sets of initial values
(four-body model) indicates a rupture of the libration. In case of
1983 SA the libration ends after 600 yr; a takes values around 4.6 AU.
1983 XF leaves the resonance near +400 yr and ends in a rather chaotic
orbit (a > 6 AU, minimum distance to Jupiter < 0.1 AU).
     We mention that in case of 1983 SA ω reaches a maximum value of 318°
and is finally reduced to 286°.

7. DISCUSSION AND CONCLUSION

Our results are surely affected by uncertainties of the starting conditions of both the massless bodies and the planets, as well as by errors
of the masses of the major bodies. Nevertheless, the following
conclusions should be sufficiently realistic. Moreover, Hahn and Rickman
(1984) found in a three-body model the same dynamical behaviour.
     Both 1983 SA and 1983 XF are in orbital resonance with Jupiter, at
least temporarily. In each case the amplitudes of the critical argument

are very large and the object passes repeatedly close to Jupiter. We note that (944) Hidalgo is the only numbered asteroid which can pass similarly close to Jupiter (0.38 AU; see Marsden, 1970).

The libration of 1983 SA is remarkable, since (279) Thule is so far the only known asteroidal librator of this type. There are indeed some conspicious dynamical differences: The maximum eccentricity of the orbit of (279) Thule is 0.14 corresponding to $Q = 4.9$ AU, the miminum distance from Jupiter is not smaller than 1.1 AU (Marsden, 1970). The critical argument oscillates around $0°$ (see Takenouchi (1962), where the reader should replace θ by $-\sigma_{4/3}$). On the other hand, the orbit of 1983 SA is highly eccentric and $Q > a_J$; $\sigma_{4/3}$ oscillates around a value of roughly $210°$ and the oscillatory behaviour is obviously disturbed. The object approaches Jupiter up to 0.6 AU according to our backward integration. The forward integration yields 0.3 AU and an entirely different type of motion is indicated for the future, where $a \approx 4.6$ AU, a value which is unknown for asteroids.

In comparing 1983 XF with asteroids, one notices that the 2/1 librators (1362) Griqua, (1921) Pala and (1922) Zulu have stable motions and cannot approach Jupiter closer than 2.0 AU (cf. Schubart, 1979). Actually, the motion of 1983 XF, characterized by a temporary libration involving close encounters with Jupiter,is rather typical for a certain mechanism of cometary 2/1 librations. A well-known example of this case is comet P/Pons-Winnecke (see Marsden (1970), Kresák (1974), and Vaghi and Rickman (1982)).

In conclusion, our investigation demonstrates that from a dynamical point of view, 1983 SA and 1983 XF must be considered as serious candidates for a cometary origin.

REFERENCES

Benest,D.: 1974, *Astron.Astrophys.* 32, 39
Bulirsch,R., Stoer,J.: 1966, *Numer.Math.* 8, 1
Hahn,G., Rickman,H.: 1984, paper submitted to *Icarus*
Kresák,L.: 1974, *Asteroids, Comets, Meteoric Matter,* Eds.: C.Cristescu
    and W.J.Klepczynski, IAU Coll. No.22, p.193
Kresák,L.: 1979, *Asteroids,* Ed. T.Gehrels, p.289
Marsden,B.G.: 1970, *Astron.J.* 75, 206
Schubart,J., Stumpff,P.: 1966, *Veröff.Astron.Rechen-Inst. Heidelberg*
    No.18
Schubart,J.: 1979, *Dynamics of the Solar System* (IAU Symp.No.81),
    Ed. R.L.Duncombe, p.207
Takenouchi,T.: 1962, *Ann. Tokyo Obs.* 7, 191
Vaghi,S., Rickman,H.: 1982, *Sun and Planetary System,* Eds. W.Fricke
    and G.Teleki, p.391

INFLUENCE OF NON-GRAVITATIONAL FORCES ON THE ORBITAL EVOLUTION OF THE SHORT-PERIOD COMETS

N.A. Belyaev and K. P. Ivanovskaya
Institute for Theoretical Astronomy
Naberezhnaya Kutuzova, 10
191187 Leningrad
USSR

ABSTRACT. This paper is concerned with the influence of non-gravitational forces on the orbital evolution of the short-period comets. This influence is variable for different comets and is especially noticeable for those comets which undergo approaches with Jupiter. When studying the dynamic evolution it is desirable to take into account the non-gravitational effects.

INTRODUCTION

An idea of the influence of the non-gravitational forces upon the cometary motion was first advanced by Encke in 1819. It had been revealed in the course of the 20th century that these forces acted upon some more comets, the number of comets whose motions experienced secular accelerations and decelerations being approximately equal. This indicated that the cause of the irregularities in cometary motion was connected with the comet's nucleus. However, consideration of the non-gravitational effects appeared to be possible without regard to the physical nature of these forces. This way of research was chosen by J.Encke, O.Backlund, M.Kamienski, A.Dubyago, G.Sitarski and other researchers of comets.

At present, there exists a fairly well developed icy nucleus model of a comet, proposed by F.Whipple and the method of taking into account the non-gravitational forces based on the above model, comprehensively presented by B. Marsden et al. in the series of their publications (Marsden, Marsden et al. 1968, 1969, 1970, 1971). There is no need to dwell upon this method at any length. It will suffice to mention that when improving cometary orbits from observations, one determines two non-gravitational parameters A1 and A2 along with the orbital elements. These two parameters define the amplitude and direction of the

A. Carusi and G. B. Valsecchi (eds.), Dynamics of Comets: Their Origin and Evolution, 371–379.
© *1985 by D. Reidel Publishing Company.*

acceleration components caused by the non-gravitational forces and they are directed along the radius-vector (parameter A1) and along the transversal (parameter A2). For A1 > 0 the acceleration is oriented away from the sun whereas for A2 > 0 - it is in the direction of increasing true anomaly.

The new method has been applied by B.Marsden for a whole series of comets and, at present, the 4th edition of Marsden's Catalogue of Cometary Orbits (1982) includes the non-gravitational parameters A1 and A2 for 34 comets.

Thus, it may be viewed upon as a real fact that 34 short-period comets experience to a greater or lesser extent the effect of the non-gravitational forces during the time interval,limited, at least, by observations of these comets. There are no reasons to think that these forces (under definite conditions) did not act prior to the comet's discovery nor is there any reason to believe they will not act in the future.

The orbital evolution of some comets has been studied with consideration of both the influence of planetary perturbations and the non-gravitational effects.

Thus, Marsden and Sekanina (1974) investigated the motion of P/Encke not only over the time interval covered by observations (from 1786), but also over the interval spanning 200 yr preceding its discovery employing the two parameters of A1= -0.0335$\cdot 10^{-8}$ and A2= -0.0180$\cdot 10^{-8}$. The authors have established that after 100 years, there is a discrepancy between the predicted and computed times of perihelion passage of 6 weeks and after a period of 200 years, this discrepancy amounts to 5 months.

Yeomans (1977) has studied P/Halley's motion backward in time to 837 of our epoch and later on in collaboration with Kiang (Yeomans and Kiang, 1981) to 1404 B.C. They have substantiated an earlier established fact that the influence of non-gravitational forces affects the time of the perihelion passage by about four days per one revolution.

Kazimirchak-Polonskaya (1968), when studying the motion of P/Wolf with account of the non-gravitational effects, arrived at conclusion that the influence of these forces on P/Wolf's orbital evolution was negligible.

This fact, however, does not allow to draw a general conclusion concerning the negligibly small influence of these forces on the motion of all 34 comets for which the numerical values of the non-gravitational forces are known. For a brief review of non-gravitational forces affecting the motions of comet, see the recent work by Marsden (1984).

METHOD OF SOLUTION

The aim of the present paper is to carry out, as far as
possible, the most careful and complete analysis on the
influence of non-gravitational forces on the dynamic evolution of short-period comets over long time periods. The
time span under consideration covers 200 yr (1800-2000).
Orbital elements for all numerical integration were taken
from the 4th edition of Marsden's Catalogue of Cometary
Orbits (1982) as well as values of the non-gravitational
parameters A1 and A2. Equations of cometary motion were
integrated by Cowell's quadrature method with consideration of the fourth order differences. Perturbations by nine
planets (Mercury-Pluto) were taken into account. We have
employed the system of masses adopted by the IAU in 1964.
       The non-gravitational parameters A1 and A2 chosen as
the starting ones were considered for the whole interval
of integration to be constant. That is, it has been assumed
that the non-gravitational forces remained unchanged in
time over the span of 200 yr.
       Of the 80 comets, observed in two or more appariti-
ons, the non-gravitational parameters are known only for
34. Among these are the comets which have constant A1 and
A2, the comets which have changing values but retain the
sign of A2 and, finally, those for which A2 has changed in
quantity and in sign.
       Therefore, for convenience of the study 34 comets
have been divided into three groups in accordance with
the above mentioned peculiarities; these groups are presented in Table I. To the first group (with constant A1
and A2) belong 13 comets; to the second group (A1 and A2
change in absolute value) - 6 (figures indicate how often
the parameters changed) and into the third group were included 15 comets (numerator indicates the total number of
the parameter changes and denominator - the number of variations with the change of a sign).
       Before passing to a presentation of the results, we
should like to point out the difficulties involved in solving the problem of the dynamic evolution of cometary orbits with full account of both the planetary perturbations
and the non-gravitational effects. If a comet was observed
in a sufficiently great number of apparitions, then, as a
rule, all observations in these apparitions cannot be linked with a required accuracy by one system of elements and
one is forced to link the apparitions by groups.
       But in such cases it is up to the researcher which
orbit should be chosen as a starting one, if he is not
willing to proceed by utilizing the first, with regard to
time, apparition for backward integration and the last
one - for forward integration in time. The preferable
technique would be a study based on one system of elements.

TABLE I

Comets experiencing non-gravitational effects

A1, A2: Constants	Variations without a change of sign	Variations with a change of sign
1. Arend (1951 X)	1. Biela (1772)-2 $^{1/}$	1. Brooks 2 (1889 V.)-4/2 $^{2/}$
2. Daniel (1909 IV)	2. Borelly (1905 II)-2	2. Brorsen (1846 III)-2/1
3. Jackson-Neujmin (1936 IV)	3. Halley (-239)-2	3. Comas-Sola (1927 III)-2/1
4. Johnson (1949 II)	4. Honda-Mrkos-Pajdusakova (1948 XII)-2	4. D'Arrest (1851 II)-7/6
5. Olbers (1815)	5. Schaumasse (1911 VII)-2	5. Encke (1786 I)-18/10
6. Perrine-Mrkos (1896 VII)	6. Schwassmann-Wachmann 2 (1929 I)-3	6. Faye (1843 III)-4/1
7. Reinmuth 17 (1928 I)		7. Finlay (1866 VII)-4/1
8. Tempel-S-Swift (1869 III)		8. Forbes (1929 II)-2/1
9. Tempel-Tuttle (1366)		9. Giacobini-Zinner (1900 III)-4/1
10. Whipple (1933 IV)		10. Grigg-Skjellerup (1902 II)-3/1
11. Wirtanen (1947 XIII)		11. Kopff (1906 IV)-4/1
12. Wolf (1884 III)		12. Pons-Winnecke (1819 III)-5/4
13. Wolf-Harrington (1924 IV)		13. Tempel 2 (1873 II)-4/2
		14. Tuttle (1790 II)-3/2
		15. Tuttle-Giacobini-Kresak (1858 III)-2/1

1/ Figure indicates how many times the non-gravitational parameters change; 2/ Numerator has the same meaning as in 1/; denominator denotes the number of A1 and A2 variations that include a sign change for A2.

In the presence of several orbits and a pair of non-gravitational parameters the problem is reduced to the choice of
the starting system of elements but for the second and the
third groups from table I this choice is complicated by
selection of the non-gravitational parameters. For the solution of the first problem we have chosen the way of Yeomans and Kiang (1981).

    As a criterion we have taken the differences (ΔT)
between the perihelion passage moments in all observed apparitions of the comet derived by numerical integration
(Tc) and those values of T, given by B.Marsden (1982). Ne-
edless to say, this was a labour-consuming work, entailing
repeated numerical integrations within the range of the
observed apparitions which for some comets (for instance,
P/Pons-Winnecke 1819 III = 1976 XIV) amounted to more than
150 yr.

    We had no clear-cut criterion for evaluating ΔT va-
lues derived in the computing process, as good or bad. However, on the whole, we were adhering to two principles:
firstly, not to employ any sharp deviations of ΔT; secondly, all the ΔT values for a given comet should be minimized in absolute values.

    Sometimes, it may be reached by a selection of the
system of elements.[1] But more often we also used the selection of non-gravitational effects.

    Table II presents for some comets those values of A1
and A2 which are given in B.Marsden's Catalogue of Cometary Orbits and which have been selected for studying the
dynamic evolution of cometary orbits. It is apparent from
this Table, that sometimes the most optimal values turned
out to be one of the A1 and A2 pairs, given by B.Marsden
in his Catalogue, sometimes A1 taken from one group and A2
from the other one, and, finally, for some of the comets
we had to select A1 and A2 artificially so that the ΔT
values would be minimized.

    DISCUSSION AND CONCLUSIONS

Table III presents a partial summary of our preliminary
study of the non-gravitational parameters for the. comets
indicated in column 2. In column 4 of the Table are given
the maximum values of the ΔT discrepancies from all the
observed apparitions derived with the highest possible (in

----

[1] For instance, for P/Pons-Winnecke ΔT = $5\overset{d}{.}7$ if the initial orbit is taken as the system of elements in 1875 and
ΔT = $1\overset{d}{.}3$ if the initial orbit is taken as the 1939 system
of elements.

TABLE II

Examples of Comets with variable values of non-gravitational parameters

Comet	Style	Intervals of constants A1 and A2	A1	A2
P/Biela	1	1805–1833	+0.28	−0.0250
		1826–1846	+0.39	−0.0254
		1832–1852	+0.36	−0.0260
		1800–2000	+0.39	−0.0254
P/Tempel 2	2	1873–1915	+0.08	+0.0021
		1915–1956	−0.04	+0.0012
		1930–1967	−0.04	+0.0008
		1956–1978	+0.08	+0.0022
		1800–2000	−0.01	+0.0008

TABLE III

Influence of non-gravitational effects on the accuracy of the perihelion passage moments of some comets

NN	Comet	Epoch	T_{max} (A1, A2≠0)	T_{max} (A1, A2=0)
1	P/Biela (1772)	$1832^{1/}$	$0^{d}.99$	$18^{d}.38$
2	P/Wolf-Harrington (1924 IV)	1952	0.03	5.83
3	P/Giacobini-Zinner (1900 III)	1959	0.72	2.65
4	P/Tempel-Swift (1869 III)	1891	0.10	1.55
5	P/Kopff (1906 IV)	1951	0.24	0.74
6	P/Forbes (1929 II)	1961	0.03	0.44
7	P/Wirtanen (1947 XIII)	1967	0.04	0.27

[1/] Epoch denotes an apparition of a comet using whose elements one gets the best results of numerical integration.

the above given sense) non-gravitational effects. In column 5 of the Table for comparison purposes are indicated the highest possible values of ΔT, derived without consideration of non-gravitational effects. Thus, this Table gives a partial proof for the necessity of taking into account the non-gravitational forces when studying the evolution of cometary orbits. The more comprehensive answer is provided by Tables IV and V. Table IV illustrates the

TABLE IV

Influence of non-gravitational effects on the evolution of the cometary orbits in the absense of the close approaches to Jupiter ($\Delta y_{min} > 0.33AU$)

1800	$\tilde{\pi}$	Ω	i	e	q	P	ΔT_{max}
P/Johnson			A1=0.78		A2= -0.0266		
A1,A2=0	317°	127°	15°	0.35	2.35AU	6.9yr	0^d.01
A1,A2≠0	317	127	15	0.35	2.34	6.9	0.03
Discre-pancies	–	–	–	–	0.01	–	0.02
P/Jackson-Neujmin			A1=0.8		A2=-0.45		
A1,A2=0	3°	184°	14°	0.65	1.49AU	8.6yr	7^d.02
A1,A2≠0	2	180	11	0.64	1.56	9.1	0.01
Discre-pancies	+1	+4	+3	+0.01	-0.07	-0.5	+7.01

influence of the non-gravitational effects upon the orbital evolution of P/Johnson and P/Jackson-Neujmin whose motions did not reveal any evidence of the close approaches with Jupiter back to 1800. For each comet, integrations have been run back to 1800 to provide orbital elements for cases both with and without consideration of the non-gravitational forces, the third line representing discrepancies between the elements. Table V, using the same scheme, presents data for two comets which experienced close approaches with Jupiter - P/Schassmann-Wachmann 2 and P/Honda-Mrkos-Pajdusakova. The data presented above and these examples show that non-gravitational forces can produce substantial perturbations in cometary orbital elements and, to a certain extent, affect the final results when studying the dynamic evolution of cometary orbits. This influence may be rather appreciable even in the absence of close approaches of the comet to Jupiter under investiga-

TABLE V

Influence of non-gravitational effecs on the evolution of cometary orbits in presence of close approaches ($\Delta q_{min} < 0.33$ AU)

1800	π	Ω	i	e	q	P	ΔT_{max}

P/Schwassmann-Wachmann 2　A1=1.02, A2=-0.1801
1926(0.18)

	π	Ω	i	e	q	P	ΔT_{max}
A1,A2=0	106°	123°	1°	0.20	3.52AU	5.23yr	$1\overset{d}{.}27$
A1,A2≠0	104	121	0.6	0.19	3.59	5.25	0.15
Differencies	-2	-2	-0.4	-0.01	+0.07	+0.02	-1.12

P/Honda-Mrkos-Pajd.　A1=0.27,　A2=-0.0420,
1876 (0.08)　1935 (0.079)

	π	Ω	i	e	q	P	ΔT_{max}
A1,A2=0	52°	271°	3°	0.68	1.22AU	7.51yr	$0\overset{d}{.}99$
A1,A2≠0	53	258	11	0.69	1.09	6.86	0.15
Differencies	+1	-13	+8	+0.01	-0.13	-0.65	-0.84

tion but become very notable in the presence of such approaches. However, the extent of this influence on the final results, as is obvious from Tables IV and V, may be variable and, therefore, it is difficult to predict the result in advance. The general conclusion of the present investigation is as follows: if at all possible then one ought to take account of the non-gravitational forces when studying the dynamic evolution of cometary orbits. Our results point out that it is necessary to determine the average values of the non-gravitational parameters A1 and A2 over the time span under consideration if a researcher is facing the problem of their choice.

A complete account of the results obtained in the present investigation took place when studying orbital evolution of all short-period comets over the 200 yr time interval (1800-2000).

REFERENCES

Kazimirchak-Polonskaya, E.I.(1967). Trudy ITA, 12, 86.
Marsden, B.G. (1968). Astron J. 73, 367.
Marsden, B.G. (1969). Astron. J. 74, 720.
Marsden, B.G. (1970). Astron. J. 75, 75.

Marsden, B.G. (1982). Catalogue of Cometary Orbits, Fo-
    urth Edition, Cambridge.
Marsden, B.G. (1985). See this volume.
Marsden, B.G.,Sekanina, Z. (1971). Astron. J. 76, 1135.
Marsden, B.G.,Sekanina, Z. (1974). Astron. J. 79, 413.
Marsden, B.G.,Sekanina, Z. and Yeomans, D.K. (1973).
    Astron. J. 78. 211.
Yeomans, D.K. (1977). Astron. J. 82, 435.
Yeomans, D.K., Kiang, T. (1981). Mon.Not.Roy.Astron.Soc.
    197, No. 2, 633.

ON THE NONGRAVITATIONAL EFFECTS IN THE COMET ENCKE MOTION

Yu.V. Batrakov and Yu.A. Chernetenko
Institute for Theoretical Astronomy
Naberezhnaya Kutuzova, 10
191187 Leningrad
USSR

ABSTRACT.   A factor depending on geomagnetic index was
introduced in the B. Marsden's formula for the nongravitational force. The orbit of the comet Encke covering seven
apparitions (1895-1914) obtained with this modified formula shows better residuals than that obtained with the non-
modified one.

For some comets (P/Encke, P/Giacobini-Zinner and others)
the nongravitational forces are subjected to irregular
changes. Whipple and Sekanina (1979) have explained such
changes by the precession of the spin axis of the oblate
comet nucleus. However, some other causes for such changes
cannot be excluded.
    Backlund (1910) was the first to suggest the possibi-
lity of a connection between the changes of the mean motion of P/Encke and the solar activity. In Figure 1 Wolf
numbers W (Loginov, 1976) and the mean daily motion changes Δn obtained by Dubjago's method (Dubjago, 1956) are
plotted. The discontinuity of the graph of Δn is explained
by the absence of observations at the comet's return in
1944. The values of Δn and their errors are given in Table I. The greatest value of Δn corresponds to 1898, if
the two observations which were made at this apparition
can be relied upon.
    The comparison of the W and Δn graphs shows that the
greatest Δn correspond to the fallings or to the minima
of the solar 11-year cycle. It is known that at the fallings of W the high speed recurrent solar wind streams
are developed and become numerous. It is natural to suppo-
se that these streams can influence the comet motion due
to changes of the outflow of matter from the comet's nucleus. This supposition comes to mind when one observes the
correlation of the outbursts of the brightness of some comets and the high speed solar wind streams (Andrienko et

A. Carusi and G. B. Valsecchi (eds.), Dynamics of Comets: Their Origin and Evolution, 381–385.
© 1985 by D. Reidel Publishing Company.

TABLE I

The changes of the mean daily motion of P/Encke

Year	Δn $(1°10^{-6}/day)$	Year	Δn $(1°10^{-6}/day)$
1891	+16.9 ± 0.2	1931	+10.5 ± 0.2
1895	+14.6 ± 0.2	1934	+ 6.7 ± 0.2
1898	+17.3 ± 0.3	1937	+ 7.3 ± 1.5
1901	+11.9 ± 0.3	1951	+ 5.3 ± 0.1
1904	+12.2 ± 0.2	1954	+ 5.5 ± 0.1
1908	+12.3 ± 0.2	1957	+ 3.7 ± 0.1
1911	+12.8 ± 0.4	1961	+ 3.1 ± 0.1
1914	+10.5 ± 0.4	1964	+ 4.3 ± 0.2
1918	+10.7 ± 0.2	1967	+ 3.0 ± 0.2
1921	+10.6 ± 0.1	1970	+ 2.6 ± 0.1
1924	+ 9.1 ± 0.1	1973	+ 3.1 ± 0.1
1928	+ 8.2 ± 0.1	1977	+ 2.5 ± 0.1

al.,1981). The high speed streams influence the terrestrial magnetic field. So, it is natural to take one of the geomagnetic indexes as the measure of the intensity of the corpuscular stream as well as the measure of the change of the nongravitational forces which can be caused by this stream. As such a measure we have taken the daily values of the aa index given by Sutorik and Cruickshank (1977) and obtained from the A_p index. We use only those values of the aa index which correspond to the corpuscular recurrent streams picked by these authors.

For the transverse component of the nongravitational acceleration we took the expression

$$F_2 = (A_2 + A_2'\tau)(1 + \chi aa)g(\tau), \qquad (1)$$

where τ is the time in years from the initial epoch 1900.0, $g(\tau)$ is the variation of the nongravitational acceleration with heliocentric distance τ (Marsden et al., 1973). aa is the geomagnetic index, the values of which range from 0 to 5. It deserves mentioning that geomagnetic indexes must be taken for the time moments at which the corpuscular streams reach the comet. Let $t + \Delta t$ be the time moment at which the corpuscular stream touches the Earth and some geomagnetic disturbance is registered by observer and, in the same way, t be the time moment when the stream touches the comet. For Δt we have the following formula (Dobrovol'skij, 1964)

$$\Delta t = \frac{\Delta \lambda}{\overline{\omega}} + \frac{\Delta \tau}{V} \ , \tag{2}$$

where $\Delta \lambda$ is the difference of the heliographic longitudes of the comet and of the Earth, $\Delta \tau$ is the difference of the heliocentric distances of the comet and of the Earth, $\overline{\omega}$ is the angular rotation rate of the Sun, V is the speed of the corpuscular stream. We used the values $\overline{\omega}$ =13°4/day, V =700km/s. For the radial and normal components of the nongravitational acceleration we took the usual form

$$F_1 = A_1\, g\,(\tau) \ , \tag{3}$$
$$F_3 = A_3\, g\,(\tau) \ ,$$

because the attempts to determine the secular coefficients A_1', A_3' and parameters in F_1 and F_3 analogous to χ have failed.

The seven apparitions of P/Encke (1895-1914) were linked taking into account the aa geomagnetic index (orbit (A)) and without this index (orbit (B)). The orbits were obtained by the Everhart's numerical integration method (Everhart, 1974) with the 1 day step. This step was chousen because the aa index values were given at 1 day intervals. In all 109 observations of the comet were used for the orbit computation. The results for the nongravitational parameters A_1, A_2, A_3, A_2', χ are summarised in Table II and the orbital elements are:

Epoch 1901 Sept. 13.0 E.T.

Orbit (A)	Orbit (B)
T = 1901 Sept. 15.9618 E.T.	T = 1901 Sept. 15.9614 E.T
ω = 183°9886 ⎫	ω = 183°9855 ⎫
Ω = 335.4932 ⎬ 1950.0	Ω = 335.4955 ⎬ 1950.0
i = 12.8993 ⎭	i = 12.8987 ⎭
e = 0.846 047	e = 0.846 046
q = 0.341 612 a.u.	q = 0.341 620 a.u.

The mean residual for the orbit (A) is smaller than that one for the orbit (B). In this sense the orbit (A) is better than the orbit (B). The comparison of $\delta_{(A)}$ and $\delta_{(B)}$ according to the Fisher's criterium showed the real significance of the χ parameter in the case. However the orbit (A) is not completely satisfactory, because there are small systematic trends in the 1895 and 1904 residuals and the maximum 1898 residual amounts to 12". Besides, the mean residual for 1918 observations not included into the orbit determination for the orbit (A) is greater than that for the orbit (B). Further investigations are certainly

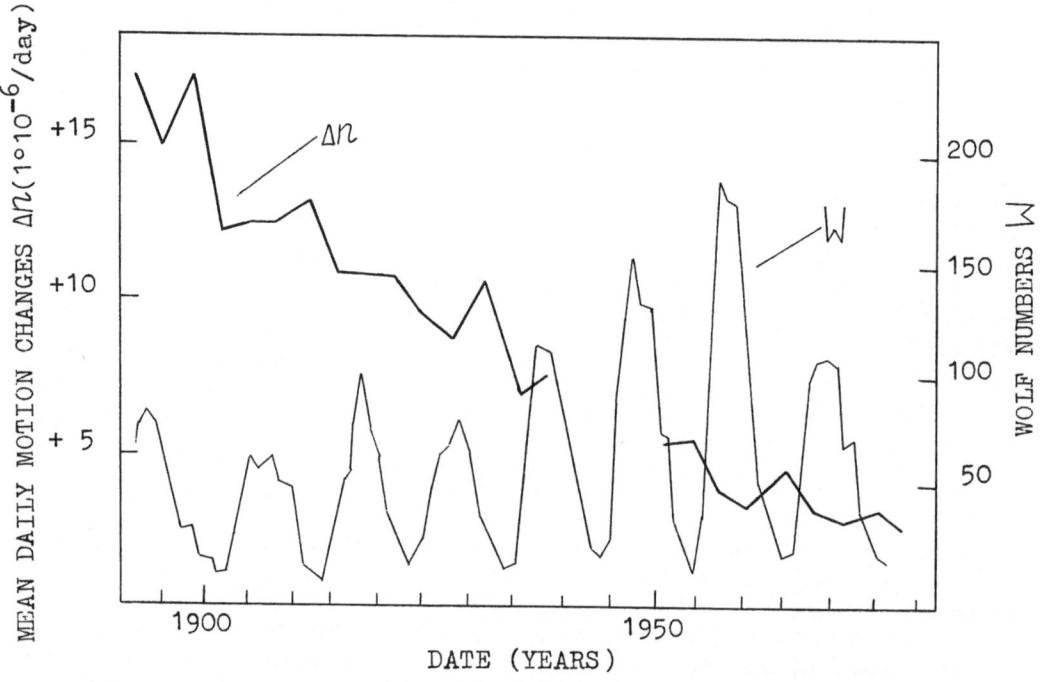

Figure 1. Mean daily motion changes Δn and Wolf numbers W during 1891 - 1977.

needed. If the parameter χ makes a real effect it means that the interaction of the high speed solar wind stream with the comet results in the rapid change of the nongravitational forces up to 2-4 times of their nominal value.

TABLE II

Nongravitational parameters for P/Encke

Orbit	Mean residual 1895-1914	$A_1 10^8$	$A_2 10^8$	$A_3 10^8$	$A_2' 10^{11}$	χ	Mean residual 1891 1918	
(A)	3".7	-0.23 $\pm$ 2	-0.01846 $\pm$ 13	-0.01 $\pm$ 12	$+0.197$ $\pm$ 14	$+2.5$ $\pm$ 2	16"	9"
(B)	4.5	-0.28 $\pm$ 2	-0.01922 $\pm$ 8	$+0.37$ $\pm$ 13	$+0.219$ $\pm$ 14	$-$	28	7

The results presented in Table II allow us to draw the conclusion that the inclusion of the geomagnetic index into Marsden's nongravitational force formula makes a non-negligible reduction of the residuals of P/Encke.

The authors wish to thank Drs. G.I. Ol', A.I. Ol' (Institute for Arctic and Antarctic, Leningrad) and Dr. L.I. Miroshnichenko (IZMIRAN, Moscow) for the data kindly placed at their disposal and for the useful advices.

References

Andrienko, D.A., Vashchenko, V.N.: 1981, Komety i korpuskuljarnoe isluchenie Solnza, Moscow.

Backlund, O.: 1910, Mon. Not. Roy. Astron. Soc., 70, 429.

Dobrovol'skij, O.V.: 1964, Komety, Moscow.

Dubjago, A.D.: 1956, Byull. astron. obs. im. V.P. Engelgardta, 116, kn. 6, N 32.

Everhart, E.: 1974, Celest. Mech., 10, 35.

Loginov, V.F.: 1976, Katalog indeksov solnechnoj i geomagnitnoj aktivnosti, Obninsk.

Marsden, B.G., Sekanina, Z., and Yeomans, D.K.: 1973, Astron. J., 78, 211.

Sutorik, J., and Cruickshank, C.M.: 1977, A division of the aa indices into six classes based on the Ap index, 1868-1976. NOAA Tech. Repr. ERL 389 - SEL 38.

Whipple, F.L. and Sekanina, Z.: 1979, Astron. J., 84, 1894.

SECTION VII

COMET P/HALLEY AND FUTURE MISSIONS TO COMETS

THE DYNAMICAL HISTORY OF COMET HALLEY

Donald K. Yeomans
Jet Propulsion Laboratory
California Institute of Technology
Pasadena, California 91109
U.S.A.

ABSTRACT. The history of the attempts to predict the motion of comet
Halley is outlined and the importance of the so-called nongravitational
forces acting upon this comet is emphasized. Recent orbital work of the
International Halley Watch Astrometry Network is reviewed. Comet
Halley's transverse nongravitational parameter is positive and nearly
constant with time suggesting that the comet is in direct rotation without precession of the spin pole. The nongravitational effects are consistent with the vaporization of water ice from the comet's nucleus and
long term integrations suggest that the comet has been in its present
orbit for at least 16,000 years and probably much longer.

I. THE HISTORY OF COMET HALLEY THROUGH THE 1909-1911 APPARITION

1.1. The Prediction of Future Perihelion Passage Times

Since 240 B.C., Chinese observers have documented a nearly unbroken
record of scientifically useful observations of comet Halley (Ho Peng
Yoke, 1964; Ho Peng Yoke and Ang Tian-Se, 1970). After the probable 240
B.C. apparition, only the 164 B.C. return went unrecorded by the Chinese
and with the exception of occasional Korean and Japanese sightings,
useful comet Halley observations made outside of China were virtually
nonexistant for over a millennium thereafter. Beginning with the cometary observations of the Florentine physician and astronomer, Paolo
Toscanelli (1397-1482), quantitative and accurate cometary positions
became available throughout the West (Celoria, 1921). However, the
necessary theory for representing a comet's motion was not available
until the publication of Isaac Newton's PRINCIPIA in 1687. Newton (1687)
outlined a semi-analytic orbit determination theory and used the comet of
1680 as an example. While Newton never applied the method to another
comet, Edmond Halley began what he termed "a prodigious deal of calculation " and applied Newton's method to determine the parabolic orbits for
two dozen well observed comets (Halley, 1705). Struck by the similarity
in the orbital elements for the comets observed in 1531, 1607 and 1682,
Halley suggested that these three apparitions were due to the same
comet, and that it might be expected again in 1758. Halley's subsequent

A. Carusi and G. B. Valsecchi (eds.), Dynamics of Comets: Their Origin and Evolution, 389–398.

calculations indicated that a close Jupiter approach in 1681 would cause
an increase in the length of the next period. Halley then revised his
earlier prediction and suggested in a publication appearing after his
death (Halley, 1759) that the comet that was to bear his name would
return again in late 1758 or early 1759.

To refine Halley's prediction, Clairaut (1758) used a modified
version of his analytic lunar theory to compute the perturbations on the
comet's orbital period due to the effects of Jupiter and Saturn over the
interval 1531-1759. Noting that calculations over the intervals 1531-
1607 and 1607-1682 predicted the 1682 perihelion passage time to within
one month, Clairaut stated that his mid-April 1759 prediction should be
good to a similar accuracy. The actual time of perihelion passage in
1759 was March 13.1 (Unless otherwise stated, all times are given in
U.T.). Beginning with Clairaut's work in 1758, all subsequent work to
1910 on the perturbed motion of comet Halley was based upon the variation
of elements technique (Lagrange, 1783). The various works differed only
in how many perturbing planets were included, how many orbital elements
were allowed to vary, and how many times per revolution the reference
ellipse was rectified by adding the perturbations in elements. Until
after the 1909-1911 apparition, no attempt was made to link the observations of two or more apparitions into one orbital solution.

In anticipating the 1835 return, Damoiseau (1820) computed the
perturbative effects of Jupiter, Saturn and Uranus on comet Halley over
the interval 1682-1835. Since the actual time of perihelion passage in
1835 was November 16.4, Damoiseau's initial prediction of November 17.15
was remarkable. However, Damoiseau (1829) later added the perturbations
due to the earth and revised his prediction to November 4.81. De
Pontecoulant considered the perturbative effects of Jupiter, Saturn and
Uranus over the interval 1682-1835 as well as the earth's perturbative
effects near the 1759 time of perihelion passage. His predictions for
the 1835 perihelion passage times were successively, November 7.5, November 13.1, November 10.8 and finally November 12.9 (de Pontecoulant
1830,1834,1835). The most complete work leading up to the 1835 return
was undertaken by O.A. Rosenberger. After a complete reduction of available observations, Rosenberger recomputed an orbit for the 1759 and 1682
apparitions (Rosenberger 1830a, 1830b). Rosenberger (1834,1835) computed
the effect on all the orbital elements from the perturbations of the
seven known planets over the 1682-1835 interval. Assuming the comet's
motion was unaffected by a resisting medium, Rosenberger's prediction for
the 1835 perihelion passage time was November 12.0. Lehmann(1835) also
investigated the motion of comet Halley over the 1607-1835 interval
taking into account the perturbative effects of Jupiter, Saturn and
Uranus. However, his perihelion passage prediction was late by more than
10 days.

In an effort to anticipate the next apparition of comet Halley,
de Pontecoulant (1864) took into account the perturbative effects of
Jupiter, Saturn and Uranus before predicting May 24.36, 1910 as the next
time of perihelion passage. The actual time of perihelion passage turned
out to be April 20.18. Cowell and Crommelin began their work with preliminary calculations to see if de Pontecoulant's prediction was approximately correct (Cowell and Crommelin 1907a, 1907b, 1907c, 1908c). Their

computations used the variation-of-elements technique, included perturbations by all the planets from Venus to Neptune (except Mars) and predicted a return to perihelion on April 8.5. Cowell and Crommelin (1910) then began a new study on the comet's motion by using numerical integration whereby the perturbed rectangular coordinates are obtained directly at each time step. This time they computed the perturbations from Venus through Neptune and used a time step that varied from 2 to 256 days. They predicted a 1910 perihelion passage time of April 17.11. The 1909 recovery of the comet required that their prediction be corrected by 3 days and they then revised their work by reducing the time steps by one half, carrying an additional decimal place and correcting certain errors in the previous work (Cowell and Crommelin, 1910). Their post recovery prediction was then revised to April 17.51 and they concluded that a least 2 days of the remaining discordance was due to causes other than errors in the calculations or errors in the planetary positions and masses. We note here that the best predictions for the 1835 perihelion passage time by Rosenberger and de Pontecoulant as well as the 1910 prediction by Cowell and Crommelin were too early by 4.4, 3.5 and 2.7 days respectively. As pointed out in Section II, this is just would one would expect since none of these predictions included the effects of the so-called nongravitational forces.

1.2. The Identification of Early Comet Halley Apparitions

Until the 20th century, all attemps at identifying ancient apparitions of comet Halley were done by either determining orbits directly from the observations or by stepping back in time at roughly 76 year intervals and testing the observations with an approximate orbit of comet Halley. Pingre (1783-84) confirmed the suspicion of Halley (1705) by showing that the comet of 1456 was an earlier apparition of comet Halley. Biot (1843) pointed out that an orbit by Burckhardt (1804) for the comet of 989 closely resembled that of comet Halley and Laugier (1843,1846) correctly identified as comet Halley the comets seen by the Chinese in 451, 760 and in the Autumn of 1378. Laugier (1842) also noted that four of the five parabolic orbital elements for the comet seen in 1301 were close to those of comet Halley. By stepping backward in time at roughly 76-77 year intervals and analysing European and Chinese observations, Hind (1850) attempted to identify comet Halley apparitions from 11B.C. to 1301. Approximate perihelion passage times were often determined directly from the observations and an identification was suggested if Halley-like orbital elements could satisfy existing observations. Although many of Hind's identifications were correct, he was seriously in error for his suggested perihelion passage times in 1223,912,837,608,373 and 11 B.C.

Using a variation of elements technique, Cowell and Crommelin (1907d) began the first effort to actually integrate the comet's equations of motion backward in time. They assumed that the orbital eccentricity and inclination were constant with time and the argument of perihelion and the longitude of the ascending node changed uniformly with time – their rates being deduced from the values computed over the 1531-1910 interval. By using Hind's (1850) times of perihelion passage or by computing new values from the observations, they deduced preliminary

values of the orbital semi-major axis for the perturbation calculations.
The motion of the comet was accurately carried back to 1301 by taking
into account first order perturbations in the comet's period from the
effects of Venus, Earth, Jupiter, Saturn, Uranus and Neptune. Using
successively more approximate perturbation methods, Cowell and Crommelin
(1907d, 1908a-e) carried the motion of the comet back to 239 B.C. At
this stage, their integration was in error by nearly 1.5 years in the
perihelion passage time and they adopted a time of May 15, 240 B.C., not
from their integration, but rather from their consideration of the observations themselves. After a complete and careful analysis of the European and Chinese observations, Kiang (1971) used the variation of elements technique to investigate the motion of comet Halley over the 240
B.C. - 1682 A.D. interval. By determining the time of perihelion passage
time directly from the observations and considering the perturbations
from all nine planets on the other orbital elements, Kiang traced the
motion of comet Halley for nearly two millennia. Hasegawa (1979) also
empirically determined perihelion passage times for comet Halley. For
each apparition from 1378 to 240 B.C., he computed several ephemerides
using Kiang's (1971) orbital elements, except for the perihelion passage
times which were chosen to make the best fit with the observations.
Attempts to represent the ancient observations of comet Halley using the
numerical integration of the comet's gravitational and nongravitational
accelerations are presented in the next Section II.

II. NONGRAVITATIONAL FORCES AND COMET HALLEY

Beginning with the work of Bessel (1835,1836), it became clear that the
motion of comet Halley was influenced by more than the solar and planetary gravitational accelerations. Michielsen (1968) pointed out that
perihelion passage time predictions that had been based upon strictly
gravitational perturbation calculations required a correction of +4.4
days over the past several revolutions. Kiang (1971) determined a mean
correction of +4.1 days. In an attempt to account for this 4 day discrepancy between the actual period of comet Halley and that computed
using perturbations from the known planets, some unorthodox solutions
have been proposed. Brady (1972) suggested the influence of a massive
trans-Plutonian planet and Rasmusen (1967) adjusted the ratio of the
sun:Jupiter mass ratio from the accepted value of 1047 to 1051. Both of
these suggested solutions must be rejected because they would produce
effects on the motion of the known planets that are not supported by
observation. Rasmusen (1981) derived an 1986 perihelion date of February 5.46 from a fit to the observations in 1835 and 1910 and then added
+3.96 days to yield a 1986 perihelion passage time prediction of February
9.42. Brady and Carpenter (1967) first suggested a 1986 perihelion
passage time of Feb. 5.37 based upon a "trial and error" fit to the
observations during the 1835 and 1910 returns. Brady and Carpenter
(1971) then introduced an empirical secular term in the radial component
of the comet's equations of motion. Although this device had the unrealistic effect of decreasing the solar gravity with time, it did allow an
accurate 1986 perihelion passage time prediction of Feb. 9.39. It is now

clear that the actual 1986 perihelion passage time (Feb. 9.44) was accurately predicted by both Rasmusen (1981) and Brady and Carpenter (1971).
However if the orbit of the comet is to be accurately computed throughout
a particular apparition or if the comet's motion is to be traced back to
ancient times, the mathematical model used to represent the obvious
nongravitational forces must be based upon a realistic physical model and
not upon empirical mathematical devices.

In introducing the icy conglomerate model for a cometary nucleus,
Whipple (1950,1951) recognized that comets may undergo substantial perturbations due to reactive forces or rocket-like effects acting upon the
cometary nucleus itself.   In an effort to accurately represent the
motions of many short periodic comets, Marsden (1968,1969) began to model
the nongravitational forces with a radial and transverse term in the
comet's equations of motion.   Marsden et al (1973) modified the nongravitational force terms to represent the vaporization flux of water ice as a
function of heliocentric distance.   The cometary equations of motion are
written;

$$\frac{d^2\vec{r}}{dt^2} = -\mu\frac{\vec{r}}{r^3} + \frac{\partial R}{\partial \vec{r}} + A1\ g(r)\hat{r} + A2\ g(r)\ \hat{T}$$

where $g(r) = \alpha (r/r_0)^{-m}(1 + (r/r_0)^n)^{-k}$

The acceleration is given in astronomical units/(ephemeris day)^2, μ is the
product of the gravitational constant and the solar mass, while R is the
planetary disturbing function.   The scale distance r_0 is the heliocentric
distance where reradiation of solar energy begins to dominate the use of
this energy for vaporizing the comet's nuclear ices.   For water ice, r_0 =
2.808 AU and the normalizing constant α = 0.111262. The exponents m,n,k
equal 2.15, 5.093 and 4.6142 respectively.   The nongravitational acceleration is represented by a radial term, A1 g(r) and a transverse term,
A2 g(r), in the equations of motion.   If the comet's nucleus were not
rotating, the outgassing would always be preferentially toward the sun
and the resulting nongravitational acceleration would act only in the
antisolar direction.   However the rotation of the nucleus, coupled with a
thermal lag angle (Θ) between the nucleus subsolar point and the point
on the nucleus where there is maximum outgassing, introduces a transverse
acceleration component in either the direction of the comet's motion or
contrary to it - depending upon the nucleus rotation direction.   The
radial unit vector ($\hat{r}$) is defined outward along the sun-comet vector,
while the transverse unit vector ($\hat{T}$) is directed normal to $\hat{r}$ in the orbit
plane and in the direction of the comet's motion.   An acceleration component normal to the orbit plane is certainly present for most comets but
its periodic nature makes detection difficult in these computations
because we are solving for an average nongravitational acceleration
effect over three or more apparitions.   While the nongravitational acceleration term g(r) was originally established for water ice, Marsden et al
(1973) have shown that if the Bond albedo in the visible range equals the
infrared albedo, then the scale distance r_0 is inversely proportional to
the square of the vaporization heat of the volatile substance.

Using observations of comet Halley over the 1607-1911 interval,

Yeomans (1977) used a least squares differential correction process to solve for the six initial orbital elements and the two nongravitational parameters A1 and A2. Different values for the scaling distance were tried with the result that r_0 = 2.808 AU was the optimum input value. This suggests that the outgassing causing the nongravitational forces acting on comet Halley are consistent with the vaporization of water ice. This result is a general one for nearly all comets for which nongravitational force parameters have been determined. The positive sign for the determined value of A2 for comet Halley indicates that the comet's nucleus is rotating in a direct sense - in the same direction as the orbital motion. Yeomans (1977) integrated the motion of comet Halley back to 837, and forward to predict a perihelion passage time of 1986 Feb. 9.66.

Brady and Carpenter (1971) were the first to apply direct numerical integration to the study of comet Halley's ancient apparitions. Using their empirical secular term to represent the nongravitational effect, they initiated their integration with an orbit that was determined from the 1682 through 1911 observations and integrated the comet's motion back to 240 B.C. in one continuous run. Because their integration was tied to no observational data prior to 1682, their early perihelion dates diverged from the dates Kiang (1971) had determined directly from the Chinese observations. Using Brady and Carpenter's (1971) orbit for comet Halley, Chang (1979) integrated the comet's motion back to 1057 B.C. However, this integration was not based upon any observations prior to 1909 nor were nongravitational effects taken into account.

Yeomans and Kiang (1981) began their investigation of comet Halley's past motion with an orbit based upon the 1759, 1682 and 1607 observations and numerically integrated the comet's motion back to 1404 B.C. Planetary and nongravitational perturbations were taken into account at each half day integration step. In nine cases, the perihelion passage times calculated by Kiang (1971) from Chinese observations were redetermined and the unusually accurate observed perihelion times in 837, 374 and 141 A.D. were used to constrain the computed motion of the comet. The dynamic model, including terms for nongravitational effects, successfully represented all the existing Chinese observations of comet Halley. This model assumed the comet's nongravitational forces remained constant with time; hence it seems that the comet's spin axis has remained stable, without precessional motion, for more than two millennia. Also implied is the relative constancy, over two millennia, of comet Halley's ability to outgas. This latter result is consistent with the comet's nearly constant intrinsic brightness over roughly the same interval (Broughton, 1979). From the list of Halley's orbital elements given by Yeomans and Kiang (1981) from 1404 B.C. to 1910 A.D., one can make a crude estimate of Halley's minimum dynamic age. The heliocentric distance to the comet's descending node increased from 0.85 AU in 1910 to 1.74 AU in 1404 B.C. If this rate of increase continued back into the distant past then the comet would not have crossed the ecliptic plane near Jupiter's orbit until 14,300 B.C. If Jupiter happened to be near during this nodal crossing, then perhaps comet Halley was captured into its current orbit configuration. Hence in 1986, comet Halley will have been in its current orbit for at least 16,000 years and probably much longer.

III. RECENT ORBITAL WORK ON COMET HALLEY

The recovery of comet Halley on October 16, 1982 at Mt. Palomar showed the comet's image to be only 9 arc seconds away from the ephemeris position provided by Yeomans (Jewitt et al, 1982). At this writing there have been additional accurate astrometric positions provided by astronomers at Kitt Peak Observatory in Arizona, the Canada-France-Hawaii Telescope in Hawaii and from the European Southern Observatory at La Silla, Chile. Recovered at a distance of more than 11 AU from the sun, the comet showed no obvious activity and the initial observational accuracy is not limited by the uncertainty of the comet's center of mass within an extensive coma. The initial astrometric positions of comet Halley are generally accurate to within 1 arc second with a series of 25 positions from La Silla in late January 1984 achieving a heretofore unrealizable root mean square accuracy of less than 0.5 arc seconds in both right ascension and declination.

There are also efforts underway to improve the accuracy of the older data. Morley (1983) has used the SAO star catlog to improve upon the positions taken at Cordoba during the last apparition, West and Schwehm (1983) have remeasured some of the Heidelberg plates and Bowell (1982) has begun to measure some Lowell Observatory plates that were never used for astrometric positions before.

Within the Astrometry Network of The International Halley Watch, the computer software for cometary orbit determination has been improved somewhat. Incoming observations times in UTC are reduced to ephemeris time, the observatory's coordinates are assigned and the right ascension and declination are corrected for the small effects of elliptic aberration. Once verified and weighted, the observations are stored in reverse chronological order on the master data file for use by the orbit determination program. This latter program takes into account the comet's nongravitational perturbations, as well as the planetary perturbations at each time step. The local error allowed at each time step can be input and the time steps of the numerical integration vary to limit the local error to the input tolerance. The partial derivatives of the observables are numerically integrated along with the comet's equations of motion. To be consistent with the reference frames used by the various flight projects to comets Halley and Giacobini-Zinner, the comet's equations of motion also include general relativistic effects by means of the parameterized space-time metric of the Eddington-Robertson-Schiff formalism. Currently this program uses a batch processed, weighted least squares technique for the orbit determination. The program can store and use a priori information matrices and map covariance matrices to specified epochs. For example, the improved orbit determination program was used to establish a prediction for the 1986 perihelion passage time based upon a new fit to the data from the 1759,1835 and 1910 returns. If this program had been available prior to the comet's recovery, the predicted time of perihelion passage would have been 1986 Feb. 9.52. At this writing the most recent orbit for comet Halley is based upon 751 observations over the interval from August 21, 1835 to March 4, 1984. The weighted RMS residual is 1.94" and the orbital elements are given below;

Epoch	1986 Feb. 19.0 E.T.
Perihelion	1986 Feb. 9.43881 E.T.
q (AU)	0.5870992
e	0.9672724
w	111.84657
node	58.14397
I	162.23932
A1	0.1471
A2	0.0155

The angular elements are referred to the ecliptic plane and the mean equinox of 1950.0 and the nongravitational parameters are given in units of 10^{-8} AU/(ephemeris day)^2 .

Alternate nongravitational force models have been tried in an effort to improve upon the existing model developed by Marsden et al (1973). Gas production rates computed by Divine(1982) were evaluated at each integration step using a comet centered rocket-like thrust direction as denoted by Sekanina (1981). Thus this new model allowed for a comet outgassing at a rate that followed the visual light curve and was asymmetric with respect to perihelion. In addition the thermal lag angle (Θ), spin pole inclination(I), and the direction of the comet's subsolar point at perihelion(ϕ) were variables in the model testing procedure. The attempted solutions proved to be insensitive to input values of ϕ and although the final solutions were not completely satisfactory, the optimum values for the spin pole inclination and thermal lag angle were approximately 30 and 5 degrees respectively. No combination of the input variable values could improve upon the existing nongravitational force model of Marsden et al (1973). It seems likely that additional improvements in the solutions using this alternate nongravitational force model will have to await information on the spin pole axis orientation expected from the Halley flight projects in March 1986.

The work described in this paper was carried out by the Jet Propulsion Laboratory, California Institute of Technology, under contract with the National Aeronautics and Space Administration.

References:

Bessel, F.W.(1835) Astron. Nachr. 13:3-6.
Bessel, F.W.(1836) Astron. Nachr. 13:345-350.
Biot, E. (1843) Conn. des Temps for 1846, Additions. p.69.
Bowell, E. (1982) Minor Planet Circular 6841.
Brady, J.L. (1972) Publ. Astron. Soc. Pac. 84:314-322.
Brady, J.L. and Carpenter, E. (1967) Astron. J. 72:365-369.
Brady, J.L. and Carpenter, E. (1971) Astron. J. 76:728-739.
Broughton, R.P. (1979) J. R. Astron. S. Can. 73:24-36.
Burckhardt, J.C. (1804) Monatliche Korrespondenz 10:162.
Celoria, G. (1921) Pubbl. Osservatoria di Brera No.55.
Chang, Y.C. (1979) Chin. Astron. 3:120-131.

Clairaut, A.C. (1758) J. Scavans (Jan.1759) 41:80-96.
Cowell, P.H. and Crommelin, A.C.D. (1907a) Mon. Not. R. Astr. Soc.
     67:174.
Cowell, P.H. and Crommelin, A.C.D. (1907b) Mon. Not. R. Astr. Soc.
     67:386-411,521.
Cowell, P.H. and Crommelin, A.C.D. (1907c) Mon. Not. R. Astr. Soc.
     67:511-521.
Cowell, P.H. and Crommelin, A.C.D. (1907d) Mon. Not. R. Astr. Soc.
     68:111-125.
Cowell, P.H. and Crommelin, A.C.D. (1908a) Mon. Not. R. Astr. Soc.
     68:173-179.
Cowell, P.H. and Crommelin, A.C.D. (1908b) Mon. Not. R. Astr. Soc.
     68:375-378.
Cowell, P.H. and Crommelin, A.C.D. (1908c) Mon. NOt. R. Astr. Soc.
     68:379-395.
Cowell, P.H. and Crommelin, A.C.D. (1908e) Mon. Not. R. Astr. Soc.
     68:665-670.
Cowell, P.H. and Crommelin, A.C.D. (1910) Publ. Astron. Gesellschaft,
     No.23.
Damoiseau, M.C.T. (1820) Memorie Della Reale Accademia Delle Scienze di
     Torino 24:1-76.
Damoiseau, M.C.T. (1829) Conn. des Temps for 1832, Additions, pp.25-34.
de Pontecoulant, G. (1830) Conn. des Temps for 1833, Additions,
     pp.104-113.
de Pontecoulant, G. (1834) Conn. des Temps for 1837, Additions,
     pp.102-104.
de pontecoulant, G. (1835) Mem. Presentes Divers Savans Acad. R. Sci.
     6:875-947.
de Pontecoulant, G. (1864) Comptes Rendus 58:825-828,915.
Divine, N. (1982) Revised Light Curve and Gas Production Rate for Comet
     Halley, JPL Interoffice Memorandum 5137-82-101, dated June 15, 1982.
Ho Peng Yoke (1964) Vistas Astr., 5:127.
Ho Peng Yoke and Ang Tian-Se (1970) Oriens Extremus 17:63-99.
Halley, E. (1705) Astronomiae Cometicae Synopsis, Oxford, 6 pp.
Halley, E. (1749) Tabulae Astronomicae, London.
Hasegawa, I. (1979) Publs. Astron. Soc. Japan 31:257
Hind, J.R. (1850) Mon. Not. R. Astr. Soc. 10:51.
Jewitt,D.C., Danielson,G.E., Gunn,J.E., Westphal,J.A., Schneider,D.P.,
     Dressler,A., Schmidt,M., and Zimmerman,B.A. (1982) I.A.U. Circular
     3737.
Kiang,T. (1971) Mem. R. Astron. Soc. 76:27-66.
Lagrange, J.L. (1783) Mem. Acad. Berlin 1783, pp.161-224.
Laugier (1842) Comptes Rendus 15:949.
Laugier (1843) Comptes Rendus 16:1003.
Laugier (1846) Comptes Rendus 23:183.
Lehmann, J.W.H. (1835) Astron. Nachr. 12:308-400.
Marsden,B.G. (1968) Astron. J. 73:367-379.
Marsden,B.G. (1969) Astron. J. 74:720-734.
Marsden,B.G., Sekanina,Z. and Yeomans,D.K. (1973) Astron. J. 78:211-225.
Michielsen,H.F. (1968) J. Spacecr. Rockets 5:328-334.
Morley,T.A. (1983) Giotto Study Note No. 46. dated Nov. 1983.

Newton, I. (1687) Philosophiae Naturalis Principia Mathematica   London:
    Book 3.
Pingre, A.G. (1783-84) Cometographie. Paris.
Rasmusen, H.Q. (1967) Publ. og Mindre Medd. fra Kobenhavns Observatorium
    No.194.
Rasmusen,H.Q. (1981) Fourth Expected Return of Comet Halley: Elements and
    Ephemerides 1981 to 1985. dated July 1981.
Rosenberger, O.A. (1830a) Astron. Nachr. 8:221-250.
Rosenberger, O.A. (1830b) Astron. Nachr. 9:53-68.
Rosenberger, O.A. (1834) Astron. Nachr. 11:157-180.
Rosenberger, O.A. (1835) Astron. Nachr. 12:187-194.
Sekanina,Z. (1981) Ann. Rev. Earth and Plan. Sci. 9:113-145.
West,R.M. and Schwehm,G. (1983) personnal communication dated Dec.13,
    1983.
Whipple,F.L. (1950) Astrophys. J. 111:375-394.
Whipple,F.L. (1951) Astrophys. J. 113:464-474.
Yeomans,D.K. (1977) Astron. J. 82:435-440.
Yeomans,D.K. and Kiang,T. (1981) Mon. Not. R. Astr. Soc. 197:633-646.

DISCUSSION

L. Kresak: How do you explain the irregular variationsof the A1
parameter as compared with the conspicuous stability of A2?

D.K. Yeomans: Because of its secular effect in adding or subtracting
orbital energy, the transverse nongravitational parameter (A2) is very
well determined. On the other hand, the radial parameter (A1) often has
an error nearly equal to the determined value itself.

J.A. Fernandez: The fact that you could fit the observed positions of
comet Halley to your computed orbit for about 2000 years: would this
indicate that random impulses - by outburst for instance - have not
played a significant role in the dynamical evolution of comet Halley?

D.K. Yeomans: Yes I think that is an accurate statement. In fact we
have conducted a covariance analysis in an effort to assess the effects
of unmodeled, stochastic nongravitational forces upon the motion of comet
Halley. For reasonable values of these stochastic effects, the long term
motion of the comet is affected very little.

J. Lissauer: How can you determine that Halley's orbit did not cross
the ecliptic near Jupiter for at least 16,000 years if Halley was
strongly perturbed by earth in 1404 B.C.?

D.K. Yeomans: In September 1404 B.C. our calculations suggest that the
computed position of comet Halley came within 0.04 AU of the earth. A
similar earth close approach took place in 837 A.D. Because of the
uncertain initial conditions prior to the earth close approach and the
lack of ancient Chinese observations to constrain the comet's motion, we
could not continue our integration back prior to 1404 B.C. However, the
uniformly increasing distance of the descending node from 1910 back to
1404 B.C. gave us confidence in our extrapolation of this rate back even
further in time. The comet could not have crossed the ecliptic plane at
Jupiter's distance more recently than 14,000 B.C. and hence the comet
must have been in its present orbit for at least 16,000 years.

THE PAST ORBIT OF COMET HALLEY AND ITS METEOR STREAM

A. Hajduk
Astronomical Institute of the Slovak Academy of Sciences
842 28 Bratislava
Czechoslovakia

ABSTRACT. The present paper studies the structural features of the meteor streams associated with Comet Halley deduced from the observations of its meteor showers, as check points of orbital elements in a deeper history of the comet orbit. Libration of the argument of perihelion of the comet and the corresponding displacement of the nodes, as recognized in the distribution of condensations within the stream, allows to estimate the maximum lifetime of the comet in the inner Solar System at about 2×10^5 years.

The present study is based on three main sources of information: The orbital elements of Comet Halley calculated by Yeomans and Kiang (1981); the libration of the argument of perihelion of the comet calculated by Kozai (1979), and the observed structural features of the associated meteor stream, as represented by the shell model of McIntosh and Hajduk (1983).

The orbital motion of P/Halley has been numerically integrated by Yeomans and Kiang back to 1404 B.C. over a total time span covering 45 revolutions. The change of orbital elements is rather small during this interval, except in the argument of perihelion (with the corresponding changes in the position of nodes) and to a lesser extent in the period (or semi-major axis). However, the changes in the period are neither periodical, nor systematic and hence attempts to use them for a reconstruction of the history of the comet's orbit (Kamienski, 1961) remained unsuccessful.

The variations of the nodal longitude and of the corresponding nodal distances are shown in Fig. 1, where the dots represent the 45 values from Yeomans and Kiang (1981) for each revolution. The solid lines represent the extrapolation of the nodal motion both backwards and forwards, within the libration limits $(-18°, +82°)$ given by Kozai (1979). The half cycle from Ω_{min} to Ω_{max} takes 140 revolutions with a corresponding change of the nodal distance from about 0.8 AU up to 4.2 AU. Each libration cycle produces a shell of orbits. The location of the present orbit of P/Halley within the present cycle is plotted in Fig. 7 by McIntosh and Hajduk (1983).

399

A. Carusi and G. B. Valsecchi (eds.), Dynamics of Comets: Their Origin and Evolution, 399–403.
© *1985 by D. Reidel Publishing Company.*

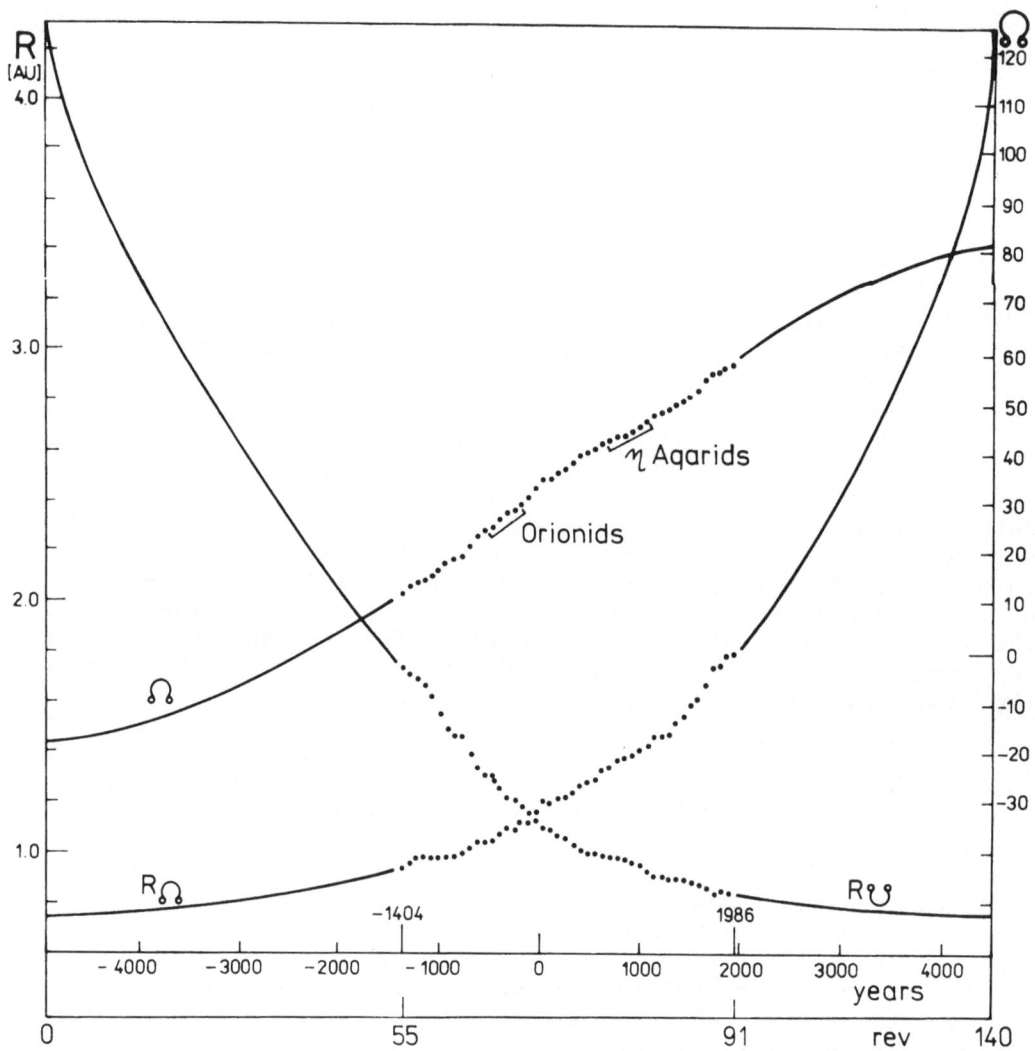

Figure 1. The nodal motion of the orbit of Comet Halley over the present libration cycle. ☊ is the heliocentric longitude of the ascending node, R☊ and R☋ are the heliocentric distances of the ascending and descending node, respectively. The dots represent values from Yeomans and Kiang (1981) for 45 revolutions.

The above data may satisfactorily serve to reconstruct the present cycle of the comet motion, but the data can give us little information about past (or future) cycles of libration, since the perturbations by Jupiter become much stronger at nodal distances near the libration limits when the orbits of Jupiter and the comet approach one another to within 0.25 AU. Kozai (1979) also lists wider changes in the other orbital elements near the libration limits.

We can use the cometary debris to reconstruct past cycles of the comet motion. According to the shell model of the cometary debris of P/Halley (McIntosh and Hajduk, 1983) the observed stable zones of increased particle density in the meteor stream correspond to the shells of particle orbits from different libration cycles of the comet. The shell is confined to a strip, the boundaries of which subtends an angle of 25° with the line of apsides. The spectrum of perturbations on particles spread out along the cometary orbit (Yabushita, 1972) causes a quasi-diffusion process, connected with the thickening of the shell, forming a belt. A typical belt of particle orbits, corresponding to one complete cycle has a width of 0.44 AU perpendicular to the comet's orbital plane and a thickness of 0.044 AU within it, at r = 1 AU.

Let us identify the observed zones of higher spatial density within the two Halley meteor showers (i.e. Orionids and Eta Aquarids) with the belts corresponding to different cycles. Then the Earth's longitudes at the time of observed particle concentrations may serve as check points for the motion of the comet in the past cycles (at the nodal distance of 1 AU and with the position of nodes at the particular Earth's longitudes). The three most recent cycles are reconstructed in this way in Fig. 2. Dots represent values for the present cycle calculated by Yeomans and Kiang. They intersect the Earth's orbit (R_{Ω} = 1.0) at the beginning of the shower periods of Eta Aquarids and Orionids at solar longitudes of 40° and 200°, respectively. The position of concentrations (belts) at the Earth's distance is indicated in Fig. 2 by open circles. Most pronounced are the central double belts at $45^{\circ}-47^{\circ}$ and $206^{\circ}-208^{\circ}$, respectively.

The three pairs of lines defined by the nodes of the belts at 1 AU and by a little different libration limits (shifted by the difference between the belts) may represent the check points for the past three cycles. Some theoretical and observational arguments support this idea. The present belt is less pronounced in observations. The comet has intersected the Earth's orbit only a couple of revolutions ago and, consequently, the belt has not yet developed enough in its width and thickness. Moreover the nodes of the present comet orbit are far from the Earth's orbit. The most pronounced central belt theoretically corresponds to the width and thickness developed in 400-500 revolutions, whereas the third belt is more diffuse with higher mass loss over 650-800 revolutions.

An analysis of the particle size distribution in the belts indicates a higher proportion of larger particles in the older belts than in the most recent one. Other structural features of

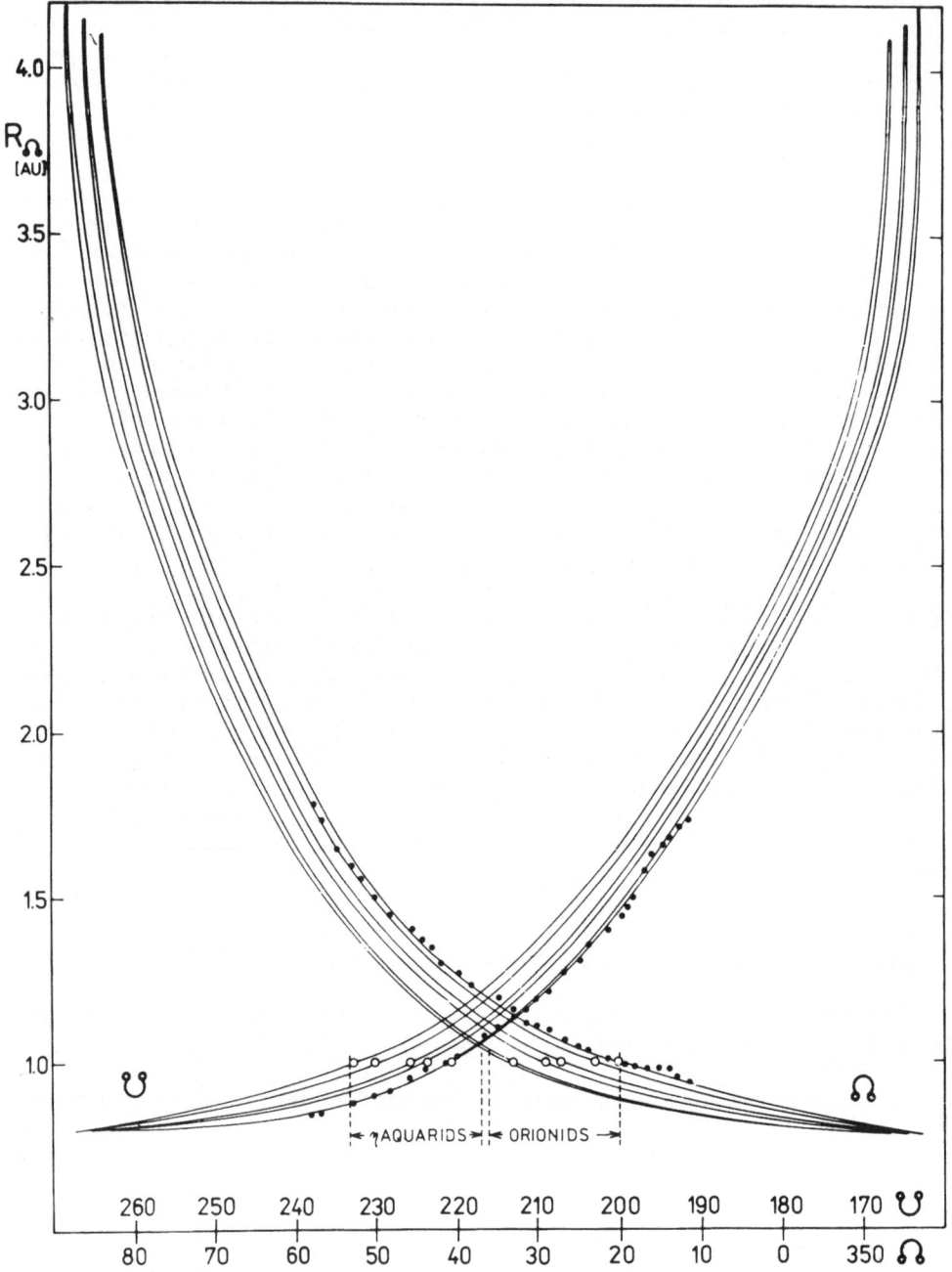

Figure 2. Three libration cycles of Comet Halley, reconstructed from meteor shower data and theoretical considerations. Circles: positions of particle concentrations at Earth's distance. Other designations as in Fig. 1.

the stream, as stream filaments and activity changes, can also be better explained by the shell model of the stream than by a classical toroidal model. For details see McIntosh and Hajduk (1983).

It can be concluded that the structure of the observed meteor streams permits us to trace at least some characteristics of the orbit of comet Halley over a time span of about 800 revolutions. There is no observational evidence for an earlier orbital history, but the sequence of three full cycles gives some possibility to extrapolate the orbital evolution further backwards in time. If we assume that the change of Ω min and Ω max corresponds to the observed difference between the cycles, then the nodal distance of Ω max has reached the orbit of Jupiter 8-10 cycles ago, giving a possible maximum lifetime of the comet in the inner Solar System of about 2 800 revolutions, or 200 000 years. Of course, the probability of a strong gravitational influence by Jupiter increases rapidly with the approach of nodes to the planet's orbit, which circumstance may considerably shorten the estimated maximum lifetime in the inner Solar System, starting with the capture of Comet Halley by Jupiter.

REFERENCES

Kamieński, M.: 1961, Acta Astronomica 11, 223-229.
Kozai, Y.: 1979, IAU Symp. 81, 231-237.
McIntosh, B.A. and Hajduk, A.: 1983, Mon. Not. Roy. Astron. Soc. 205, 931-943.
Yabushita, S.: 1972, Astron. Astrophys. 20, 205-214.
Yeomans, D.K. and Kiang, T.: 1981, Mon. Not. Roy. Astron. Soc. 197, 633-646.

DISCUSSION

D.K. Yeomans: Have you made an effort to include the ancient Chinese observations of the Orionid and Eta Aquarid meteor showers in your analysis ?

A. Hajduk: Yes. The ancient records collected by Imoto and Hasegawa (1958, Smithson. Contr. Astrophys., 2, 131) report conspicuous meteor showers at solar longitudes (L_S) around the time of intersection of the Earth's orbit with that of the comet (between 401 and 934 A.D. for the Eta Aquarids at $L_S \sim 41°$, with intersection between 530 and 607 A.D.; for the Orionids the intersection is too far back in time, between 836 and 763 B.C.). More details can be found in my recent paper in Asteroids, Comets, Meteors, Uppsala 1983, 425-429.

THE INTERNATIONAL COMETARY EXPLORER (ICE) MISSION TO COMET GIACOBINI-ZINNER (G/Z)

J. C. Brandt, R. W. Farquhar, S. P. Maran, M. B. Niedner
and T. von Rosenvinge
NASA-Goddard Space Flight Center
Greenbelt, MD 20771 USA

ABSTRACT. The ICE spacecraft will pass through the tail of
P/Giacobini-Zinner on September 11, 1985, to make in situ
measurements of particles, fields, and waves that will contribute
significantly to the knowledge of plasma tails and other aspects of
the cometary/solar wind interaction. By obtaining data on the
downstream side of G/Z, the ICE will complement the later upstream
measurements obtained by the Comet Halley probes.

COMET GIACOBINI-ZINNER

This short period comet was discovered by M. Giacobini at Nice in
1900 and rediscovered in 1913 by E. Zinner at Bamberg. The
discovery and recovery history of the comet is summarized in Table 1.
The perihelion distance of 1.03 a.u. occurs near the ecliptic plane,
making G/Z well suited to spacecraft encounters during its intervals
of maximum nuclear and atmospheric activity. The orbital
eccentricity and inclination are 0.71 and 31.9°, respectively.
Typically, the plasma tail of G/Z is observed to develop beginning
at a heliocentric distance of about 1.7 a.u. According to
conventional photographic observations at previous apparitions, the
plasma tail can attain a length exceeding 500,000 km. The coma,
also observed photographically, reaches a typical diameter of about
50,000 km (see Figure 1).

G/Z is associated with the Draconid (Giacobinid) meteor
showers. A recent analysis by D. K. Yeomans (private communication)
indicates that there is a limited possibility that a Giacobinid
shower will occur on October 8.5, 1985, although it is unlikely to
rival the occasional great displays of this shower. Groundbased
observers and interplanetary dust particle experimenters should
attempt to observe this shower.

The dust generated by G/Z, besides producing meteor showers,
represents a possible significant hazard to the ICE spacecraft. Of
special concern is dust-impact degradation of the solar power

A. Carusi and G. B. Valsecchi (eds.), Dynamics of Comets: Their Origin and Evolution, 405–414.

Table 1

DISCOVERY AND RECOVERY OF
PERIODIC COMET GIACOBINI-ZINNER

Apparition	Designation	First Observation	Observer/ Institution	Approximate Magnitude	Reference
1	1900III (1900c)	Dec. 20, 1900	Giacobini Observatoire de Nice	11	*A.N.* **154**, 161
2	1913V (1913e)	Oct. 23, 1913	Zinner Remeis-Sternwarte	9-10	*A.N.* **196**, 167 & 353
3	1926VI (1926e)	Oct. 16, 1926	Schwassmann Hamburger Sternwarte	14	*A.N.* **229**, 122
4	1933III (1933c)	Apr. 23, 1933	Schorr Hamburger Sternwarte	15	IAU Circ. 435
5	1940I (1939I)	Oct. 15, 1939	Van Biesbroeck Yerkes Observatory	15	IAU Circ. 797
6	1946V (1946c)	May 29, 1946	Jeffers Lick Observatory	17	IAU Circ. 1046
7	1959VIII (1959b)	May 8, 1959	Roemer U.S. Naval Observatory	20	IAU Circ. 1677
8	1966I (1965g)	Sept. 17, 1965	Roemer & Lloyd U.S. Naval Observatory	20	IAU Circ. 1923
9	1972VI (1972d)	Mar. 11, 1972	Roemer & McCallister University of Arizona	19	IAU Circ. 2390
10	1979III (1978h)	Apr. 30, 1978	Shao & Schwartz Harvard College Observatory	20-21	IAU Circ. 3216
11	1985_ (1984e)	Apr. 3, 1984	Djorgovski & Spinrad Univ. of California, Berkeley Will & Belton Kitt Peak National Observatory	23	IAU Circ. 3937

General References: (1) "Liste Générale Des Comètes De L'Origine A 1948," by M.F. Baldet, *Annuaire du Bureau des Longitudes (1950)*.

(2) *Catalogue of Cometary Orbits*, 4th Ed., by B.G. Marsden, Smithsonian Astrophysical Observatory (1982).

arrays, which could reduce the power available to operate the spacecraft and payload during the cometary intercept. Studies of this hazard, based on a G/Z dust model prepared by N. Divine at NASA-Jet Propulsion Laboratory, are underway at NASA-GSFC so that operating procedures can be developed to deal with it.

The early recovery of G/Z by ground-based observers was of special interest to the ICE mission flight dynamicists. In the event that the observed orbit had differed significantly from the expected orbit of G/Z, early detection of G/Z would have been crucial to carefully schedule operation of the on-board propulsion capability so as to successfully effect the flyby. Fortunately, the first recovery photographs (Figure 2), obtained on April 3, 1984, by S. Djorgovski, H. Spinrad, G. Will, and M. Belton with the 4-m Mayall telescope at Kitt Peak National Observatory, revealed that

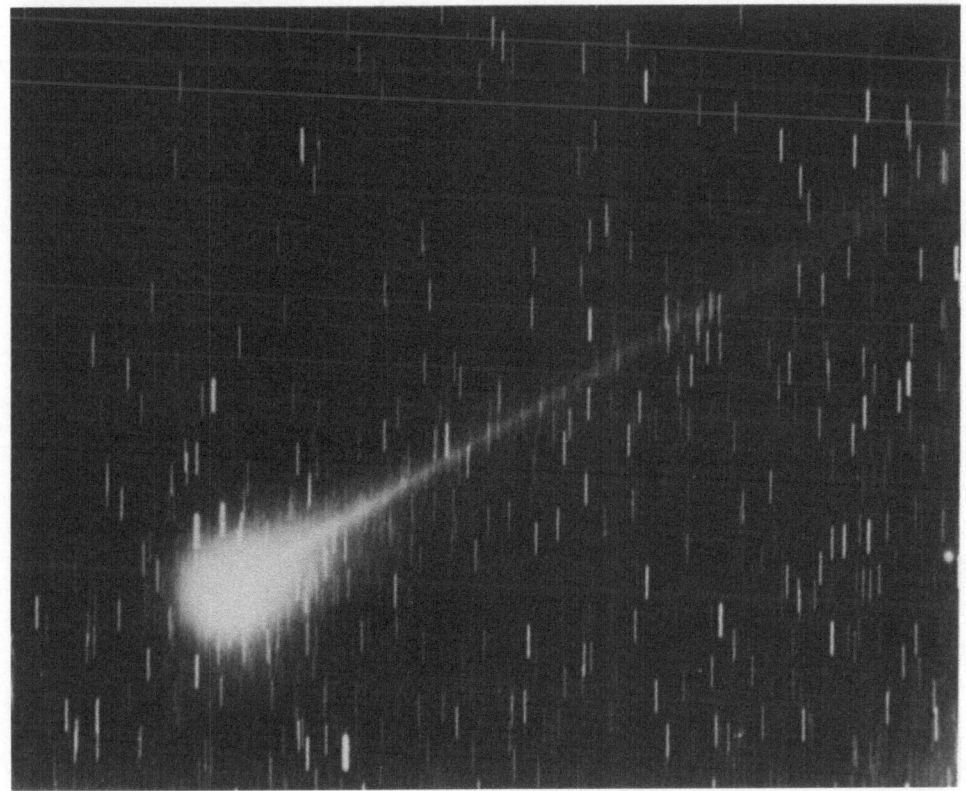

Figure 1. Photograph of Comet Giacobini—Zinner, obtained by E. Roemer
on October 26, 1959. (Official U.S. Navy photograph).

G/Z was within 10,000 km of the anticipated position. Current
projections (May 1984) indicate that if no further use were made of
the propulsion system (last exercised for trajectory adjustment in
November 1983), ICE would nevertheless pass through the nominal
location of the G/Z tail at about 100,000 km from the nucleus.
Ground commands will, however, target the tail intercept at a lesser
distance from the nucleus, to be proposed by the ICE science team
for approval by NASA. Should a scientific requirement exist, flight
controllers could navigate ICE to within about 500 km of the
nucleus, its approximate positional uncertainty near encounter. The
intercept will occur with the spacecraft travelling from south to
north in the rest frame of the comet.

The present, tentative ICE targeting strategy study utilized a
baseline tail model (Figure 3) which assumes a gas production rate
of 2.3×10^{28} molecules/sec at perihelion (derived by N. Divine
from the measured brightness of G/Z at a previous apparition). The
model suggests a plasma tail width of about 5000 km at 55,000 km
from the nucleus, consistent with the measured size on a photograph
of G/Z obtained by E. Roemer on October 26, 1959.

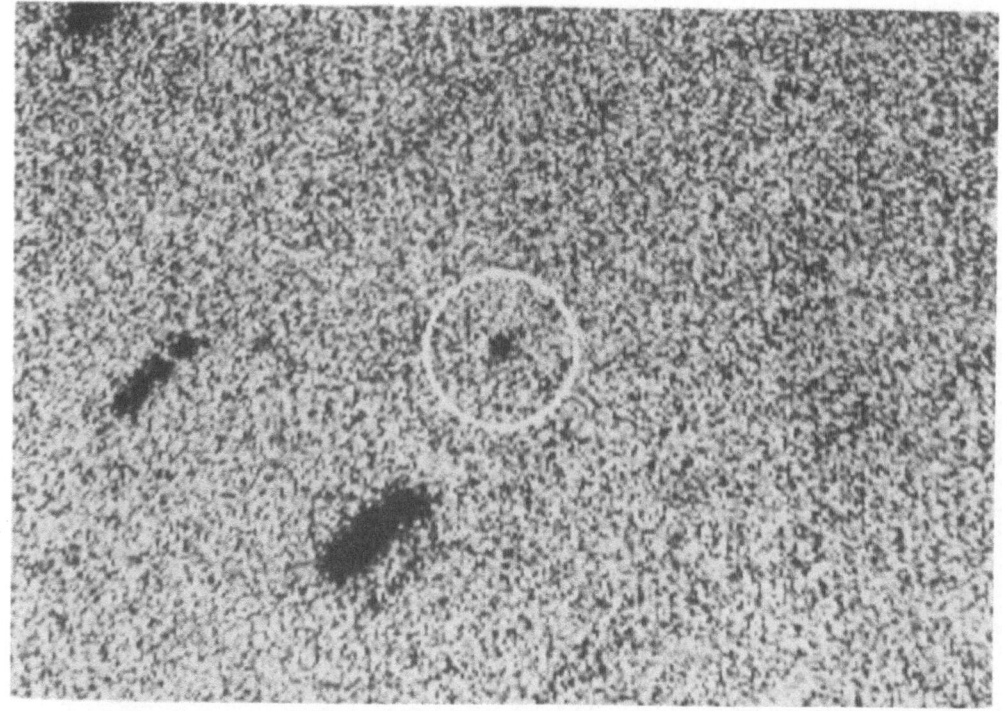

Figure 2. Recovery photograph of Comet Giacobini-Zinner, obtained by
S. Djorgovski, H. Spinrad, G. Will, and M. Belton on April 3, 1984.
(Kitt Peak National Observatory photograph).

SPACECRAFT AND PAYLOAD

The ICE mission objective is to provide _in situ_ data on the
interaction between the solar wind and a cometary atmosphere. This
will be accomplished through the intercept of the tail of G/Z.
Secondary objectives include the support of the various planned
intercepts of Comet Halley through measuring solar wind phenomena
upstream of P/Halley, following the G/Z tail intercept. ICE will
pass 0.93 a.u. upstream of P/Halley on October 31, 1985 and 0.21
a.u. upstream of P/Halley on March 28, 1986.
     The ISEE-3 spacecraft, launched August 12, 1978 on a Delta
rocket, was renamed ICE by NASA, effective December 22, 1983, when
the spacecraft made a close lunar swingby, passing only 120 km above
the lunar surface to obtain the gravitational assist necessary to
depart from the Earth-Moon system (Figure 4) and travel onward to
intercept G/Z. Previously, ISEE-3 operated in a halo orbit about
the sunward Lagrangian point of the Earth-Moon system. In 1983,
ISEE-3 accomplished an extended mission to make the first extensive
survey of the distant geotail. Its survey, conducted to distances
as great as 237 earth radii in the tail, was accomplished through a
series of lunar swingby maneuvers. The study of particle, field,
and wave phenomena in the geotail, besides the scientific

COMET GIACOBINI-ZINNER AT PERIHELION

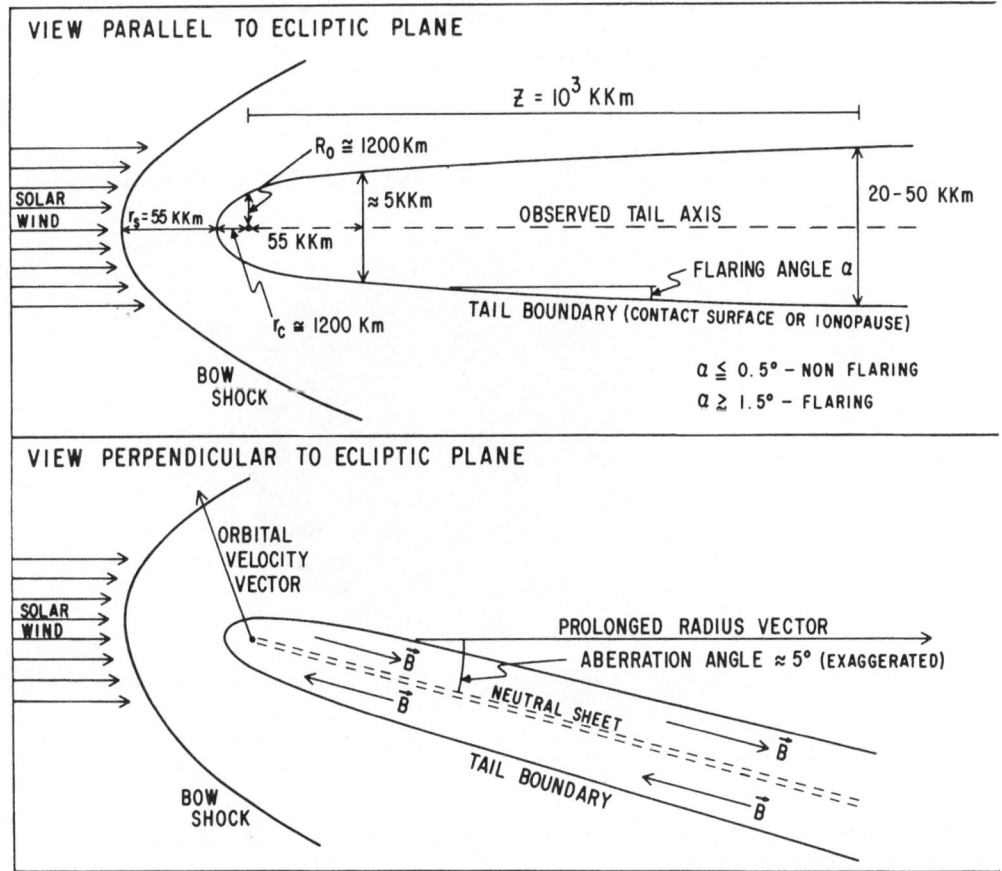

VIEW PARALLEL TO ECLIPTIC PLANE

$\bar{Z} = 10^3$ KKm

$R_0 \cong 1200$ Km

$r_s = 55$ KKm

SOLAR WIND

≈ 5 KKm

OBSERVED TAIL AXIS

55 KKm

20-50 KKm

$r_c \cong 1200$ Km

FLARING ANGLE α

TAIL BOUNDARY (CONTACT SURFACE OR IONOPAUSE)

BOW SHOCK

α ≤ 0.5° – NON FLARING

α ≥ 1.5° – FLARING

VIEW PERPENDICULAR TO ECLIPTIC PLANE

ORBITAL VELOCITY VECTOR

SOLAR WIND

PROLONGED RADIUS VECTOR

$\vec{B}$

ABERRATION ANGLE ≈ 5° (EXAGGERATED)

$\vec{B}$

NEUTRAL SHEET

$\vec{B}$

TAIL BOUNDARY

$\vec{B}$

BOW SHOCK

Figure 3. Schematic diagram of baseline ion tail model for Comet Giacobini-Zinner near perihelion.

significance of the results obtained, provided an excellent rehearsal and scientific baseline for experiment operations during the single brief comet tail encounter that will occur on September 11, 1985.

A schematic of the ICE spacecraft is given in Figure 5. From current knowledge, five of the ICE instruments are considered most suited to make major contributions to cometary science through measurements obtained in the G/Z intercept:

Vector Helium Magnetometer

Obtains three, high-accuracy, triaxial measurements per second in ranges extending from $\pm$ 4γ to +1.4 gauss.

Principal Investigator: E. Smith, Jet Propulsion Laboratory.

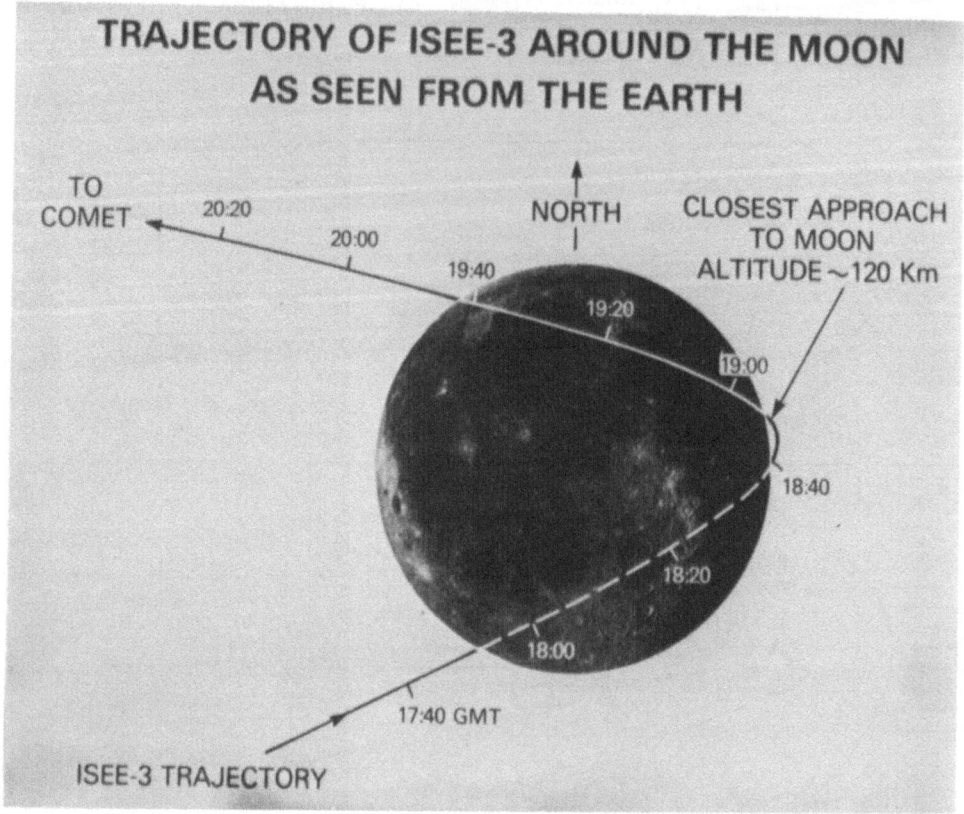

Figure 4. The orbit of ISEE-3 near the moon at the time of the
close lunar swingby on December 22, 1983. The official name of the
spacecraft became ICE as it departed from the vicinity of the moon.

Plasma Wave Experiment

Samples 16 channels in electric field in one second in the frequency
range 20 Hz to 10^5 Hz.
    Principal Investigator: F. Scarf, TRW.

Radio Wave Experiment

Samples the range 30 kHz to 1 MHz in 12 steps with Δf = 3 kHz.
Samples the range 40 kHz to 2 MHz in 12 steps with Δf = 10 kHz.
Each scan takes 56 seconds.
    Principal Investigator: J.-L. Steinberg, Observatoire de Meudon.

Plasma Electron Experiment

Scans in 16 steps covering the energy range 5 eV to 1500 eV.
Sampling takes 24 seconds.

   Principal Investigator: S. Bame, Los Alamos National
Laboratory.

Plasma Ion Experiment

Has velocity selector. Determines M/Q in 25 steps for the range 4
to 50 and velocity in 25 steps for the range 20 km/sec to 200
km/sec. Complete set of measurements takes 15 minutes.
   Principal Investigator: K. Ogilvie, NASA-Goddard Space Flight
Center.

ISEE-3 SPACECRAFT

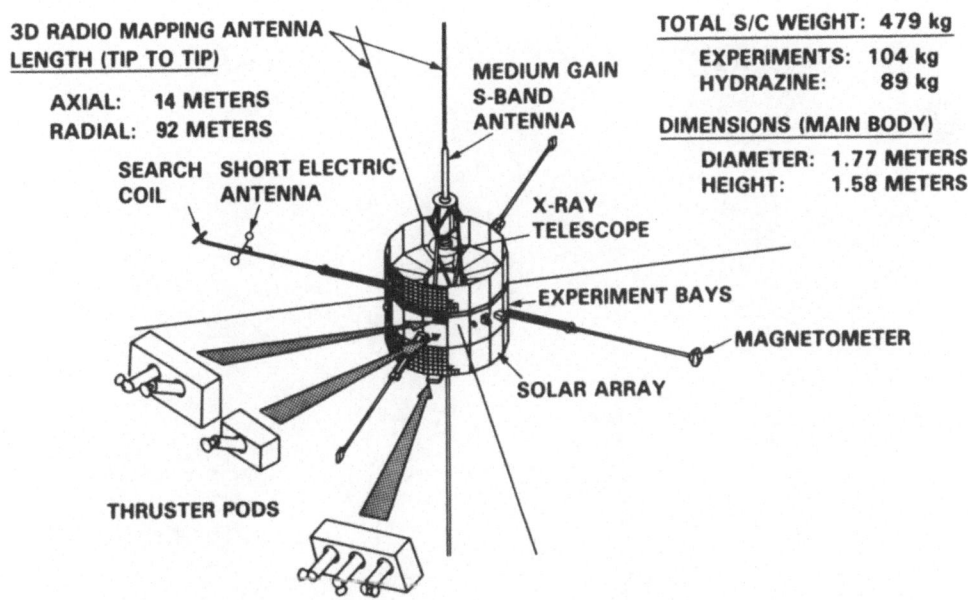

Figure 5. Schematic diagram of the International Cometary Explorer,
as built under the name of ISEE-3.

   Other experiments are intended to measure <u>Energetic Protons</u>, <u>X
Rays</u>, <u>Low Energy to High Energy Cosmic Rays</u>, <u>Cosmic Ray Electrons</u>,
and <u>Gamma Ray Bursts</u>. These experiments do not seem as likely to
produce cometary data as those described above, other than upper
limits. However, our detailed knowledge of cometary physics and
structures is sufficiently insecure that these instruments should be

operated in a serendipitous mode during the cometary intercept,
provided that sufficient operating power is available. ICE has no
imaging nor dust experiments.

The ICE spacecraft measurements will be supplemented with
ground-based measurements of G/Z obtained by arrangement with the
International Halley Watch and concentrated in September, 1985.
Astrometric and orbital work is being carried out in collaboration
with Dr. Yeomans.

MISSION TARGETING

It is thought that a cometary plasma tail is formed from
interplanetary magnetic field lines that are captured and frozen
into the cometary ionosphere and which entrain plasma that is
accelerated in the antisunward direction. In the simplest model,
the tail is organized into two magnetic lobes of opposite field
direction, separated by a current sheet (Figure 6). Directly in the

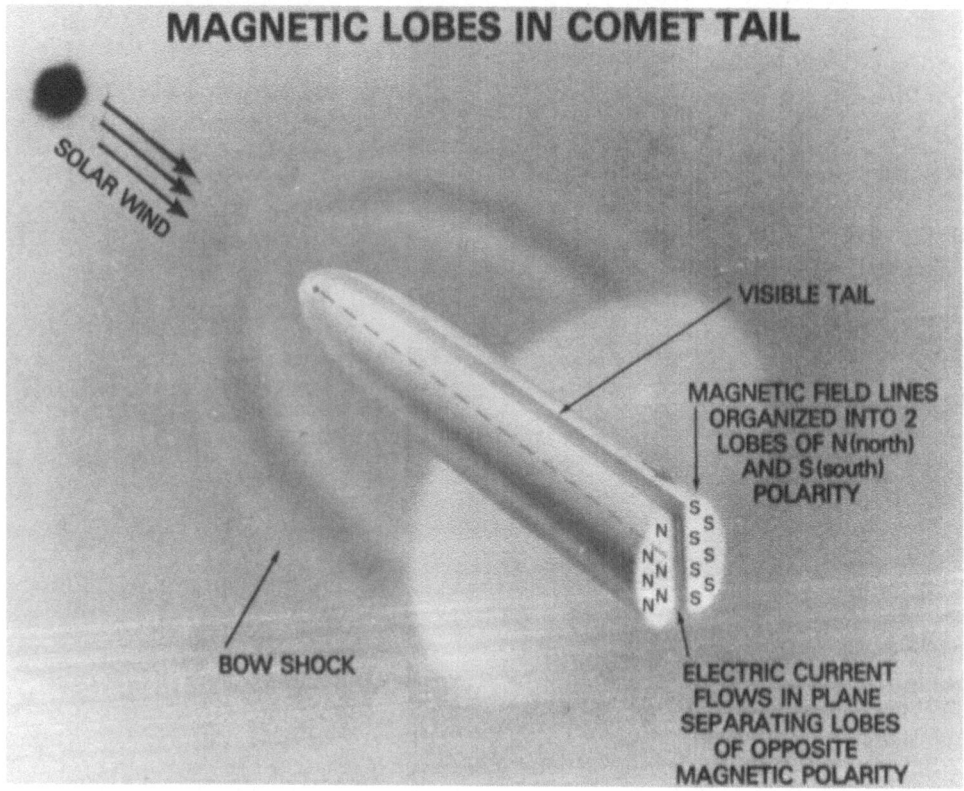

Figure 6. Schematic diagram of simple ion tail model with two
magnetic lobes.

wake of the nucleus, however, conditions may be chaotic and it is
thought that the tail may not be organized as just described closer
than a few thousand km to the nucleus. This sets a lower bound to
the distance from the nucleus at which the ICE science team would
prefer that the intercept occur. At very large distances from the
nucleus, the observed tendency of a plasma tail to wag back and
forth in response to solar wind velocity variations would lower the
probability of a successful tail intercept. Therefore an intercept
distance in the range of about 5000 km to 15,000 km may be
desirable. The plasma tail at that location may be obscured by the
coma on photographs taken from the ground, so that a measurement of
the tail width by ICE would be especially interesting. From such a
measurement, an independent (but model−dependent) derivation of the
gas production rate can be made. (Specifically, an accurate
determination of the width of the contact surface at the flank
should be possible without model−dependent assumptions). The dust
hazard will also be considered in making the final targeting
decision. In any case, ICE will surely probe the bow shock and
contact discontinuity regions predicted by current theory of the
cometary/solar wind interaction, if they exist on roughly the
anticipated scales. A precise intercept of the tail could allow
investigation of the current sheet that divides the two magnetic
lobes in the simple plasma tail model, assuming the model is
applicable. Even if precise targeting is not possible, the
existence of the current sheet should be inferred from the detection
of waves generated there (Figure 7).

As already mentioned, the targeting accuracy for ICE is
potentially about 500 km. Tracking data will allow very precise
calculation of the spacecraft location, so that the dominant term in
the targeting accuracy actually attained will be the uncertainty in
the location of the comet. G/Z is an active comet with irregular
brightness variations. Non−gravitational forces on G/Z increased
during 1900 to 1965. Therefore, an accurate dynamical model may be
necessary to achieve the targeting accuracy goal. The development
of such a model may allow the identification of one or more
quantities that can be measured by the instruments on ICE. We
invite the attention of dynamicists and modellers to the possibility
of expanding their methodology for studying nongravitational
phenomena in comets using astrometric observations via the
incorporation of in situ measurements.

CONCLUSION

In situ measurements of a comet's tail region and other features
associated with the comet's interaction with the solar wind should
be of interest to all comet scientists. In particular, the ICE
tailward measurements are complementary to the sunward measurements
to be made by the missions to Halley's Comet.

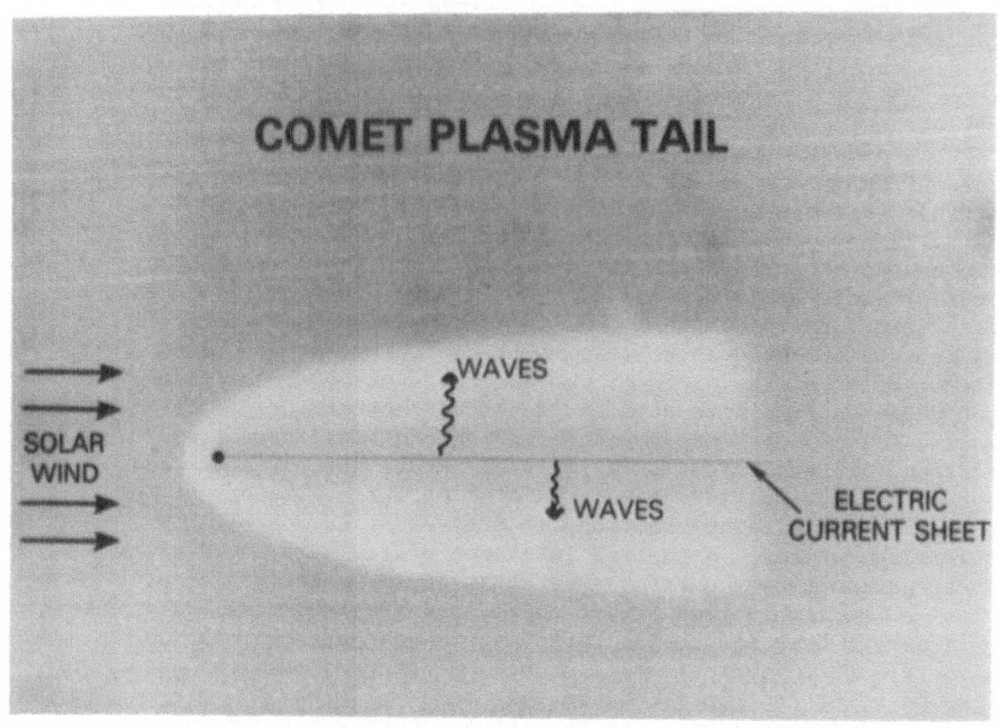

Figure 7. Schematic diagram indicates presence of waves that are
thought to be generated in the ion tail of a comet and which may be
detectable as ICE passes through the tail of Comet Giacobini-Zinner.

REFERENCES

Ogilvie, K.W., von Rosenvinge, T., and Durney, A.C. 1977,
    "International Sun-Earth Explorer: A Three-Spacecraft Program,"
    Science, 198 (No. 4313, Oct. 14), p. 131.
Farquhar, R.W., and Dunham, D.W. 1983, "A Late Entry in the Great
    Comet Chase," Astronautics and Aeronautics, 21 (No. 9, Sept.),
    p. 50.
Comet Subcommittee of the ISEE Science Working Team, June 1982
    (Revised May 1983), Report entitled "Intercept of Giacobini-
    Zinner by ISEE-3."
Brandt, J.C., Farquhar, R.W., Maran, S.P., Niedner, M.B., and
    Von Rosenvinge, T. 1983, "The Third International Sun-Earth
    Explorer (ISEE-3) Mission to Comet Giacobini-Zinner (G/Z),
    Bull. Am. Astron. Soc., 15 (No. 4), p. 960.

THE SELECTION OF COMETS FOR FUTURE SPACE MISSIONS

D.K. Yeomans
Jet Propulsion Laboratory
California Institute of Technology
Pasadena, California 91109
U.S.A.

ABSTRACT.   The criteria used to select a short period comet for possible future rendezvous space missions are stated and the selection process is outlined. For the time period 1900 - 2000, several candidate comets offer opportunities for spacecraft rendezvous. Two of the best candidates are periodic comets Kopff and Wild 2.

1. INTRODUCTION

The scientific objectives of future space missions to comets are generally agreed upon (NAS-ESF, 1983). These objectives include the following:
1. Determine the chemical composition and physical structure of the comet's nucleus and characterize the nucleus as a function of heliocentric distance.
2. Characterize the coma through the identification of parent molecules and the processes by which these molecules are transformed in the inner coma and observe the changes in coma activity as a function of time and heliocentric distance.
3. Determine the composition and distribution of coma grains and study their interactions with the coma plasma.
4. Study the interaction of the cometary plasma with the solar wind.

The International efforts underway to study comets Halley and Giacobini-Zinner from flyby spacecraft will go a long way in addressing the above scientific objectives (Reinhard, 1982). Following its last lunar swingby in December 1983, the International Cometary Explorer (ICE) spacecraft was targetted toward a September 11, 1985 flyby of periodic comet Giacobini-Zinner. This spacecraft, with its complement of plasma instruments, will fly through the ion tail of this comet and hence concentrate on the fourth scientific objective listed above. The ability to redirect an earth orbital spacecraft (whose major mission goals had already been attained) to fly by one of the very few short period comets with a visible ion tail was a fortunate circumstance. There are plans for a flotilla of five spacecraft to fly by comet Halley during the interval March 6-14, 1986. The Soviet Union will send the first of its two VEGA

A. Carusi and G. B. Valsecchi (eds.), Dynamics of Comets: Their Origin and Evolution, 415–421.

spacecraft past Halley on March 6 at a distance of 10,000 km on the sunward side. The second identical VEGA spacecraft will follow in a similar flyby trajectory on March 9. The Japanese spacecraft, Planet A and MS-T5 will pass closest to comet Halley on March 8 at a distance larger than 100,000 km. The European Space Agency's Giotto spacecraft will fly through Halley's inner coma (< 1000 km sunside) around midnight (G.M.T.) on March 13. The extraordinary interest in space observations of comet Halley is understandable when one realizes that this comet is the only one displaying the full range of cometary phenomena and having a predictable orbital path. When taken together with the coordination of ground based observations by The International Halley Watch, the space observations of comets Halley and Giacobini-Zinner should address many of the scientific objectives for cometary missions.

While the scientific return from the various planned flyby missions to comets Halley and Giacobini-Zinner will return a vast amount of valuable data on the tail plasma and the coma's gas and dust components, the primary scientific objective (to study the nucleus) will only be addressed in a preliminary fashion. An intensive study of a comet's nucleus will require a space mission to rendezvous or fly alongside a comet for a period of time, to orbit the nucleus itself and to study its chemical composition and structure over a far longer time than is afforded by a fast flyby space mission. The remainder of this paper will discuss the target selection process that has been undertaken for future rendezvous missions to comets.

2.   TARGET SELECTION CRITERIA FOR COMET RENDEZVOUS MISSIONS

In selecting a group of comets that are attractive candidates for a future rendezvous mission, the following criteria were used;

1. The target comet's orbital motion should be well understood.
2. The comet should exhibit both quiescent and active stages and it should be possible to rendezvous with the comet well before it becomes active.
3. Near perihelion, the comet should have a relatively high gas production rate.
4. A good observational history should exist for each comet.
5. During the rendezvous phase of the mission, the comet should be easily observable from the ground.
6. The orbit of the target comet should be such that it does not place unnecessary cost burdens upon the launch vehicle, spacecraft or ground operations.

An obvious criterion for a rendezvous target comet is that its orbital motion be well understood. There have been cases where periodic comets have poorly known orbital motions (ie. comet Westphal) so that the spacecraft navigation to the comet would be quite difficult, if not impossible. For this same reason, one apparition comets are a poor choice despite the fact that they are often the most active comets. In general, three apparitions of a comet are necessary before a comet's orbit becomes well known (Marsden, 1968).

The study of the target comet's nucleus is the primary scientific objective, and since the nucleus should be observed evolving from its inactive to active phases, the target comet should have both a quiescent and active phase. Very close measurements of a comet's nucleus are required to satisfy the scientific objectives so that these most important observations should be made first and when the comet's dust and gas environment is least hostile - when the comet is inactive. In addition, the navigation of a spacecraft in orbit about a comet's nucleus will require a knowledge of the mass of the nucleus and a mass determination derived from Doppler tracking data is most easily effected in the absence of gas and dust drag forces acting upon the spacecraft (Yeomans et al, 1980).

In order to study the parent and daughter species in the cometary coma, the spacecraft instruments must be supplied with a sufficiently dense gas and dust environment to make meaningful measurements. Hence the target comet should be relatively active near perihelion. In Table I, candidate comets are listed in order of their maximum magnitude which is usually the comet's apparent magnitude at perihelion (reduced to 1 AU from the earth). M. Festou has shown that the comet's maximum magnitude is a good indicator of the comet's maximum gas production rate (NAS-ESF, 1983).

Each candidate comet should have a fairly good history of observations so that environmental gas and dust models can be prepared for the comet. These models are necessary for determining instrument sensitivity requirements and the requirements for shielding against dust and electromagnetic radiation.

For each candidate comet listed in Table I, the most current orbit has been integrated forward to determine the times of perihelion passage in the 1990-2000 period. Full planetary and nongravitational perturbations were taken into account at each time step and an ephemeris was generated for extended intervals on either side of perihelion. In most cases, the initial conditions were taken from the catalog of B.G. Marsden (1982). All those comets listed in Table I have good ground based viewing opportunities surrounding at least one of the perihelion passages. Good ground based viewing conditions during the rendezvous phase of the cometary mission will allow the correlation of close up spacecraft measurements of the nucleus and inner coma with the ground based observations of the outer coma regions. The combination of these two data sets will allow the total science return to be larger than the sum of the individual efforts.

In Table I, each candidate comet is listed in order of its maximum visual magnitude, reduced to one AU from the earth. Comet Halley is included for comparison. The perihelion distance(q), orbital inclination(I) and number of observed apparitions are then given followed by future perihelion passage times and the ground based viewing opportunities on either side of the given perihelion. Those periodic comets that do not have good ground based viewing near their perihelion passages in the 1990-2000 interval are not included nor are those comets whose maximum magnitude gets no brighter than 12. The comets listed in Table I satisfy the first five of the six selection criteria. For comets in highly inclined orbits (I >25 degrees), the energy requirements for a rendezvous spacecraft are prohibitive. Ideally, the rendezvous comet target should have a low inclination so that major orbital plane changes in the spacecraft trajectory are not necessary. Hence, comets with large inclinations such as Borrelly and Giacobini-Zinner were not given further consideration. The most attractive opportunities for a comet rendezvous mission are given below. These comets and their perihelion times of interest are given in order of their maximum brightness.

Encke (May 1997)
Kopff (July 1996)
Wild 2 (May 1997)
Honda-Mrkos-Pajdusakova (Dec. 1995)
d'Arrest (July 1995)
Tempel 2 (Sept. 1999)
Tempel 1 (July 1994)
Churyumov-Gerasimenko (Jan. 1996)
Tuttle-Giacobini-Kresak (July 1995)

It should be noted that the 1997 apparition of comet Encke is only fair in terms of ground based viewing opportunities and comet Wild 2 has only been observed at two apparitions to date. However, because of its relatively high gas production rate, Wild 2 has been retained as a possible mission target. During a recent study, the above list of comets was studied in light of their ease of access for a rendezvous spacecraft (Yen, 1983). Comet Tempel 1 was not considered because it would have required a launch date considered to be too early for current advanced mission planning. The current injected spacecraft mass was assumed to be 2787 kg, the launch vehicle was assumed to be NASA's space shuttle with a Centaur G' booster and the arrival of the spacecraft at the comet was constrained to be at least 100 days prior to perihelion. Opportunities to fly by one or two asteroids enroute to the comet rendezvous were also considered important in the final selection process. Preliminary studies soon indicated that the shuttle launch capability was inadequate for effecting a rendezvous with comet Encke (during its 1997 apparition) and comet d'Arrest (during its 1995 apparition). The field of acceptable target candidates was then reduced to Kopff(1996), Wild 2(1997), Honda-Mrkos-Pajdusakova(1995), Tempel 2(1999), Churyumov-Gerasimenko(1996) and Tuttle-Giacobini-Kresak(1995). Additional studies were then conducted on these six comets in an effort to determine their relative merits for future rendezvous missions (Yen, 1984). Of these six comets, comet Kopff is the brightest one near perihelion and hence the most physically attractive rendezvous target.

Under the stated assumptions, Table II presents the rendezvous opportunities for the three comets of most interest for a comet rendezvous mission in the mid 1990's. After each comet's name is given the perihelion passage time (T), the arrival date(AD) given in number of days before perihelion, the first day of the launch period (LD), the flight time to the comet in years for an optimum ballistic trajectory (FTC), the flight time, from the earth, until both the comet and spacecraft reach perihelion (FTP), the launch energy required (C3), the Shuttle/Centaur G'

TABLE I: Potential Targets For A Comet Rendezvous Mission

Comet	Last Apparition	Max. Mag.	q (AU	I (Deg.)	No. of Appar.	Perihelion Passage	Grd. Based Viewing
Halley	1982i	3.0	0.59	162	29	1986 Feb. 9	Fair
Encke	1980XI	7.1	0.33	12	52	1987 Jul.17	Poor
						1990 Oct.28	Good
						1994 Feb. 9	Good
						1997 May 23	Fair
						2000 Sep.29	Poor
Kopff	1982k	8.2	1.58	5	12	1983 Aug.10	Exc.
						1990 Jan.20	Poor
						1996 Jul. 2	Exc.
Wild 2	1983s	9.1	1.49	3	2	1984 Aug.20	Poor
		9.4	1.58			1990 Dec.17	Good
		9.4	1.58			1997 May 7	Exc.
Honda-Mrkos-Pajdusak.	1980I	9.3	0.54	4	6	1985 May 24	Poor
						1990 Sep.13	Exc.
						1995 Dec.26	Exc.
d'Arrest	1982VII	9.3	1.29	19	14	1982 Sep.14	Exc.
						1989 Feb. 4	Poor
						1995 Jul.27	Exc.
Tempel 2	1982d	9.7	1.38	12	17	1983 Jun. 1	Good
		10.0	1.48			1988 Sep.17	Good
		10.0	1.48			1994 Mar.17	Poor
		10.0	1.48			1999 Sep. 8	Good
Tempel 1	1982j	10.9	1.49	11	7	1983 Jul.10	Exc.
			1.50			1989 Jan. 4	Poor
			1.49			1994 Jul. 3	Exc.
Churyumov-Gerasimenko	1982VIII	11.1	1.30	7	3	1982 Nov.12	Exc.
						1989 Jun.19	Poor
						1996 Jan.18	Exc.
Tuttle-Giacobini-Kresak	1978XXV	11.2	1.12	10	6	1984 Aug.28	Fair
		11.1	1.07			1990 Feb. 8	Good
		11.1	1.06			1995 Jul.29	Fair

injected mass maximum (Mo), the post launch delta V required to acomplish
rendezvous (ΔVPL1), and the delta V remaining for enroute asteroid flybys
and post rendezvous operations of the spacecraft around the comet
itself (ΔVPL2). From Table II, it appears that the fuel supply (ΔVPL2)
for asteroid flybys and post rendezvous maneuvers is marginal for comet
Honda-Mrkos-Pajdusakova. Clearly the comet Kopff 1996 opportunity is the
most desirable in terms of available fuel at rendezvous and arrival time
at the comet. Although without a long history of observations, comet
Wild 2, during its 1997 apparition, is also attractive in terms of the
ease of spacecraft rendezvous.

TABLE II: Comet Rendezvous Mission Parameters

Comet	T	AD	LD	FTC yrs.	FTP yrs.	C3 $(km/s)^2$	Mo kg.	ΔVPL1 km/s	ΔVPL2 km/s
Kopff	1996 Jul. 2	−890	1990 Jul. 4	3.6	6.0	81.1	2532	1.80	0.63
		−100	1991 Jul.13	4.7	5.0	76.0	2787	2.19	0.48
Wild 2	1997 May 7	−845	1991 Mar. 9	3.9	6.2	76.0	2787	1.99	0.60
		−341	1992 Mar.22	4.2	5.2	76.0	2787	2.40	0.19
Honda-Mrkos-Pajdusakova	1995 Dec.26	−100	1990 Nov.14	4.8	5.1	76.1	2767	2.64	−0.07
		− 50		5.0	5.1	76.0	2787	2.58	−0.01

Recently the comet Kopff 1996 opportunities were recommended for the
proposed first rendezvous mission to a comet by NASA's Mariner Mark II
spacecraft project. The recommended mission included a flyby of one or
two main belt asteroids with a flyby of C-type asteroid 772 Tanete being
an attractive opportunity enroute to the comet Kopff rendezvous (see
Figure 1).

The work described in this paper was carried out by the Jet Propulsion
Laboratory, California Institute of Technology, under contract with the
National Aeronautics and Space Administration.

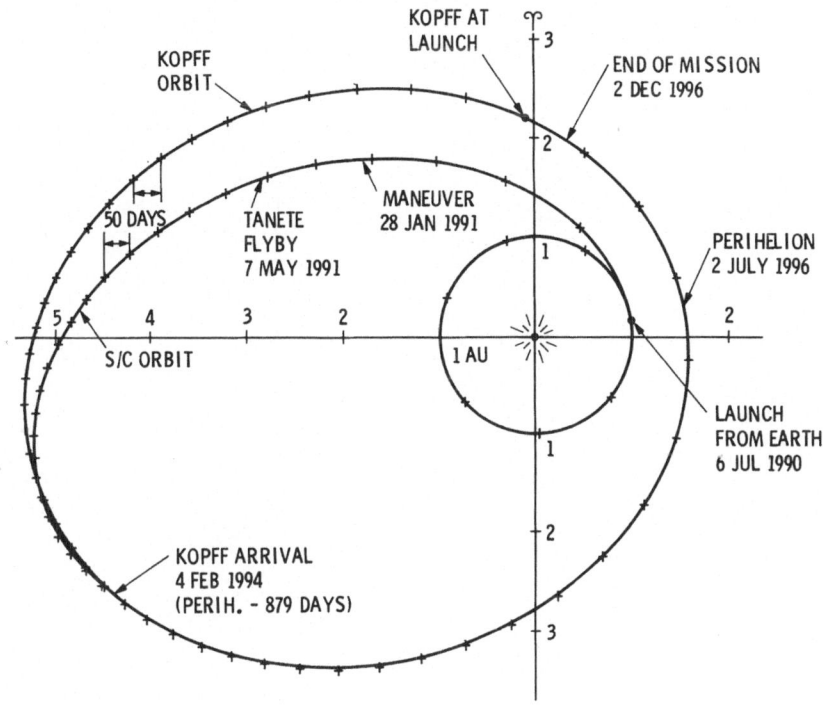

Figure 1. Orbit diagram of Kopff rendezvous with enroute flyby of Asteroid 772 Tanete.

References:

Marsden, B.G.: 1968, "Comets and Nongravitational Forces", Astronomical Journal, Vol. 73, No.5, Part 1, June 1968, pp.367-379.

Marsden, B.G.: 1982, "Catalogue of Cometary Orbits" Minor Planet Center, Central Bureau for Astronomical Telegrams.

NAS-ESF Joint Working Group on Planetary Exploration; Report of the Primitive Bodies Study Team, October, 1983.

Reinhard, R.: 1982, "Space Missions to Halley's Comet and Related Activities" ESA Bulletin, No.29, February 1982, pp.68-83.

Yen, C.L.: 1983, "Ballistic Comet Rendezvous Mission Opportunities", presentation to NASA's Comet Rendezvous Science Working Group, October 11, 1983.

Yen, C.L.: 1984, "Mariner Mark II Comet Rendezvous Mission Performance Update",presentationto NASA's Comet Rendezvous Science Working Group, Jan. 11, 1984.

Yeomans, D.K., Ananda, M., Sjogren, W.L., and Wood, L.J.: 1980, "Cometary Mass Determination" The Journal of the Astronautical Sciences, Vol. 29, No.1, Jan.-Mar. 1980, pp.19-33.

SUBJECT INDEX

NAME INDEX

Abt, H.A. 67
A'Hearn, M.F. 167
Aksnes, K. 340
Aliev, S. 299
Allen, C.W. 17, 141
Alvarez, L.W. 28, 40, 84
Alvarez, W. 84
Ananda, M. 421
Andrews, J.N. 29
Andrienko, D.A. 385
Ang Tian-Se 397
Arnold, V.I. 225
Asaro, F. 84
Astapovic, I.S. 126
Atreya, S.K. 168
Augustyniak, W.M. 169
Aumann, H.H. 170

Babadzhanov, P.B. 126, 148
Backlund, O. 385
Bailey, M.E. 28, 40, 67, 85,
    95, 166, 277, 299, 316
Bardwell, C.M. 168, 340
Barker, E.S. 167
Batrakov, Yu.N. 201
Beichman, C.A. 170
Beintema, D.A. 170
Belyaev, N.A. 126, 253, 277
Bender, D.F. 310
Benest, D. 166, 299, 370
Benettin, G. 225
Bessel, F.W. 396
Bettis, D.G. 201, 234, 235
Bhatt, H.C. 40
Bien, R. 166, 299
Biermann, L. 17, 67, 85, 111,
    166
Binzel, R.P. 166, 167, 177
Biot, E. 126, 396
Black, W. 235
Boclet, D. 28
Bogart, R.S. 67

Boggess, N. 170
Bohm-Vitense, E. 28
Bonté, Ph. 28
Bohor, B.F. 28, 85
Bowell, E. 167, 299, 396
Boyce, J.M. 182
Bradley, J.P. 28
Brady, J.L. 351, 396
Brandt, J.C. 414
Briggs, R.E. 299
Brin, G.D. 167, 170
Broughton, R.P. 396
Brouwer, D. 126, 202, 225
Brown, R.H. 167
Brown, W.L. 167, 169
Brownlee, D.E. 28
Buckley, R.J. 253
Buffoni, L. 310
Bulirsch, R. 201, 370
Burns, J.A. 126, 178
Buckhardt, J.C. 396
Burton, W.B. 29, 40, 68
Butcher, J.C. 201
Byl, J. 67, 111

Cameron, A.G.W. 67, 85, 95, 167
Campbell, D.B. 299
Capaccioni, F. 167, 177
Capps, R.W. 169
Cardelli, J. 28
Carpenter, E. 351, 396
Carusi, A. 167, 201, 213, 234, 253,
    254, 277, 299, 339
Catullo, V. 177
Celoria, G. 310, 396
Cerroni, P. 167, 177
Chang, Y.C. 396
Chapman, C.R. 8, 167
Chebotarev, G.A. 67, 167, 241, 254,
    277
Chernykh, N.S. 255
Clairaut, A.C. 397

429

DISCUSSION INDEX

LIST OF PARTICIPANTS

M.E. Bailey   Astronomy Centre, University of Sussex, Falmer,
                        BRIGHTON, BN1 6JQ, United Kingdom.

M.A. Barucci   IAS – Reparto Planetologia, Viale Università 11,
                        00185 ROME, Italy.

D. Benest   Observatoire de Nice, Le Mont-Gros, F-06300 NICE,
                        France.

R. Bianchi   IAS – Reparto Planetologia, Viale Università 11,
                        00185 ROME, Italy.

R. Bien   Astronomisches Rechen-Institut, D-6900 HEIDELBERG,
                        Federal Republic of Germany.

J.C. Brandt   NASA, Goddard Space Flight Center, GREENBELT, MD
                        20771, U.S.A.

R. Burchi   Osservatorio Astronomico Collurania, 64100 TERAMO,
                        Italy.

F. Capaccioni   IAS – Reparto Planetologia, Viale Università 11,
                        00185 ROME, Italy.

A. Carusi   IAS – Reparto Planetologia, Viale Università 11,
                        00185 ROME, Italy.

R. Casacchia   IAS – Reparto Planetologia, Viale Università 11,
                        00185 ROME, Italy.

V. Celebonovic   Cubrina 5A, 11000 BEOGRAD, Yugoslavia.

P. Cerroni   IAS – Reparto Planetologia, Viale Università 11,
                        00185 ROME, Italy.

S.V.M. Clube   Royal Observatory, Blackford Hill, EDINBURGH EH9 3HJ
                        United Kingdom.

A. Coradini   IAS – Reparto Planetologia, Viale Università 11,
                        00185 ROME, Italy.

M. Coradini   IAS – Reparto Planetologia, Viale Università 11,
                        00185 ROME, Italy.

A.H. Delsemme   Department of Physics and Astronomy, University of
                        Toledo, TOLEDO, Ohio 43606, U.S.A.

M. Di Martino   Osservatorio Astronomico di Torino, 10025 PINO TO-
                        RINESE, Italy.

G. Dulinski   Institute of Astronomy, N. Copernicus University,
                        ul. Chopina 12/18, 87-100 TORUN, Poland.

D.W. Dunham   (X-603) System Science Division, Computer Sciences
                        Corporation, 8728 Colesville Road, SILVER SPRING,
                        MD 20910, U.S.A.

R. Dvorak   Astronomisches Institut, University of Graz, Uni-
                        versitatsplatz 5, A-8010 GRAZ, Austria.

E. Everhart   Physics Department, University of Denver, DENVER, CO

                              80208, U.S.A.

P. Farinella   Istituto di Matematica dell'Università, Piazza dei
                    Cavalieri 2, 56100 PISA, Italy.

C. Federico   IAS - Reparto Planetologia, Viale Università 11,
                    00185 ROME, Italy.

J.A. Fernández   Observatorio do Valongo, U.F.R.J. Ladeira Pedro An-
                    tonio 43, 20.080 RIO DE JANEIRO, Brazil.

E. Flamini   IAS - Reparto Planetologia, Viale Università 11,
                    00185 ROME, Italy.

G. Forti   Osservatorio Astrofisico di Arcetri, Largo E. Fermi
                    5, 50125 FIRENZE, Italy.

C. Froeschlé   Observatoire de Nice, B.P. 252, 06007 NICE, France.

M. Fulchignoni   IAS - Reparto Planetologia, Viale Università 11,
                    00185 ROME, Italy.

J.M. Greenberg   Huygens Laboratory, Postbus 9504, 2300 RA LEIDEN,
                    Holland.

R. Greenberg   Planetary Science Institute, 2030 E. Speedway, Suite
                    201, TUCSON, AZ 85719, U.S.A.

S. Grudzinska   Institute of Astronomy, N. Copernicus University,
                    ul. Chopina 12/18, 87-100 TORUN, Poland.

A. Hajduk   Astronomical Institute of SAV, 84228 BRATISLAVA,
                    Czechoslovakia.

D.W. Hughes   Department of Physics, The Hicks Building, The Uni-
                    versity, SHEFFIELD S10 2TN, United Kingdom.

A. Kitov   Observatoire de Paris, 18 Villa Marguerite, 92130
                    ISSY-LES-MOULINEAUX, France.

L. Kresák   Astronomical Institute of SAV, 84228 BRATISLAVA,
                    Czechoslovakia.

M. Kresáková   Astronomical Institute of SAV, 84228 BRATISLAVA,
                    Czechoslovakia.

P. Lanciano   IAS - Reparto Planetologia, Viale Università 11,
                    00185 ROME, Italy.

B.A. Lindblad   Lund Observatory, Box 1107, S-22104 LUND, Sweden.

J. Lissauer   MS 243-3, NASA-Ames Research Center, MOFFETT FIELD,
                    CA 94035, U.S.A.

B. Lokanadham   Centre of Advanced Study in Astronomy, Osmania Univer-
                    sity, 500 007 HYDERABAD, India.

Rh. Lüst   Max Planck Institut für Astrophysik, Karl Schwartz-
                    childstrasse 1, 8046 GARCHING, Federal Republic of
                    Germany.

G. Magni   IAS - Reparto Planetologia, Viale Università 11,
                    00185 ROME, Italy.

B.G. Marsden   Smithsonian Astrophysical Observatory, 60 Garden
                    Street, CAMBRIDGE, Mass. 02138, U.S.A.

V.R. Matas	Astronomisches Rechen-Institut, Monchhofstrasse 12-14, 6900 HEIDELBERG, Federal Republic of Germany.
F. Mignard	C.E.R.G.A., Avenue Copernic, 06130 GRASSE, France.
A. Manara	Osservatorio Astronomico di Brera, Via Brera 28, 20121 MILANO, Italy.
A. Milani	Istituto di Matematica dell'Università, Piazza dei Cavalieri 2, 56100 PISA, Italy.
W.M. Napier	Royal Observatory, Blackford Hill, EDIMBURGH EM9 3MJ, United Kingdom.
E. Perozzi	ESOC, Robert-Bosch-Str. 5, 6100 DARMSTADT, Federal Republic of Germany.
E.M. Pittich	Astronomical Institute of SAV, 84228 BRATISLAVA, Czechoslovakia.
S. Pozio	IAS - Reparto Planetologia, Viale Università 11, 00185 ROME, Italy.
F. Remy	C.E.R.G.A., Avenue Copernic, 06130 GRASSE, France.
H. Rickman	Astronomiska Observatoriet, Box 515, S-75120 UPPSALA, Sweden.
F. Roque	Observatoire de Paris-Meudon, Place Jules Jansen, F-92190 MEUDON, France.
K.S. Russell	U.K. Schmidt Telescope Unit, Private Bag, COONABARA-BRAN, NSW 2357, Australia.
H. Scholl	Astronomisches Rechen-Institut, Munchhofstrasse 12-14, 6900 HEIDELBERG, Federal Republic of Germany.
R. Shubert	Department of Electrical Engeneering, California State University, FULLERTON, CA 92634, U.S.A.
I. Stellmacher	Bureau des Longitudes, 77 Avenue Denfert-Rochereau, 75014 PARIS, France.
G.B. Valsecchi	IAS - Reparto Planetologia, Viale Università 11, 00185 ROME, Italy.
P.R. Weissman	Jet Propulsion Laboratory, 4800 Oak Grove Drive, PASADENA, CA 91103, U.S.A.
I.P. Williams	Department of Applied Mathematics, Queen Mary College, Mile End Road, LONDON EI 4NS, United Kingdom.
S. Yabushita	Dept. of Applied Mathematics and Physics, Kyoto University, KYOTO 606, Japan.
D.K. Yeomans	Jet Propulsion Laboratory, 4800 Oak Grove Drive, PASADENA, CA 91103, U.S.A.
P.E. Zadunaisky	CNIE Observatorio Nacional de Fisica Cosmica, Av. Mitre 3100, SAN MIGUEL (Buenos Aires), Argentina.